Advanced Differential Equations

Advanced Differential Equations

Youssef N. Raffoul
Professor of Mathematics
University of Dayton
Dayton, OH, United States

ACADEMIC PRESS
An imprint of Elsevier

Academic Press is an imprint of Elsevier
125 London Wall, London EC2Y 5AS, United Kingdom
525 B Street, Suite 1650, San Diego, CA 92101, United States
50 Hampshire Street, 5th Floor, Cambridge, MA 02139, United States
The Boulevard, Langford Lane, Kidlington, Oxford OX5 1GB, United Kingdom

Notices

Knowledge and best practice in this field are constantly changing. As new research and experience
broaden our understanding, changes in research methods, professional practices, or medical treatment
may become necessary.

Practitioners and researchers must always rely on their own experience and knowledge in evaluating
and using any information, methods, compounds, or experiments described herein. In using such
information or methods they should be mindful of their own safety and the safety of others, including
parties for whom they have a professional responsibility.

To the fullest extent of the law, neither the Publisher nor the authors, contributors, or editors, assume
any liability for any injury and/or damage to persons or property as a matter of products liability,
negligence or otherwise, or from any use or operation of any methods, products, instructions, or ideas
contained in the material herein.

Library of Congress Cataloging-in-Publication Data
A catalog record for this book is available from the Library of Congress

British Library Cataloguing-in-Publication Data
A catalogue record for this book is available from the British Library

ISBN: 978-0-323-99280-0

For information on all Academic Press publications
visit our website at https://www.elsevier.com/books-and-journals

Publisher: Katey Birtcher
Editorial Project Manager: Rafael G. Trombaco
Production Project Manager: Prem Kumar Kaliamoorthi
Designer: Margaret Reid

Typeset by VTeX

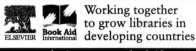

*To my wonderful wife Nancy, who is
the wind beneath my wings.*

Contents

Student Resources

For the Partial Solutions Manual, visit the companion site:
https://www.elsevier.com/books-and-journals/book-companion/9780323992800

Preface

Differential equations are widely used by mathematicians, physicists, engineers, biologists, chemists, and scientists who work in relevant fields. They encounter the use of differential equations in the study of Newton's law of cooling, Maxwell's equations, Newton's laws of motion, fluid dynamics equations, equations in plasma dynamics, equations in stellar dynamics, Hook's law, Schrödinger's equation, acoustic wave equation, equations in chemical kinetics, equations in thermodynamics, Einstein's equations for general relativity, population models, epidemics, and so on. For many years, the author has been encouraged by the graduate students at the University of Dayton to write a concise and reader-friendly book on the subject of advanced differential equations. So this book grew out of lecture notes that the author has been constantly revising and using for a graduate course in differential equations. The book should serve as a two-semester graduate textbook in exploring the theory and applications of ordinary differential equations and differential equations with delays. It is intended for students who have basic knowledge of ordinary differential equations and real analysis. While writing this book, the author tried to balance rigor and presenting the most difficult material in an elementary format by adopting easier and friendlier notations that make the book accessible to a wide audience. It was the author's main intention to provide many examples to illustrate the theory conveyed in the theorems. The author made every effort to include contemporary topics such as the use of *fixed point theory* in several places to prove the existence and uniqueness, various notions of stability, and the existence of positive periodic solutions on Banach spaces. What makes the book appealing and distinguished from other books is the addition of Chapters 8 and 9 on delay differential equations with advanced topics. The author is convinced that any student who completes the whole book, especially Chapters 8 and 9, should be ready to carry on with meaningful research in delay differential systems.

Much of the pedagogical and mathematical development of this book is influenced by the author's style of presentation. The literature on differential equations is vast and well established, and some of the ideas found their way into this book.

Since stability is the central part of this book, namely by the Lyapunov method, we must mention some history. Lyapunov functions are named after Alexander Lyapunov, a Russian mathematician, who in 1892 published his book *The General Problem of Stability of Motion*. Lyapunov was the first to consider the modifications necessary in nonlinear systems to the linear theory of stability based on linearizing near a point of equilibrium. His work, initially published in Russian and then translated to French, received little attention for many years. Interest in Lyapunov stability started suddenly during the Cold War period when his method was found to be applicable to the stability of aerospace guidance systems, which typically contain strong nonlinearities not treatable by other methods. More recently, the concept of the Lyapunov exponent related to Lyapunov's first method of discussing stability has received wide interest in connection with chaos theory.

Chapter 1 deals with various introductory topics, including variation of parameters formula, metric spaces, and Banach spaces.

In Chapter 2 we introduce Gronwall's inequality that we make use of to prove the uniqueness of solutions. We introduce theorems on the existence and uniqueness of solutions, their dependence on initial data, and their continuation on maximal interval.

In Chapter 3 we introduce systems of differential equations. We briefly discuss how the existence and uniqueness theorems of Chapter 2 are extended to suit systems. Then we develop the notion of the fundamental matrix as a solution and utilize it to write solutions of non-homogeneous systems so that they can be analyzed.

Stability theory is the central part of this book. Chapters 4–8 are totally devoted to stability. In Chapter 4 we are mainly concerned with the stability of linear systems via the variation of parameters. The chapter also includes a nice section of Floquet theory with its application to Mathieu's equation.

Chapters 5 and 6 are deeply devoted to the study of the stability of linear systems, near-linear systems, perturbed systems, autonomous systems in the plane, and stability by linearization. Chapter 5 is ended with the study of Hamiltonians and gradient systems. We begin Chapter 6 by looking at stability diagrams in scalar equations and move into the study of bifurcations as it naturally arises while looking at stability. Bifurcation occurs when the dynamics abruptly change as certain parameters move across certain values. We end Chapter 6 by considering stable and unstable manifolds, which then delves into the Hartman–Grobman theorem. The theorem says that the behavior of a dynamical system in the domain near a hyperbolic equilibrium point is qualitatively the same as the behavior of its linearization near this equilibrium point, where hyperbolicity means that no eigenvalue of the linearization has a zero real part.

Chapter 7 delves deeply into the stability of general systems using Lyapunov functions. We prove general theorems regarding the stability of autonomous and non-autonomous systems by assuming the existence of such Lyapunov function. We touch on the notion of ω-limit set and its correlation to Lyapunov functions. The chapter is concluded with a detailed discussion on exponential stability.

Chapter 8 is solely devoted to the study of delay differential equations. It contains recent development in the research of delay differential equations. We begin the chapter by pointing out how basic results from ordinary differential equations are easily extended to delay differential equations. We introduce the *method of steps* and show how to piece together a solution. The transition of moving from ordinary differential equations to delay differential equations was made simple through the extension of Lyapunov functions to Lyapunov functionals. Then we move on to a whole new concept, *fixed point theory*. The use of fixed point theory alleviates some of the difficulties that arise from the use of Lyapunov functionals when studying stability. Later on, we apply fixed point theory to the study of stability and the existence of positive periodic solutions of neutral differential equations and neutral Volterra integro-differential equations, respectively. We end the chapter with the use of Lyapunov functionals to obtain necessary conditions for the exponential stability of Volterra integro-differential equations with finite delay.

Chapter 9 deals with current research concerning the use of a *new variation of parameters formula*. The objective is to introduce a new method for inverting first-order ordinary differential equations with time-delay terms to obtain a new variation of parameters formula that we use to study the stability, boundedness, and periodicity of general equations in ordinary and delay differential equations.

A combination of Chapters 1–3, 6, and 7 can be used to deliver a course on nonlinear systems for engineers.

The author has not attempted to give the historical origin of the theory, except in very rare cases. This resulted in the situation that not every reference listed in References is mentioned in the text or the body of the book.

Exercises play an essential learning tool of the course and accompany each chapter. They range from routine calculations to solving more difficult problems to open-ended ones. Students must read the relevant material before attempting to do the exercises.

I am indebted to Dr. Mohamed Aburakhis, who fully developed all the codes for all the figures in the book. I like to thank the hundreds of graduate students at the University of Dayton whom the author taught for the last 20 years and who helped the polishing and refining of the lecture notes, most of which have become this book.

A heartfelt appreciation to Jeff Hemmelgarn from the University of Dayton for carefully reading the whole book and pointing out many typos.

Youssef N. Raffoul
University of Dayton
Dayton, Ohio
June 2021

Preliminaries and Banach spaces

We briefly discuss basic topics of ordinary differential equations and provide examples that illustrate the need for a comprehensive and systematic theory of differential equations. In addition, we introduce metrics and Banach spaces, which we will use throughout the book.

1.1 Preliminaries

Let $\mathbb{R}$ denote the set of real numbers, let I be an interval in $\mathbb{R}$, and consider a function $x : I \to \mathbb{R}$. We say the function x has a derivative at a point $t^* \in I$ if the limit

$$\lim_{t \in I,\, t \to t^*} \frac{x(t) - x(t^*)}{t - t^*}$$

exists as a finite number. In this case, we adopt the notation

$$x'(t^*) = \lim_{t \in I,\, t \to t^*} \frac{x(t) - x(t^*)}{t - t^*},$$

where $x'(t^*)$ is the instantaneous rate of change of the function x at t^*. If t^* is one of the endpoints of the interval I, then the above definition of derivative becomes that of a one-sided derivative. If $x'(t^*)$ exists at every point $t^* \in I$, then we say that x is differentiable on I and write $x'(t)$. Throughout the book, we might use $\frac{dx}{dt}$ to indicate $x'(t)$. Similarly, if $x'(t)$ has a derivative function, then we call it the second derivative of the function $x(t)$ and denote it by $x''(t)$. For higher-order derivatives, we use the notations

$$x'''(t), x^{(4)}(t), \ldots, x^{(n)}(t), \quad \text{or} \quad \frac{d^n x}{dt^n}.$$

The *order of a differential equation* is defined by the highest derivative present in the equation. An *nth-order ordinary differential equation* is a functional relation of the form

$$F\left(t, x, \frac{dx}{dt}, \frac{d^2 x}{dt^2}, \frac{d^3 x}{dt^3}, \ldots, \frac{d^n x}{dt^n}\right) = 0, \; t \in \mathbb{R} \tag{1.1.1}$$

Advanced Differential Equations. https://doi.org/10.1016/B978-0-32-399280-0.00007-3

between the independent variable t and the dependent variable x, and its derivatives

$$\frac{dx}{dt}, \frac{d^2x}{dt^2}, \frac{d^3x}{dt^3}, \dots, \frac{d^n x}{dt^n}.$$

Loosely speaking, by a solution of (1.1.1) on an interval I, we mean a function $x(t) = \varphi(t)$ such that

$$F\big(t, \varphi(t), \varphi'(t), \dots, \varphi^{(n)}(t)\big)$$

is defined for all $t \in I$ and

$$F\big(t, \varphi(t), \varphi'(t), \dots, \varphi^{(n)}(t)\big) = 0$$

for all $t \in I$. If we require, for some initial time $t_0 \in \mathbb{R}$, a solution $x(t)$ to satisfy the initial conditions

$$x(t_0) = a_0, \ \frac{dx}{dt}(t_0) = a_1, \ \frac{d^2x}{dt^2}(t_0) = a_2, \ \dots, \ \frac{d^{n-1}x}{dt^{n-1}}(t_0) = a_{n-1}, \qquad (1.1.2)$$

for constants a_i, $i = 0, 1, 2, \dots, n-1$, then (1.1.1) along with (1.1.2) is called an initial value problem (IVP).

Following the notation of (1.1.1), a first-order differential equation takes the form

$$F\Big(t, x, \frac{dx}{dt}\Big) = 0.$$

Hence, if we assume that we can solve for $\dfrac{dx}{dt}$, then we have

$$x'(t) = f(t, x)$$

for some function f that satisfies certain continuity conditions.

Let $x : I \to \mathbb{R}$ be a function. A differentiable function $z : I \to \mathbb{R}$ is called an *antiderivative* of the function x on the interval I if

$$z'(t) = x(t) \ \text{ for all } \ t \in I.$$

The set of all antiderivatives of x is denoted by

$$\int x(t)dt$$

and called the indefinite integral of the function x. When we calculate the indefinite integral $\int t^2 dt$, we in fact solve the first-order differential equation $x'(t) = t^2$. The family of its solutions is given by $t^3/3 + c$, where c is any constant. Thus we may write the solution as $x(t) = t^3/3 + c$, which is a one-parameter family of solutions, the same as the family of all the antiderivatives of t^3. Now, if we impose an initial condition on the differential equation, say $x(t_0) = x_0$, for some initial time

t_0 and real number x_0, then the constant c is uniquely determined by the relation $x_0 = t_0^3/3 + c$. In this case the differential equation has the unique solution given by $x(t) = t^3/3 + x_0 - t_0^3/3$. However, without imposing the condition $x(t_0) = x_0$, the differential equation would have infinitely many solutions given by $x(t) = t^3/3 + c$.

Differential equations play an important role in modeling the behavior of physical systems such as falling bodies, vibration of a mass on a spring, and swinging pendulum. To illustrate the need for the theoretical study of differential equations and in particular nonlinear ones, we examine a few examples.

Consider the first-order differential equation

$$x'(t) = h(t)g(x), \; x(t_0) = x_0, \; t \geq t_0,$$

where $h, g : \mathbb{R} \to \mathbb{R}$ are continuous.

1. If $g(x_0) = 0$, then $x(t) = x_0$ is a solution.
2. In a region where $g(x) \neq 0$, we can divide by $g(x)$ so that

$$\frac{x'(t)}{g(x(t))} = h(t).$$

Separating the variables and then integrating both sides from t_0 to t give

$$\int_{t_0}^t \frac{x'(s)ds}{g(x(s))} = \int_{t_0}^t h(s)ds.$$

Using the transformation $u = x(s)$ with $x(t_0) = x_0$, we arrive at

$$\int_{x_0}^{x(t)} \frac{du}{g(u)} = \int_{t_0}^t h(s)ds.$$

If for some function G, we have $\dfrac{dG}{dx} = \dfrac{1}{g}$, then the above expression implies that

$$G(x(t)) - G(x(t_0)) = \int_{t_0}^t h(s)ds$$

or

$$x(t) = G^{-1}\left(G(x_0) + \int_{t_0}^t h(s)ds\right),$$

provided that the inverse of G exists. Note that the right side of the above expression depends on the initial time t_0 and the initial value x_0. Therefore to emphasize the dependence of solutions on the initial data, we may write a solution $x(t)$ in the form $x(t) = x(t, t_0, x_0)$.

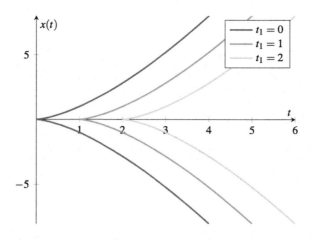

FIGURE 1.1

This example shows infinitely many solutions.

In the next example, we illustrate the existence of more than one solution. Consider the differential equation

$$x'(t) = \frac{3}{2}x^{1/3}(t), \; x(0) = 0, \; t \in \mathbb{R}.$$

It is clear that $x(t) = 0$ is a solution. Hence we may consider a solution $x_1(t) = 0$ and let

$$x_2(t) = \begin{cases} 0 & \text{for } t \leq 0, \\ t^{3/2} & \text{for } t > 0, \end{cases}$$

which is also a continuous and differentiable solution. Likewise, for $t_1 > 0$, we have

$$x_3(t) = \begin{cases} 0 & \text{for } t \leq t_1, \\ (t - t_1)^{3/2} & \text{for } t > t_1. \end{cases}$$

Continuing in this way, we see that the differential equation has infinitely many solutions. Similarly, if x is a solution, then $-x$ is also a solution (see Fig. 1.1).

In the next example, we show that solutions may escape (become unbounded) in finite time. The differential equation

$$x'(t) = x^3(t), \; x(t_0) = x_0 > 0, \; t \geq t_0$$

has the solution

$$x(t) = \sqrt{\frac{1}{\frac{1}{x_0^2} + 2(t_0 - t)}}.$$

We can easily see that the solution is only valid for $t < t_0 + \dfrac{1}{2x_0^2}$ and becomes unbounded (escapes) as t approaches $t_0 + \frac{1}{2x_0^2}$ from the left.

The most important application in engineering problems is Newton's law

$$m\frac{d^2x(t)}{dt^2} = F[t, x(t), \frac{dx(t)}{dt}]$$

for the position $x(t)$ of a particle with mass m acted on by a force F, which may be a function to time t, the position $x(t)$, and the velocity $\frac{dx(t)}{dt}$. For example, if the force is only due to gravity, then we have the second-order differential equation

$$m\frac{d^2x(t)}{dt^2} = -mg,$$

which has the solution

$$x(t) = -\frac{1}{2}gt^2 + c_1 t + c_2,$$

where c_1 and c_2 are constants that can be uniquely determined by specifying the position and velocity of the particle at some instant of time.

In the next example, we consider the problem of *leaky bucket*.

Example 1.1. We have a bucket with no water flowing into it and having a hole in the bottom. If Q is the volumetric flow rate, then $Q_{in} - Q_{out} = Q_{stored}$. Since $Q_{in} = 0$, we obtain $Q_{stored} = -Q_{out}$. Let $h(t)$ be the height of the water in the bucket at time t, and assume that the initial height at time $t = 0$ is h_0. Let $v(t)$ be the velocity of the leaked water (flow velocity). The volumetric flow rate Q_{stored} can be calculated by multiplying the velocity by the area of the bucket, that is, $Q_{stored} = A_{bucket}\frac{dh(t)}{dt}$. Similarly, Q_{out} is the flow velocity multiplied by the area of the hole A_{hole}. It follows that $Q_{out} = A_{hole}v(t)$. For fluids of height $h(t)$, the velocity of water coming out at the bottom is $v(t) = \sqrt{2gh(t)}$. By rearranging the terms, we arrive at the first-order differential equation in h given by

$$\frac{dh(t)}{dt} = -k\sqrt{h}, \quad h(0) = h_0, \tag{1.1.3}$$

where $k = \dfrac{A_{hole}}{A_{bucket}}\sqrt{2g} > 0$. By separating the variables in (1.1.3) and then integrating, we arrive at the solution

$$h(t) = \left(\sqrt{h_0} - \frac{kt}{2}\right)^2. \tag{1.1.4}$$

Note that the solution $h(t)$ given by (1.1.4) decreases from the initial height h_0. Moreover, at time $t^* = \frac{2\sqrt{h_0}}{k}$, the water is completely drained (by gravity), $h(t^*) = 0$, and the bucket will remain empty or the height will remain zero after $t^* = \frac{2\sqrt{h_0}}{k}$. Therefore we may write the solution as

$$
h(t, 0, h_0) = \begin{cases} \left(\sqrt{h_0} - \frac{kt}{2}\right)^2, & 0 \le t \le \frac{2\sqrt{h_0}}{k}, \\ 0, & t > \frac{2\sqrt{h_0}}{k}. \end{cases}
$$

Another interesting application is the projectile problem that we analyze in the next section.

1.2 Escape velocity

Let M and R be the mass and radius of the Earth, respectively. We are interested in finding the smallest initial velocity for a mass m to exit the Earth's gravitational field, the so-called *escape velocity*. We assume that no external forces are acting on the system other than the gravitational force. Newton's universal gravitational law states that the force between two massive bodies is proportional to the product of the masses and inversely proportional to the square of the distance between them, where the mass of each body can be considered as concentrated at its center. For a mass m with position x above the surface of the Earth, the force F on the mass is given by

$$
F = -G \frac{Mm}{(R+x)^2},
$$

where G is the proportionality constant in the universal gravitational law. The minus sign means the force on the mass m points in the direction of decreasing x. By Newton's second law, force $F = m\frac{d^2x}{dt^2}$ (mass times acceleration). When $x = 0$, that is, at the Earth's surface, the gravitational force equals $-mg$, where g is the gravitational constant. Therefore

$$
\frac{GM}{R^2} = g, \ \ g \approx 9.8 \, \text{m/s}^2.
$$

We have the second-order differential equation

$$
\begin{aligned}
\frac{d^2x}{dt^2} &= -G \frac{M}{(R+x)^2} \\
&= -\frac{g}{(1 + \frac{x}{R})^2},
\end{aligned}
\tag{1.2.1}
$$

where the radius of the Earth is known to be $R \approx 6350 \, \text{km}$. We transform Eq. (1.2.1) into a first-order differential equation in terms of the velocity v by noting that $\frac{d^2x}{dt^2} = \frac{dv}{dt}$. If we write $v(t) = v(x(t))$, that is, considering the velocity of the mass m as a function of its distance above the Earth. Then using the chain rule, we have

$$\frac{dv}{dt} = \frac{dv}{dx}\frac{dx}{dt}$$
$$= v\frac{dv}{dx},$$

since $v = \frac{dx}{dt}$. As a consequence, (1.2.1) becomes the first-order differential equation

$$\frac{dv}{dx} = -\frac{g}{(1 + \frac{x}{R})^2}\frac{1}{v}.$$

Suppose the mass is shot vertically from the Earth's surface with initial velocity $v(x = 0) = v_0$. Separating the variables and integrating both sides, we obtain

$$\int v \, dv = -g \int \frac{dx}{(1 + \frac{x}{R})^2},$$

which gives

$$\frac{v^2}{2} = \frac{gR^2}{R + x} + c$$

for some constant c. Using the initial velocity condition, we find

$$c = -gR + \frac{v_0^2}{2}.$$

Substituting c back into the solution and simplifying, we arrive at the solution

$$v^2 = v_0^2 - \frac{gRx}{R + x}.$$

The escape velocity is defined as the minimum initial velocity v_0, such that the mass can escape to infinity. Therefore $v_0 = v_{escape}$ when $v \to 0$ as $x \to \infty$. Taking the limit, we have

$$0 = v_0^2 - \lim_{x \to \infty}\frac{gRx}{R + x} = v_0^2 - 2gR,$$

or

$$v_{escape}^2 = 2gR.$$

With $R \approx 6350\,\text{km}$ and $g = 127{,}008\,\text{km/h}^2$, we get $v_{escape} = \sqrt{2gR} \approx 40{,}000\,\text{km/h}$. In contrast, the muzzle velocity of a modern high-performance rifle is $4300\,\text{km/h}$, which is not enough for a bullet shot into the sky to escape from Earth's gravity.

Now we formally attempt to qualitatively analyze differential equations.

Definition 1.2.1. Let D be an open subset of $\mathbb{R}^2$, and let $f : D \to \mathbb{R}$ be a continuous function. Let $(t_0, x_0) \in D$. We say that $x(t) = x(t, t_0, x_0)$ is a solution of

$$x' = f(t, x), \quad x(t_0) = x_0, \tag{1.2.2}$$

on an interval I if $t_0 \in I$, $x : I \to \mathbb{R}$ is differentiable, $(t, x(t)) \in D$ for $t \in I$, $x'(t) = f(t, x(t))$ for $t \in I$, and $x(t_0) = x_0$.

Definition 1.2.2. A solution $x(t)$ of (1.2.2) is said to be bounded on the interval $I = [0, \infty)$ if for any $t_0 \in [0, \infty)$ and $r > 0$, there exists a *positive number* $\alpha(t_0, r)$ depending on t_0 and r such that $|x(t, t_0, x_0)| \leq \alpha(t_0, r)$ for all $t \geq t_0$ and x_0 such that $|x_0| < r$. It is uniformly bounded if α is independent of the initial time t_0.

Definition 1.2.3. Let $x(t)$ and $y(t)$ be solutions of (1.2.2) with respect to initial conditions x_0 and y_0, respectively. The solution $x(t)$ is then said to be stable if for every $\varepsilon > 0$, there exists $\delta = \delta(\varepsilon, t_0) > 0$ such that

$$|x(t) - y(t)| < \varepsilon \text{ whenever } |x_0 - y_0| < \delta.$$

Consider the linear differential equation

$$x'(t) = 1, \ x(t_0) = x_0. \tag{1.2.3}$$

It is easy to check that $x(t) = x_0 + (t - t_0)$ is the solution of (1.2.3). If $y(t)$ is another solution with $y(t_0) = y_0$, then we have $y(t) = y_0 + (t - t_0)$. For any $\varepsilon > 0$, let $\delta = \varepsilon$. Then

$$|x(t) - y(t)| = |x_0 + (t - t_0) - y_0 - (t - t_0)| = |x_0 - y_0| < \varepsilon$$

whenever $|x_0 - y_0| < \delta$. Hence the solution $x(t)$ is stable but clearly unbounded. This simple example shows that the notion of a solution being unbounded does not automatically imply that the same solution is unstable with respect to another solution starting at a different initial point. In Chapters 6 and 7 we will discuss boundedness and stability in more detail. The previous examples illustrate the need for a coherent theory for addressing the following issues:

1. Existence and uniqueness.
2. Boundedness and stability.
3. The dependence of solutions on the initial data.

1.3 Applications to epidemics

The *law of mass action* is a useful concept that describes the behavior of a system that consists of many interacting parts, such as molecules, that react with each other, or viruses that are passed along from a population of infected individuals to non-immune ones. The law of mass action was first derived for chemical systems but subsequently found wide use in epidemiology and ecology. To describe the law of mass action, we assume that m substances $s_1, s_2, \ldots, s_m$ together form a product with concentration p. Then the law of mass action states that $\frac{dp}{dt}$ is proportional to the product of the m concentrations s_i, $i = 1, \ldots, m$, that is,

$$\frac{dp}{dt} = ks_1 s_2 \ldots s_m.$$

Suppose we have a homogeneous population of fixed size divided into two groups. Those who have the disease are called infective, and those who do not have the disease are called susceptible. Let $S = S(t)$ be the susceptible portion of the population, and let $I = I(t)$ be the infective portion. Then by assumption, we may normalize the population and have $S + I = 1$. We further assume that the dynamics of this epidemic satisfies the law of mass action. Hence, for positive constant λ, we have the nonlinear differential equation

$$I'(t) = \lambda S I. \tag{1.3.1}$$

Let $I(0) = I_0$, $0 < I(0) < 1$, be a given initial condition. By substituting $S = 1 - I$ into (1.3.1) it follows that

$$I'(t) = \lambda I (1 - I), \quad I(0) = I_0. \tag{1.3.2}$$

If we can solve (1.3.2) for $I(t)$, then $S(t)$ can be found from the relation $I + S = 1$. We separate the variables in (1.3.2) and obtain

$$\frac{dI}{I(1 - I)} = \lambda dt.$$

Using partial fraction on the left side of the equation and then integrating both sides yield

$$\mathrm{Ln}|I| - \mathrm{Ln}|1 - I| = \lambda t + c,$$

or for positive constant c_1, we have

$$I(t) = \frac{c_1 e^{\lambda t}}{1 + c_1 e^{\lambda t}}.$$

Applying $I(0) = I_0$ gives the solution

$$I(t) = \frac{I_0 e^{\lambda t}}{1 - I_0 + I_0 e^{\lambda t}}. \tag{1.3.3}$$

Now for $0 < I(0) < 1$, the solution given by (1.3.3) is increasing with time as expected. Moreover, using L'Hospital's rule, we have

$$\lim_{t \to \infty} I(t) = \lim_{t \to \infty} \frac{I_0 e^{\lambda t}}{1 - I_0 + I_0 e^{\lambda t}} = 1.$$

Hence the infection will grow, and everyone in the population will eventually get infected.

1.4 Metrics and Banach spaces

This section is devoted to introductory materials related to Cauchy sequences, metric spaces, contraction, compactness, contraction mapping principle, and Banach spaces. Materials in this section will be of use in several places of the book, especially in Chapters 2, 8, and 9. Throughout the book, by $C(I, \mathbb{R}^n)$, we denote the space of all continuous functions $f : I \to \mathbb{R}^n$ on an interval I, possibly infinite.

Definition 1.4.1. A pair (E, ρ) is a metric space if E is a set and $\rho : E \times E \to [0, \infty)$ such that for all y, z, and u in E, we have

(a) $\rho(y, z) \geq 0$, $\rho(y, y) = 0$, and $\rho(y, z) = 0$ implies $y = z$;
(b) $\rho(y, z) = \rho(z, y)$; and
(c) $\rho(y, z) \leq \rho(y, u) + \rho(u, z)$.

The next definition is concerned with Cauchy sequences.

Definition 1.4.2. (Cauchy sequence) A sequence $\{x_n\} \subseteq E$ is a Cauchy sequence if for each $\varepsilon > 0$, there exists $N \in \mathbb{N}$ such that $n, m > N \implies \rho(x_n, x_m) < \varepsilon$.

Complete metric spaces play a major role when showing that a fixed point belongs to the metric space of interest.

Definition 1.4.3. (Completeness of metric space) A metric space (E, ρ) is said to be complete if every Cauchy sequence in E converges to a point in E.

Definition 1.4.4. A set L in a metric space (E, ρ) is compact if each sequence in L has a subsequence with a limit in L.

Definition 1.4.5. Let $\{f_n\}$ be a sequence of real functions with $f_n : [a, b] \to \mathbb{R}$.

1. $\{f_n\}$ is uniformly bounded on $[a, b]$ if there exists $M > 0$ such that $|f_n(t)| \leq M$ for all $n \in \mathbb{N}$ and $t \in [a, b]$.
2. $\{f_n\}$ is equicontinuous at t_0 if for each $\varepsilon > 0$, there exists $\delta > 0$ such that for all $n \in \mathbb{N}$, if $t \in [a, b]$ and $|t_0 - t| < \delta$, then $|f_n(t_0) - f_n(t)| < \varepsilon$. Also, $\{f_n\}$ is equicontinuous if $\{f_n\}$ is equicontinuous at each $t_0 \in [a, b]$.
3. $\{f_n\}$ is uniformly equicontinuous if for each $\varepsilon > 0$, there exists $\delta > 0$ such that for all $n \in \mathbb{N}$, if $t_1, t_2 \in [a, b]$ and $|t_1 - t_2| < \delta$, then $|f_n(t_1) - f_n(t_2)| < \varepsilon$.

It is easy to see that $\{f_n\} = \{x^n\}$ is not an equicontinuous sequence of functions on $[0, 1]$ but each f_n is uniformly continuous.

Proposition 1.1. *[Cauchy criterion for uniform convergence] If $\{F_n\}$ is a sequence of bounded functions that is Cauchy in the uniform norm, then $\{F_n\}$ converges uniformly.*

Definition 1.4.6. A real-valued function f defined on $E \subseteq \mathbb{R}$ is said to be Lipschitz continuous with Lipschitz constant K if $|f(x) - f(y)| \le K|x - y|$ for all $x, y \in E$.

It is easy to see that the function $f(x) = x^2$ is not Lipschitz on $\mathbb{R}$. This is due to the fact that for any x and y in $\mathbb{R}$, we have that $f(x) - f(y) = |x^2 - y^2| = |x + y||x - y|$, and so there is no constant K such that $|x^2 - y^2| \le K|x - y|$. Definition 1.4.6 implies that f is globally Lipschitz since the constant K is uniform for all x and y in $\mathbb{R}$.

Remark 1.1. It is an easy exercise that a Lipschitz continuous function is uniformly continuous. Also, if each f_n in a sequence of functions $\{f_n\}$ has the same Lipschitz constant, then the sequence is uniformly equicontinuous.

Lemma 1.1. *If $\{f_n\}$ is an equicontinuous sequence of functions on a closed bounded interval, then $\{f_n\}$ is uniformly equicontinuous.*

Proof. Suppose $\{f_n\}$ is equicontinuous on $[a, b]$. Let $\varepsilon > 0$. For each $x \in K$, let $\delta_x > 0$ be such that $|y - x| < \delta_x \implies |f_n(x) - f_n(y)| < \varepsilon/2$ for all $n \in \mathbb{N}$. The collection $\{B(x, \delta_x/2) : x \in [a, b]\}$ is an open cover of $[a, b]$, so it has a finite subcover $\{B(x_i, \delta_{x_i}/2) : i = 1, \dots, k\}$. Let $\delta = \min\{\delta_{x_i}/2 : i = 1, \dots, k\}$. Then, if $x, y \in [a, b]$ with $|x - y| < \delta$, then there is some i with $x \in B(x_i, \delta_{x_i}/2)$. Since $|x - y| < \delta \le \delta_{x_i}/2$, we have $|x_i - y| \le |x_i - x| + |x - y| < \delta_{x_i}/2 + \delta_{x_i}/2 = \delta_{x_i}$. Hence $|x_i - y| < \delta_{x_i}$ and $|x_i - x| < \delta_{x_i}$. So, for any $n \in \mathbb{N}$, we have $|f_n(x) - f_n(y)| \le |f_n(x) - f_n(x_i)| + |f_n(x_i) - f_n(y)| < \varepsilon/2 + \varepsilon/2 = \varepsilon$. So $\{f_n\}$ is uniformly equicontinuous. $\square$

The next theorem gives us the main method of proving compactness in the spaces we are interested in.

Theorem 1.4.1. *[Ascoli–Arzelà] If $\{f_n(t)\}$ is a uniformly bounded and equicontinuous sequence of real-valued functions on an interval $[a, b]$, then there is a subsequence that converges uniformly on $[a, b]$ to a continuous function.*

Proof. Since $\{f_n(t)\}$ is equicontinuous on $[a, b]$, by Lemma (1.1) $\{f_n(t)\}$ is uniformly equicontinuous. Let $t_1, t_2, \dots$ be a listing of the rational numbers in $[a, b]$ (note that the set of rational numbers is countable, so this enumeration is possible). The sequence $\{f_n(t_1)\}_{n=1}^{\infty}$ is a bounded sequence of real numbers (since $\{f_n\}$ is uniformly bounded), so it has a subsequence $\{f_{n_k}(t_1)\}$ converging to a number, which we denote $\phi(t_1)$. It will be more convenient to represent this subsequence without subsubscripts, so we write f_k^1 for f_{n_k} and switch the index from k to n. So the subsequence is written as $\{f_n^1(t_1)\}_{n=1}^{\infty}$. Now, the sequence $\{f_n^1(t_2)\}$ is bounded, so it has a convergent subsequence, say $\{f_n^2(t_2)\}$, with limit $\phi(t_2)$. We continue in this way obtaining a sequence of sequences $\{f_n^m(t)\}_{n=1}^{\infty}$ (one sequence for each m), each of which is a subsequence of the previous one. Furthermore, we have $f_n^m(t_m) \to \phi(t_m)$ as $n \to \infty$ for each $m \in \mathbb{N}$. Now, consider the "diagonal" functions defined $F_k(t) = f_k^k(t)$. Since $f_n^m(t_m) \to \phi(t_m)$, it follows that $F_r(t_m) \to \phi(t_m)$ as $r \to \infty$ for each $m \in \mathbb{N}$ (in other words, the sequence $\{F_r(t)\}$ converges pointwise at each t_m). We now show that $\{F_k(t)\}$ converges uniformly on $[a, b]$ by showing that it is Cauchy in the uniform norm. Let $\varepsilon > 0$, and let $\delta > 0$ be as in the definition of uniformly equicontinuous for $\{f_n(t)\}$ applied with $\varepsilon/3$. Divide $[a, b]$ into p intervals, where $p > \frac{b-a}{\delta}$. Let ξ_j be a rational number in the jth interval for $j = 1, \ldots, p$. Recall that $\{F_r(t)\}$ converges at each of the points ξ_j, since they are rational numbers. So, for each j, there is $M_j \in \mathbb{N}$ such that $|F_r(\xi_j) - F_s(\xi_j)| < \varepsilon/3$ whenever $r, s > M_j$. Let $M = \max\{M_j : j = 1, \ldots, p\}$. If $t \in [a, b]$, then it is in one of the p intervals, say the jth. So $|t - \xi_j| < \delta$, and thus $|f_r^r(t) - f_r^r(\xi_j)| = |F_r(t) - F_r(\xi_j)| < \varepsilon/3$ for every r. Also, if $r, s > M$, then $|F_r(\xi_j) - F_s(\xi_j)| < \varepsilon/3$ (since M is the maximum of the M_i). So for $r, s > M$, we have

$$
\begin{aligned}
|F_r(t) - F_s(t)| &= |F_r(t) - F_r(\xi_j) + F_r(\xi_j) - F_s(\xi_j) + F_s(\xi_j) - F_s(t)| \\
&\leq |F_r(t) - F_r(\xi_j)| + |F_r(\xi_j) - F_s(\xi_j)| + |F_s(\xi_j) - F_s(t)| \\
&\leq \frac{\varepsilon}{3} + \frac{\varepsilon}{3} + \frac{\varepsilon}{3} = \varepsilon.
\end{aligned}
$$

By the Cauchy criterion for convergence, the sequence $\{F_r(t)\}$ converges uniformly on $[a, b]$. Since each $F_r(t)$ is continuous, the limit function $\phi(t)$ is also continuous. $\square$

Remark 1.2. The Ascoli–Arzelà theorem can be generalized to a sequence of functions from $[a, b]$ to $\mathbb{R}^n$. Apply the Ascoli–Arzelà theorem to the first coordinate function to get a uniformly convergent subsequence. Then apply the theorem again, this time to the corresponding subsequence of functions restricted to the second coordinate, getting a subsubsequence, and so on.

The next criterion, known as the Weierstrass M-test plays an important role in showing the existence of solutions.

Lemma 1.2. *(Weierstrass M-test) Let $\{f_n\}$ be a sequence of functions defined on a set E. Suppose that for all $n = 1, \ldots$, there is a constant M_n such that $|f_n(t)| \leq M_n$ for all $t \in E$. If*

$$
\sum_{n=1}^{\infty} M_n < \infty, \quad then \quad \sum_{n=1}^{\infty} f_n(t)
$$

converges absolutely and uniformly on the E.

We remark that the Weierstrass M-test can be easily generalized if the domain of the sequence of functions is a subset of Banach space endowed with an appropriate norm.

Here is an example of the Weierstrass M-test.

Example 1.2. For $n = 1, 2, \ldots$, define the sequence of functions $\{f_n\}$ on $\mathbb{R}$ by $f_n(t) = \frac{1}{t^2 + n^2}$. Then $|f_n(t)| = |\frac{1}{t^2 + n^2}| \leq \frac{1}{n^2} := M_n$ for all $t \in \mathbb{R}$ and $n \geq 1$. Since

the series $\displaystyle\sum_{n=1}^{\infty} \frac{1}{n^2}$ converges, by the Weierstrass M-test the series $\displaystyle\sum_{n=1}^{\infty} \frac{1}{t^2 + n^2}$ converges uniformly on $\mathbb{R}$. Moreover, as each term of the series is continuous and the
convergence is uniform, the sum function is also continuous. (As the uniform limit
of continuous functions is continuous.)

Here is another example with a simple twist to it.

Example 1.3. We prove that the series

$$\sum_{n=1}^{\infty} \frac{n^2 + x^4}{n^4 + x^2}$$

converges to a continuous function $f : \mathbb{R} \to \mathbb{R}$.

Let c be a positive constant. Then for all $x \in [-c, c]$, we have that

$$\left| \frac{n^2 + x^4}{n^4 + x^2} \right| \leq \frac{n^2 + x^4}{n^4} \leq \frac{1}{n^2} + \frac{c^4}{n^4} := M_n.$$

On the other hand, the series

$$\sum_{n=1}^{\infty} M_n = \sum_{n=1}^{\infty} \frac{1}{n^2} + c^4 \sum_{n=1}^{\infty} \frac{1}{n^4}$$

converges, so Weierstrass M-test implies that the series converges uniformly to
a function f on the bounded interval $[-c, c]$. Each term in the series is continuous
and since the uniform limit of continuous functions is continuous, the limit function f is continuous on $[-c, c]$ for every $c > 0$. Now since every $x \in \mathbb{R}$ lies in such
an interval for sufficiently large c, it follows that f is continuous on $\mathbb{R}$. Note that the
series does not converge uniformly on $\mathbb{R}$, so we cannot use the argument that the sum
is continuous on $\mathbb{R}$ because the series converges uniformly on $\mathbb{R}$.

Banach spaces form an important class of metric spaces. We now define Banach
spaces in several steps.

Definition 1.4.7. A triple $(V, +, \cdot)$ is said to be a linear (or vector) space over
a field F if V is a set and the following are true.

1. Properties of $+$
   a. $+$ is a function from $V \times V$ to V. Outputs are denoted $x + y$.
   b. for all $x, y \in V$, $x + y = y + x$ ($+$ is commutative).
   c. for all $x, y, w \in V$, $x + (y + w) = (x + y) + w$ ($+$ is associative).
   d. there is a unique element of V, which we denote 0, such that for all $x \in V$,
        $0 + x = x + 0 = x$ (additive identity).
   e. for each $x \in V$, there is a unique element of V, which we denote $-x$, such
        that $x + (-x) = -x + x = 0$ (additive inverse).

2. Scalar multiplication

    a. $\cdot$ is a function from $F \times V$ to V. Outputs are denoted $\alpha \cdot x$ or αx.

    b.   for all $\alpha, \beta \in F$ and $x \in V$, $\alpha(\beta x) = (\alpha\beta)x$.

    c.   for all $x \in V$, $1 \cdot x = x$.

    d.   for all $\alpha, \beta \in F$ and $x \in V$, $(\alpha + \beta)x = \alpha x + \beta x$.

    e.   for all $\alpha \in F$ and $x, y \in V$, $\alpha(x + y) = \alpha x + \alpha y$.

Commonly, the real numbers and complex numbers are fields in the above definition. For our purposes, we only consider the field of real numbers $F = \mathbb{R}$.

Definition 1.4.8. (Normed spaces) A vector space $(V, +, \cdot)$ is a *normed space* if for each $x \in V$, there is a nonnegative real number $\|x\|$, called the *norm* of x, such that for all $x, y \in V$ and $\alpha \in \mathbb{R}$,

1. $\|x\| = 0$ if and only if $x = 0$,

2. $\|\alpha x\| = |\alpha|\|x\|$,

3. $\|x + y\| \leq \|x\| + \|y\|$.

Remark 1.3. A norm on a vector space always defines a metric $\rho(x, y) = \|x - y\|$ on the vector space. Given a metric ρ defined on a vector space, it is tempting to define $\|v\| = \rho(v, 0)$. But this is not always a norm.

Definition 1.4.9. A Banach space is a complete normed vector space, that is, a vector space $(X, +, \cdot)$ with norm $\|\cdot\|$ for which the metric $\rho(x, y) = \|x - y\|$ is complete.

Example 1.4. The space $(\mathbb{R}^n, +, \cdot)$ over the field $\mathbb{R}$ is a vector space (with the usual vector addition $+$ and scalar multiplication $\cdot$), and there are many suitable norms for it. For example, if $x = (x_1, x_2, \ldots, x_n)$, then

1. $\|x\| = \max\limits_{1 \leq i \leq n} |x_i|,$

2. $\|x\| = \sqrt{\sum\limits_{i=1}^{n} x_i^2},$

3. $\|x\| = \sum\limits_{i=1}^{n} |x_i|,$

4. $\|x\|_p = \left(\sum\limits_{i=1}^{n} |x_i|^p\right)^{1/p}, p \geq 1$

are all suitable norms. Norm 2 is the Euclidean norm: the norm of a vector is its Euclidean distance to the zero vector, and the metric defined from this norm is the usual Euclidean metric. Norm 3 generates the "taxi-cab" metric on $\mathbb{R}^2$, and Norm 4 is the l^p norm.

Throughout the book, it should cause no confusion to use $|\cdot|$ instead of $\|\cdot\|$ to denote a particular norm.

Remark 1.4. Consider the vector space $(\mathbb{R}^n, +, \cdot)$ as a metric space with its metric defined by $\rho(x, y) = \|x - y\|$, where $\|\cdot\|$ is any of the norms in Example 1.4. The completeness of this metric space comes directly from the completeness of $\mathbb{R}$, and hence $(\mathbb{R}^n, \|\cdot\|)$ is a Banach space.

Remark 1.5. In the Euclidean space $\mathbb{R}^n$, compactness is equivalent to closedness and boundedness (Heine–Borel theorem). In fact, the metrics generated from any of the norms in Example 1.4 are equivalent in the sense that they generate the same topologies. Moreover, compactness is equivalent to closedness and boundedness in each of those metrics.

Example 1.5. Let $C([a,b], \mathbb{R}^n)$ denote the space of all continuous functions $f : [a,b] \to \mathbb{R}^n$.

1. $C([a,b], \mathbb{R}^n)$ is a vector space over $\mathbb{R}$.
2. If $\|f\| = \max_{a \leq t \leq b} |f(t)|$, where $|\cdot|$ is a norm on $\mathbb{R}^n$, then $(C([a,b], \mathbb{R}^n), \|\cdot\|)$ is a Banach space.
3. Let M and K be two positive constants and define

$$L = \{f \in C([a,b], \mathbb{R}^n) : \|f\| \leq M; |f(u) - f(v)| \leq K|u - v|\}.$$

Then L is compact.

Proof (of part 3). Let $\{f_n\}$ be any sequence in L. The functions are uniformly bounded by M and have the same Lipschitz constant, K. So the sequence is uniformly equicontinuous. By the Ascoli–Arzelà theorem there is a subsequence $\{f_{n_k}\}$ that converges uniformly to a continuous function $f : [a,b] \to \mathbb{R}^n$. We now show that $f \in L$. Well, $|f_n(t)| \leq M$ for each $t \in [a,b]$, so $|f(t)| \leq M$ for each $t \in [a,b]$ and hence $\|f\| \leq M$. Now fix $u, v \in [a,b]$ and $\varepsilon > 0$. Since $\{f_{n_k}\}$ converges uniformly to f, there is $N \in \mathbb{N}$ such that $|f_{n_k}(t) - f(t)| < \varepsilon/2$ for all $t \in [a,b]$ and $k \geq N$. So fixing any $k \geq N$, we have

$$
\begin{aligned}
|f(u) - f(v)| &= |f(u) - f_{n_k}(u) + f_{n_k}(u) - f_{n_k}(v) + f_{n_k}(v) - f(v)| \\
&\leq |f(u) - f_{n_k}(u)| + |f_{n_k}(u) - f_{n_k}(v)| + |f_{n_k}(v) - f(v)| \\
&< \varepsilon/2 + K|u - v| + \varepsilon/2 = K|u - v| + \varepsilon.
\end{aligned}
$$

Since $\varepsilon > 0$ was arbitrary, $|f(u) - f(v)| \leq K|u - v|$. Hence $f \in L$. We have demonstrated that $\{f_n\}$ has a subsequence converging to an element of L. Hence L is compact. $\qquad\square$

Example 1.6. Consider $\mathbb{R}$ as a vector space over $\mathbb{R}$ and define the metric $d(x,y) = \dfrac{|x - y|}{1 + |x - y|}$. For each $x \in \mathbb{R}$, we define $\|x\| = d(x, 0)$. Explain why $\|\cdot\|$ is not a norm on $\mathbb{R}$.

Example 1.7. Let $\phi : [a,b] \to \mathbb{R}^n$ be continuous, and let S be the set of continuous functions $f : [a,c] \to \mathbb{R}^n$ with $c > b$ and $f(t) = \phi(t)$ for $a \leq t \leq b$. Define $\rho(f,g) = \|f - g\| = \sup_{a \leq t \leq c} |f(t) - g(t)|$ for $f, g \in S$. Then (S, ρ) is a complete metric space but not a Banach space since $f + g$ is not in S.

Example 1.8. Let (S, ρ) be the space of continuous bounded functions $f : (-\infty, 0] \rightarrow \mathbb{R}$ with $\rho(f, g) = \|f - g\| = \sup_{-\infty < t \leq 0} |f(t) - g(t)|$. Then:

1. (S, ρ) is a Banach space.
2. The set $L = \{f \in S : \|f\| \leq 1, |f(u)f(v)| \leq |u - v|\}$ is not compact in (S, ρ).

Proof (of 2). Consider the sequence of functions defined

$$f_n(t) = \begin{cases} 0 & \text{if } t \leq -n, \\ \frac{t}{n} + 1 & \text{if } -n < t \leq 0. \end{cases}$$

Then the sequence converges pointwise to $f = 1$, but $\rho(f_n, f) = 1$ for all $n \in \mathbb{N}$. So there is no subsequence of $\{f_n\}$ converging in the norm $\| \cdot \|$ (i.e., converging uniformly) to f. □

Example 1.9. Let (S, ρ) be the space of continuous functions $f : (-\infty, 0] \rightarrow \mathbb{R}^n$ with

$$\rho(f, g) = \sum_{n=1}^{\infty} 2^{-n} \rho_n(f, g)/\{1 + \rho_n(f, g)\},$$

where

$$\rho_n(f, g) = \max_{-n \leq s \leq 0} |f(s) - g(s)|,$$

and $|\cdot|$ is the Euclidean norm on $\mathbb{R}^n$. Then:

1. (S, ρ) is a complete metric space. The distance between all functions is bounded by 1.
2. $(S, +, \cdot)$ is a vector space over $\mathbb{R}$.
3. (S, ρ) is not a Banach space because ρ does not define a norm, since $\rho(x, 0) = \|x\|$ does not satisfy $\|\alpha x\| = |\alpha| \|x\|$.
4. Let M and K be given positive constants. Then the set

$$L = \{f \in S : \|f\| \leq M \text{ on } (-\infty, 0], |f(u) - f(v)| \leq K|u - v|\}$$

is compact in (S, ρ).

Proof (of 4). Let $\{f_n\}$ be a sequence in L. It is clear that if $f_n \rightarrow f$ uniformly on compact subsets of $(-\infty, 0]$, then we have $\rho(f_n, f) \rightarrow 0$ as $n \rightarrow \infty$. Let us begin by considering $\{f_n\}$ on $[-1, 0]$. Then the sequence is uniformly bounded and equicontinuous, and so there is a subsequence, say $\{f_n^1\}$, converging uniformly to some continuous f on $[-1, 0]$. Moreover, the argument of Example 1.5 shows that $|f(t)| \leq M$ and $|f(u) - f(v)| \leq K|u - v|$. Next, we consider $\{f_n^1\}$ on $[-2, 0]$. Then the sequence is uniformly bounded and equicontinuous, and so there is a subsequence, say $\{f_n^2\}$, converging uniformly, say, to some continuous f on $[-2, 0]$.

Continuing this way, we arrive at $F_n = f_n^n$, which has a subsequence of $\{f_n\}$ converging uniformly on compact subsets of $(-\infty, 0]$ to a function $f \in L$. This proves that L is compact.   □

We leave the proof of next result to the reader.

Theorem 1.4.2. *Let $g : (-\infty, 0] \to [1, \infty)$ be a continuous strictly decreasing function with $g(0) = 1$ and $g(r) \to \infty$ as $r \to -\infty$. Let $(S, | \cdot |_g)$ be the space of continuous functions $f : (-\infty, 0] \to \mathbb{R}^n$ for which*

$$|f|_g := \sup_{-\infty < t \leq 0} \frac{|f(t)|}{|g(t)|}$$

exists. Then:

1. *$(S, | \cdot |_g)$ is a Banach space.*
2. *Let M and K be given positive constants. Then the set*

$$L = \{f \in S : \|f\| \leq M \text{ on } (-\infty, 0], |f(u) - f(v)| \leq K|u - v|\}$$

is compact in (S, ρ).

Definition 1.4.10. Let (E, ρ) be a metric space, and let $D : E \to E$. The operator or mapping D is a contraction if there exists $\alpha \in (0, 1)$ such that

$$\rho\Big(D(x), D(y)\Big) \leq \alpha \rho(x, y).$$

The next theorem is known by the name of Caccioppoli theorem, or the Banach contraction mapping principle [9]. A proof can be found in many places, such as Burton [15] or Smart [58].

Theorem 1.4.3. *(Contraction mapping principle) Let (E, ρ) be a complete metric space, and let $D : E \to E$ be a contraction operator. Then there exists a unique $\phi \in E$ such that $D(\phi) = \phi$. Moreover, if $\psi \in E$ and if $\{\psi_n\}$ is defined inductively by $\psi_1 = D(\psi)$ and $\psi_{n+1} = D(\psi_n)$, then $\psi_n \to \phi$, the unique fixed point.*

Proof. Let $y_0 \in E$ and define the sequence $\{y_n\}$ in E by $y_1 = Dy_0$, $y_2 = Dy_1 = D(Dy_0) = D^2 y_0, \ldots, y_n = Dy_{n-1} = D^n y_0$. Next, we show that $\{y_n\}$ is a Cauchy sequence. Indeed, if $m > n$, then

$$\rho(y_n, y_m) = \rho(D^n y_0, D^m y_0)$$
$$\leq \alpha \rho(D^{n-1} y_0, D^{m-1} y_0)$$
$$\vdots$$
$$\leq \alpha^n \rho(y_0, y_{m-1})$$

$$\leq \alpha^n \big\{ \rho(y_0, y_1) + \rho(y_1, y_2) + \ldots + \rho(y_{m-n-1}, y_{m-n}) \big\}$$
$$\leq \alpha^n \big\{ \rho(y_0, y_1) + \alpha \rho(y_0, y_1) + \ldots + \alpha^{m-n-1} \rho(y_0, y_1) \big\}$$
$$\leq \alpha^n \rho(y_0, y_1) \big\{ 1 + \alpha + \ldots + \alpha^{m-n-1} \big\}$$
$$\leq \alpha^n \rho(y_0, y_1) \frac{1}{1-\alpha}.$$

Thus, since $\alpha \in (0, 1)$, we have that

$$\rho(y_n, y_m) \to 0 \text{ as } n \to \infty.$$

This shows that the sequence $\{y_n\}$ is Cauchy. Since (E, ρ) is a complete metric space, $\{y_n\}$ has a limit, say y in E. Since the mapping D is continuous, we have that

$$D(y) = D(\lim_{n\to\infty} y_n) = \lim_{n\to\infty} D(y_n) = \lim_{n\to\infty} y_{n+1} = y,$$

and y is a fixed point. It remains to show that y is unique. Let $x, y \in E$ be such that $D(x) = x$ and $D(y) = y$. Then

$$0 \leq \rho(x, y) = \rho\big(D(x), D(y)\big) \leq \alpha \rho(x, y),$$

which implies that

$$0 \leq (1 - \alpha)\rho(x, y) \leq 0.$$

Since $1 - \alpha \neq 0$, we must have $\rho(x, y) = 0$, and hence $x = y$. This completes the proof. $\qquad\square$

Another form of the contraction mapping principle:

Theorem 1.4.4. *(Contraction mapping principle, Banach fixed point theorem) Let* (E, ρ) *be a complete metric space, and let* $P : E \to E$ *be such that* P^m *is a contraction for some fixed positive integer m. Then there is a unique* $x \in E$ *such that* $P(x) = x$.

1.5 Variation of parameters

In this section we develop the variation of parameters formula, which we use, in one form or another, throughout this book. Consider the nonhomogeneous differential equation

$$x'(t) = a(t)x(t) + g(t, x(t)), \ x(t_0) = x_0, \ t \geq t_0 \geq 0, \tag{1.5.1}$$

where $g \in C([0, \infty) \times \mathbb{R}, \mathbb{R})$ and $a \in C([0, \infty), \mathbb{R})$.

Multiplying both sides of (1.5.1) by the integrating factor $e^{-\int_{t_0}^t a(u)du}$ and observing that

$$\frac{d}{dt}\left(x(t)e^{-\int_{t_0}^t a(u)du}\right) = x'(t)e^{-\int_{t_0}^t a(u)du} - a(t)x(t)e^{-\int_{t_0}^t a(u)du},$$

we arrive at

$$\frac{d}{dt}\left(x(t)e^{-\int_{t_0}^t a(u)du}\right) = g(t, x(t))e^{-\int_{t_0}^t a(u)du}.$$

Integration of the above expression from t_0 to t and using $x(t_0) = x_0$ yield

$$x(t)e^{-\int_{t_0}^t a(u)du} = x_0 + \int_{t_0}^t g(s, x(s))e^{-\int_{t_0}^s a(u)du}ds,$$

from which we get

$$x(t) = x_0 e^{\int_{t_0}^t a(u)du} + \int_{t_0}^t g(s, x(s))e^{\int_s^t a(u)du}ds, \ t \geq t_0 \geq 0. \tag{1.5.2}$$

It can be easily shown that if $x(t)$ satisfies (1.5.2), then it satisfies (1.5.1). Expression (1.5.2) is known as the variation of parameters formula. Note that (1.5.2) is a functional equation in x since the integrand is a function of x. If we replace the function g with a function $h \in C([0, \infty), \mathbb{R})$, then (1.5.2) takes the special form

$$x(t) = x_0 e^{\int_{t_0}^t a(u)du} + \int_{t_0}^t h(s)e^{\int_s^t a(u)du}ds, \ t \geq t_0 \geq 0. \tag{1.5.3}$$

Another special form of (1.5.2) is that if the function $a(t)$ is constant for all $t \geq t_0$ and g is replaced with h as before, then from (1.5.3), we have that

$$x(t) = x_0 e^{a(t-t_0)} + \int_{t_0}^t e^{a(t-s)}h(s)ds, \ t \geq t_0 \geq 0. \tag{1.5.4}$$

It is easy to compute, using (1.5.4), that the differential equation

$$x'(t) = 2x(t) + t, \ x(0) = 3,$$

has the solution

$$x(t) = \frac{13}{4}e^{2t} - \frac{t}{2} - \frac{1}{4}.$$

Remark 1.6. The variation of parameters formula given by (1.5.2) is valid for

$$g \in C(\mathbb{R} \times \mathbb{R}, \mathbb{R}) \ \text{ and } \ a \in C(\mathbb{R}, \mathbb{R}).$$

Next, we consider the following application regarding the variation of parameters.

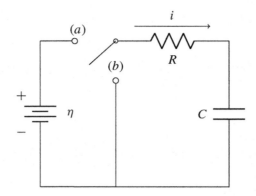

FIGURE 1.2

RC circuit.

1.5.1 **RC circuit**

Fig. 1.2 shows a resistor R and capacitor C connected in a series. The battery connected to this circuit by a switch provides an electromotive force, or *emf* force η. Initially, there is no charge on the capacitor. When the switch is flipped to (a), the battery connects, and the capacitor charges. Similarly, when the switch is flipped to (b), the battery disconnects, and the capacitor discharges with energy dissipated in the resistor. Our aim is to determine the voltage drop across the capacitor during charging and discharging.

We start by observing that the voltage drops across a capacitor V_C and resistor V_R are given by

$$V_C = \frac{q}{C}, \quad V_R = iR,$$

respectively, where C is the capacitance and R is the resistance. The current i and the charge q are related by the relation

$$i = \frac{dq}{dt}.$$

From the first equation, we have

$$\frac{dV_C}{dt} = \frac{dq}{dt} \cdot \frac{1}{C} = \frac{i}{C} = \frac{1}{C} \frac{V_R}{R},$$

which implies that

$$V_R = RC \frac{dV_C}{dt}.$$

Kirchhoff's voltage law states that the *emf* η in any closed loop is equal to the sum of the voltage drops in that loop. When the switch is thrown to (a), this gives

$$V_R + V_C = \eta.$$

Substituting V_R into $V_R + V_C = \eta$ gives the linear differential equation

$$\frac{dV_C}{dt} + V_C \frac{1}{RC} = \eta \frac{1}{RC}, \quad V_C(0) = 0.$$

An application of the variation of parameters formula gives the solution

$$V_C(t) = e^{-t/RC} \int_0^t \frac{\eta}{RC} e^{s/RC} ds,$$

or

$$V_C(t) = \eta \left(1 - e^{-t/RC} \right).$$

Thus the voltage starts at zero and rises slowly but exponentially to η with characteristic time scale given by RC. In other words,

$$V_C(t) \to \eta \text{ as } t \to \infty.$$

Now suppose the switch is thrown to (b). Then by Kirchhoff's voltage law

$$V_R + V_C = 0,$$

which results in the differential equation

$$\frac{dV_C}{dt} + V_C \frac{1}{RC} = 0, \quad V_C(0) = 0.$$

Then the solution during the discharge phase is given by

$$V_C(t) = \eta e^{-t/RC},$$

and the voltage decays exponentially to zero with the characteristic time scale given by RC.

1.6 Special differential equations

Next, we consider second-order nonlinear differential equations of the form

$$x'' = f(t, x, x'), \tag{1.6.1}$$

where $f(t, x, y)$ is a function of three variables defined in the region $D = \{(t, x, y) : a_1 < t < a_2, b_1 < x < b_2, c_1 < y < c_2\}$ with f, $\frac{\partial f}{\partial x}$, and $\frac{\partial f}{\partial y}$ continuous in D. Little is known about how to solve (1.6.1) except in certain particular cases. Two interesting and useful cases occur when (1.6.1) has one of the following forms:

$$F(t, x', x'') = 0 \tag{1.6.2}$$

or

$$F(x, x', x'') = 0. \tag{1.6.3}$$

For (1.6.2), the transformation $x' = u$ transforms the original second-order differential equation into a first-order differential equation $F(t, u, u') = 0$, which can be solved. To see this, we consider the nonlinear second-order differential equation

$$tx'' + 2x' + t = 1, \; t > 0.$$

For $x' = u$, we have the first-order differential equation $tu' + 2u + t = 1$. Using the variation of parameters formula given by (1.5.2), we arrive at

$$u = \frac{c_1}{t^2} + \frac{1}{2} - \frac{t}{3}$$

for some constant c_1. Since $x(t) = \int u(t)dt$, we arrive at the general solution

$$x(t) = -\frac{c_1}{t} + \frac{t}{2} - \frac{t^2}{6} + c_2$$

for some constant c_2.

Equations of the form (1.6.3) can be reduced to first-order equations by the substitution $x' = u$. Eq. (1.6.3) is now replaced by the equivalent system of two differential equations

$$x' = u, \quad F(x, u, u') = 0.$$

By using x as the independent variable we have

$$x'' = \frac{du}{dt} = \frac{du}{dx}\frac{dx}{dt} = \frac{du}{dx}u.$$

Hence (1.6.3) reduces to the first-order differential equation

$$F(x, u, u\frac{du}{dx}) = 0.$$

As an application, we consider the differential equation

$$x'' + (x')^2 x + x'x = 0, \; x(0) = 0, \; x'(0) = -1.$$

Then, under the mentioned substitution, we have the new first-order differential equation

$$u\frac{du}{dx} + u^2 x + ux = 0$$

or

$$\frac{du}{dx} + xu = -x.$$

By the variation of parameters, we arrive at the solution

$$u = -1 + ce^{\frac{x^2}{2}}.$$

Using $u = -1$ when $x = 0$, we have $c = 0$. Now using the transformation $x' = u$, it follows that

$$\frac{dx}{dt} = -1.$$

Using $x(0) = 0$, we arrive at the solution

$$x(t) = -t.$$

1.7 Exercises

Exercise 1.1. Show that the differential equation

$$x'(t) = \frac{3}{2} x^{1/3}(t), \ x(0) = 0, \ t \in \mathbb{R}$$

has infinitely many solutions.

Exercise 1.2. Show that the differential equation

$$x'(t) = x^{\alpha}(t), \ x(0) = 0, \ t \in \mathbb{R}$$

has infinitely many solutions for $\alpha \in (0, 1)$.

Exercise 1.3. Find the escape time of the solutions for the differential equation

$$x'(t) = x^2(t), \ x(t_0) = x_0 \neq 0, \ t \in \mathbb{R}.$$

Exercise 1.4. Show that the solution $x(t, t_0, x_0)$ of

$$x'(t) = t, \ x(t_0) = x_0 \neq 0, \ t \geq 0$$

is stable but unbounded.

Exercise 1.5. Let $x = (x_1, x_2, \ldots, x_n), x_i \in \mathbb{R}, i = 1, 2, \ldots, n$.

Show that $\|x\| = \displaystyle\sum_{i=1}^{n} x_i^2$ does not define a norm.

Exercise 1.6. Consider the following power series $L(x)$, which is also known as Euler's dilogarithm function:

$$L(x) = \sum_{n=1}^{\infty} \frac{x^n}{n^2}.$$

(a) Compute the domain of convergence for $L(x)$. Be sure to give a full analysis of the endpoints.
(b) Show that $L(x)$ uniformly converges on its entire domain of convergence. (Use Weierstrass M-test.)
(c) Explain why the $L(x)$ is continuous on its domain of convergence.

Exercise 1.7. Consider the sequence $\{f_n\}$ of functions $f_n : \mathbb{R} \to \mathbb{R}$ defined by

$$f_n(x) = \frac{nx}{\sqrt{1 + n^2 x^2}}.$$

Find the pointwise limit of this sequence as $n \to \infty$. Does the sequence converge uniformly on $\mathbb{R}$?

Exercise 1.8. Show that the space $C(I, \mathbb{R})$, where I is any compact subset of $\mathbb{R}$, endowed with the supremum norm

$$\|f\|_\infty = \sup_{t \in I} |f(t)|$$

is a Banach space.

Exercise 1.9. Show that the space Ω_T of all continuous T-periodic functions given by

$$\Omega_T = \{f \in C(\mathbb{R}, \mathbb{R}) : f(t + T) = f(T) \text{ for all } t \in \mathbb{R}\}$$

with the maximum norm

$$\|f\| = \max_{t \in [0,T]} |f(t)|$$

is a Banach space.

Exercise 1.10. Show that the space $C([0, 1], \mathbb{R})$ endowed with the L^1 norm

$$\|f\|_1 = \int_0^1 |f(t)| dt$$

is a metric space but is not complete.

Exercise 1.11. For $1 \le p < \infty$, let

$$l_p = \{a = (a_1, a_2, \ldots) : a_k \in \mathbb{R}, \sum_k |a_k|^p < \infty\}$$

and

$$||a||_p = \left(\sum_k |a_k|^p \right)^{1/p}.$$

Show that l_p endowed with $|| \cdot ||_p$ is a Banach space.

Exercise 1.12. Prove Remark 1.1.

Exercise 1.13. Find a continuous solution satisfying

$$x'(t) + x(t) = f(t), \ x(0) = 0,$$

where $f(t) = \begin{cases} 1, & 0 \leq t \leq 1, \\ 0, & t > 1. \end{cases}$

Is the solution differentiable at $t = 1$?

Exercise 1.14. Find a continuous solution satisfying

$$x'(t) + 2tx(t) = f(t), \ x(0) = 2,$$

where $f(t) = \begin{cases} t, & 0 \leq t < 1, \\ 0, & t \geq 1. \end{cases}$

Is the solution differentiable at $t = 1$?

Exercise 1.15. For the SI epidemic model of Section 1.3, use Eq. (1.3.1) and the relation $I + S = 1$ to obtain a first-order differential equation in terms of S and S'. Assume a positive condition $S(0) = 1 - I(0) := S_0$ and find the solution $S(t)$ and $\lim_{t \to \infty} S(t)$.

Exercise 1.16. Let $d > 0$, and let $f : [0, d] \to \mathbb{R}$ be continuous. Let k be any constant and suppose $x(t)$ solves the differential inequality

$$x'(t) \leq kx(t) + f(t), \ t \in [0, d].$$

Show that

$$x(t) \leq x(0)e^{kt} + \int_0^t e^{k(t-s)} f(s)ds, \ t \in [0, d].$$

Hint: There is a continuous and nonnegative function $g(t)$ such that

$$x'(t) + g(t) = kx(t) + f(t), \ t \in [0, d].$$

Exercise 1.17. Suppose a function $f : [0, d] \times \mathbb{R} \to \mathbb{R}$ is continuous and that $f(t, x)$ is bounded for all $(t, x) \in [0, d] \times \mathbb{R}$, where d is a positive constant. Suppose $x = \varphi(t)$ solves the differential equation

$$x'(t) = xf(t, x), \ x(0) = 1, \ t \in [0, d].$$

Show that there is a constant K such that $\varphi(t) \leq e^{Kt}$ for all $t \in [0, d]$.

Exercise 1.18. In Problems 1–5 solve the differential equation. If no initial conditions are given, find the general solution.

1. $t^2 x'' + t = 1, t > 0$.
2. $t^2 x'' + 2tx' = 1, t > 0$.
3. $xx'' - 2(x')^2 + 4x^2 = 0, x(1) = 1, x'(1) = 2$.
4. $\sin(x')x'' = \sin(x), x(1) = 2, x'(1) = 1$.
5. $tx'' - x' = t^2 e^t, t > 0$.

Existence and uniqueness

In Chapter 1 we laid out the basics we need to use in this and the following chapters. This chapter is devoted to the existence and uniqueness and the continuation of solutions. We limit our discussions and proofs to scalar differential equations, and in Chapter 3 we indicate how the theory can be naturally extended to vector equations.

2.1 Existence and uniqueness of solutions

This section is devoted to the existence and uniqueness of solutions of the initial value problem (IVP)

$$x' = f(t, x), \quad x(t_0) = x_0, \tag{2.1.1}$$

where we assume that $f : D \to \mathbb{R}$ is continuous and D is a subset of $\mathbb{R} \times \mathbb{R}$. In the case the differential equation (2.1.1) is linear, a solution can be found. However, in general, this approach is not feasible when the differential equation is not linear, and hence another approach must be indirectly adopted that establishes the existence of a solution of (2.1.1). For the development of the existence theory, we need a broader definition of Lipschitz condition.

Definition 2.1.1. The function $f : D \to \mathbb{R}$ is said to satisfy the *global* Lipschitz condition in x if there exists a Lipschitz constant $k > 0$ such that

$$|f(t, x) - f(t, y)| \le k|x - y| \text{ for } (t, x), (t, y) \in D. \tag{2.1.2}$$

Definition 2.1.2. The function $f : D \to \mathbb{R}$ is said to satisfy a *local* Lipschitz condition in x if for any $(t_1, x_1) \in D$, there exists a domain $D_1 \subset D$ such that $f(t, x)$ satisfies a Lipschitz condition in x on D_1, that is, there exists a positive constant k_1 such that

$$|f(t, x) - f(t, y)| \le K_1|x - y|, \text{ for } (t, x), (t, y) \in D_1. \tag{2.1.3}$$

Definition 2.1.1 can be easily extended to functions $f : D \to \mathbb{R}^n$, where $D \subset \mathbb{R} \times \mathbb{R}^n$ under a proper norm. Let $R = \{(t, x) : |t| \le a, |x| \le b\}$ be any rectangle in D. If f and $\frac{\partial f}{\partial x}$ are continuous on R, which is the case in this chapter, then f and $\frac{\partial f}{\partial x}$ are bounded on R. Therefore there exist positive constants $\mathcal{M}$ and $\mathcal{K}$ such that

$$|f(t, x)| \le \mathcal{M} \quad \text{and} \quad |\frac{\partial f}{\partial x}| \le \mathcal{K} \tag{2.1.4}$$

Advanced Differential Equations. https://doi.org/10.1016/B978-0-32-399280-0.00008-5

for all points (t, x) in R. Now for any two points (t, x_1), (t, x_2) in R, by the mean value theorem there exists a constant $\eta \in (x_1, x_2)$ such that

$$f(t, x_1) - f(t, x_2) = \frac{\partial f}{\partial x}(t, \eta)(x_1 - x_2),$$

from which it follows that

$$|f(t, x_1) - f(t, x_2) \le |\frac{\partial f}{\partial x}(t, \eta)||x_1 - x_2|$$
$$\le \mathscr{K}|x_1 - x_2|. \tag{2.1.5}$$

We have shown that if f and $\frac{\partial f}{\partial y}$ are continuous on R, then f satisfies a global Lipschitz condition on R.

Example 2.1. Consider $f(t, x) = x^{2/3}$ in the rectangle $R = \{(t, x) : |t| \le 1, |x| \le 1\}$. We claim f does not satisfy the Lipschitz condition on R. Consider the pair of points (t, x_1) and $(t, 0)$ in R where $x_1 > 0$. Then

$$\frac{|f(t, x_1) - f(t, 0)}{x_1 - 0}| = x_1^{-1/3} \to \infty \text{ as } x_1 \to 0^+,$$

and hence there exists no $\mathscr{K}$ such that (2.1.5) holds.

Remark 2.1. Consider (2.1.1) on the rectangle $D \subset \mathbb{R} \times \mathbb{R}$ defined by

$$D = \{(t, x) : |t - t_0| \le a, \ |x - x_0| \le b\},$$

where a and b are positive constants. Let

$$\tau = t - t_0, \ u = x - x_0, \text{ and } g(\tau, u) = f(\tau + t_0, u + x_0).$$

Then the function g is defined on the rectangle

$$D^* = \{(\tau, u) : -a \le \tau \le a, \ -b \le u \le b\},$$

and the IVP (2.1.1) is equivalent to the IVP problem

$$u' = g(\tau, u), \ u(0) = 0.$$

For emphasis, we restate the following definition.

Definition 2.1.3. We say that x is a solution of (2.1.1) on an interval I if $x : I \to \mathbb{R}$ is differentiable, $(t, x(t)) \in D$ for $t \in I$, $x'(t) = f(t, x(t))$ for $t \in I$, and $x(t_0) = x_0$ for $(t_0, x_0) \in D$.

In preparation for the next theorem, we observe that the IVP (2.1.1) is equivalent to

$$x(t) = x_0 + \int_{t_0}^{t} f(s, x(s))ds. \tag{2.1.6}$$

Relation (2.1.6) is an *integral equation* since it contains an integral of the unknown function x. This integral is not a formula for the solution, but rather it provides another relation satisfied by a solution of (2.1.1). We have the following definition.

Definition 2.1.4. We say $x : I \to \mathbb{R}$ is a solution of the integral equation (2.1.6) on an interval I if $t_0 \in I$, x is continuous on I, $(t, x(t)) \in D$ for $t \in I$, and (2.1.6) is satisfied for $t \in I$.

The next theorem is fundamental for the proof of the existence theorems.

Theorem 2.1.1. *Let D be an open subset of $\mathbb{R}^2$, and let $(t_0, x_0) \in D$. Then x is a solution of (2.1.1) on an interval I if and only if x satisfies the integral equation given by (2.1.6) on I.*

Proof. Let $x(t)$ be a solution of (2.1.1) on an interval I. Then $t_0 \in I$, x is differentiable on I, and hence x is continuous on I. Moreover, $(t, x(t)) \in D$ for $t \in I$, $x(t_0) = x_0$, and $x'(t) = f(t, x(t))$ for $t \in I$. Now an integration of $x'(t) = f(t, x(t))$ from t_0 to t gives (2.1.6) for $t \in I$. For the converse, if x satisfies (2.1.6) for $t \in I$, then $t_0 \in I$, and x is continuous on I. Moreover, $(t, x(t)) \in D$ for $t \in I$, and (2.1.6) is satisfied for $t \in I$. Thus $x(t)$ is differentiable on I. By differentiating (2.1.6) with respect to t, we arrive at $x'(t) = f(t, x(t))$ for all $t \in I$ and $x(t_0) = x_0 + \int_{t_0}^{t_0} f(s, x(s))ds = x_0$. This completes the proof. $\qquad\square$

A graduate textbook in differential equations would be incomplete without the statement and proof of Picard's local existence and uniqueness using successive approximations. In the next section we prove similar theorems on Banach spaces using fixed point theory.

We note that Picard's local existence and uniqueness theorem and Cauchy–Peano existence theorem are widely used, and variant proofs of the theorems can be found in [17], [29], [31], [54], and [70].

The heart of proving our next results is the construction of a sequence of functions that converge to a limit function that satisfies the IVP (2.1.1), although the members of the sequence individually do not.

Theorem 2.1.2 addresses the three basic issues:

(1) Members of the constructed sequence $\{x_n\}$ exist for all time.

(2) The sequence converges, and the limiting function satisfies the integral equation (2.1.6) and hence the IVP (2.1.1).

(3) The limiting function, which is a solution of (2.1.1), is unique.

Theorem 2.1.2. *(Picard's local existence and uniqueness) Let $D \subset \mathbb{R} \times \mathbb{R}$ be defined as*

$$D = \{(t, x) : |t - t_0| \leq a, \ |x - x_0| \leq b\},$$

where a and b are positive constants. Assume that $f \in C(D, \mathbb{R})$ and f satisfies the Lipschitz condition (2.1.2). Let

$$M = \max_{(t,x) \in D} |f(t, x)| \tag{2.1.7}$$

and

$$h = \min\{a, \frac{b}{M}\}. \tag{2.1.8}$$

Then the IVP (2.1.1) has a unique solution, denoted by $x(t, t_0, x_0)$, on the interval $|t - t_0| \leq h$ and passing through (t_0, x_0). Furthermore,

$$|x(t) - x_0| \leq b \ for \ |t - t_0| \leq h.$$

Proof. Before we begin the proof, we recommend to look at Fig. 2.1.

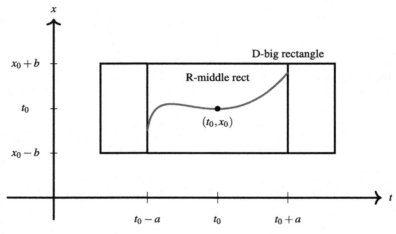

FIGURE 2.1

Interval of existence.

First, we note that since f is continuous on D, condition (2.1.7) is automatically implied. The idea of the proof is to construct a sequence of functions $\{x_n\}$ that converges uniformly to a unique x on the interval $|t - t_0| \leq h$ and is a solution of (2.1.1). We begin by successively defining a sequence of functions $\{x_n\}$ for $|t - t_0| \leq h$ by setting

$$x_0(t) = x_0,$$

$$x_1(t) = x_0 + \int_{t_0}^{t} f(s, x_0(s))ds,$$

$$\vdots$$

$$x_n(t) = x_0 + \int_{t_0}^{t} f(s, x_{n-1}(s))ds. \tag{2.1.9}$$

According to (2.1.7) and (2.1.8), we have that

$$|x_1(t) - x_0| \leq hM \leq b \ \text{for} \ |t - t_0| \leq h.$$

Therefore $\int_{t_0}^{t} f(s, x_1(s)) ds$ can be defined for $|t - t_0| \leq h$, and hence

$$|x_2(t) - x_0| \leq hM \leq b.$$

Similarly, we may show that $x_3(t), \ldots, x_n(t)$ are well defined on $|t - t_0| \leq h$. Thus

$$|x_n(t) - x_0| = |\int_{t_0}^{t} f(s, x_{n-1}(s)) ds| \leq M|t - t_0|. \tag{2.1.10}$$

Next, we show that $\{x_n\}$ converges uniformly to a function, say x, on $|t - t_0| \leq h$. Note that

$$x_n(t) = x_0(t) + [x_1(t) - x_0(t)] + [x_2(t) - x_1(t)] + \cdots + [x_n(t) - x_{n-1}(t)],$$

or

$$x_n(t) = x_0(t) + \sum_{j=0}^{n-1} [x_{j+1}(t) - x_j(t)].$$

Taking the limit if it exists, we have

$$\lim_{n \to \infty} x_n(t) = x_0(t) + \lim_{n \to \infty} \sum_{j=0}^{n-1} [x_{j+1}(t) - x_j(t)]. \tag{2.1.11}$$

Our next task is to compute $|x_{j+1}(t) - x_j(t)|$. Using (2.1.9) and then (2.1.2) followed by (2.1.10) give

$$|x_2(t) - x_1(t)| \leq \int_{t_0}^{t} |f(s, x_1(s)) - f(s, x_0(s))| ds$$

$$\leq K \int_{t_0}^{t} |x_1(s) - x_0(s)| ds$$

$$\leq KM \int_{t_0}^{t} |s - t_0| ds = KM \int_{t_0}^{t} (s - t_0) ds$$

$$= KM \frac{(t - t_0)^2}{2}, \quad |t - t_0| \leq a.$$

In a similar fashion, we obtain

$$|x_3(t) - x_2(t)| \leq \int_{t_0}^{t} |f(s, x_2(s)) - f(s, x_1(s))| ds$$

$$\leq K \int_{t_0}^{t} |x_2(s) - x_1(s)| ds$$

$$\leq K \int_{t_0}^{t} KM \frac{(s-t_0)^2}{2} ds$$

$$= MK^2 \frac{(t-t_0)^3}{3!}, \quad |t-t_0| \leq a.$$

Continuing this way, we arrive at

$$|x_j(t) - x_{j-1}(t)| \leq MK^{j-1} \frac{(t-t_0)^j}{j!}. \tag{2.1.12}$$

To complete the induction argument, we assume that (2.1.12) holds for j and show that it holds for $j+1$. Using (2.1.12), we arrive at

$$|x_{j+1}(t) - x_j(t)| \leq \int_{t_0}^{t} \left| f(s, x_j(s)) - f(s, x_{j-1}(s)) \right| ds$$

$$\leq K \int_{t_0}^{t} \left| x_j(s) - x_{j-1}(s) \right| ds$$

$$\leq K \int_{t_0}^{t} MK^{j-1} \frac{(s-t_0)^j}{j!} ds$$

$$= MK^j \frac{(t-t_0)^{j+1}}{j!(j+1)} = MK^j \frac{(t-t_0)^{j+1}}{(j+1)!}, \quad |t-t_0| \leq a.$$

Next, we substitute (2.1.12) into (2.1.11) and get

$$\lim_{n \to \infty} x_n(t) = x_0 + \frac{M}{K} \lim_{n \to \infty} \sum_{j=0}^{n-1} \frac{|K(t-t_0)|^{j+1}}{(j+1)!}$$

$$\leq x_0 + \frac{M}{K} \lim_{n \to \infty} \sum_{j=0}^{n-1} \frac{(Kh)^{j+1}}{(j+1)!}$$

$$\leq x_0 + \frac{M}{K} \left(e^{Kh} - 1 \right) < \infty.$$

Hence, by the Weierstrass M-test, $\{x_n(t)\}$ converges uniformly, say to a function $x(t)$. Next, we show that $x(t)$ is a solution of (2.1.1) on D. We have

$$|x(t) - x_0| = |x(t) - x_n(t) + x_n(t) - x_0|$$

$$\leq |x(t) - x_n(t)| + |x_n(t) - x_0|$$

$$\leq |x(t) - x_n(t)| + M|t - t_0|$$

for every fixed $t \in (t_0 - h, t_0 + h)$. As a result,

$$|x(t) - x_0| = \lim_{x \to \infty} |x(t) - x_n(t)| + M|t - t_0|$$

$$= 0 + M|t - t_0|$$
$$\leq Mh \leq b$$

for $t \in (t_0 - h, t_0 + h)$. So $x(t) \in D$. If the sequence $\{x_n(t)\}$ converges uniformly and $\{x_n(t)\}$ is continuous on the interval $|t - t_0| \leq h$, then

$$\lim_{n \to \infty} \int_{t_0}^{t} f(s, x_n(s))ds = \int_{t_0}^{t} \lim_{n \to \infty} f(s, x_n(s))ds$$

by the uniform continuity of f. As a consequence,

$$x(t) = \lim_{n \to \infty} x_{n+1}(t)$$

$$= x_0 + \lim_{n \to \infty} \int_{t_0}^{t} f(s, x_n(s))ds$$

$$= x_0 + \int_{t_0}^{t} \lim_{n \to \infty} f(s, x_n(s))ds$$

$$= x_0 + \int_{t_0}^{t} f(s, \lim_{n \to \infty} x_n(s))ds$$

$$= x_0 + \int_{t_0}^{t} f(s, x(s))ds,$$

that is,

$$x(t) = x_0 + \int_{t_0}^{t} f(s, x(s))ds, \quad |t - t_0| \leq h. \tag{2.1.13}$$

The integrand $f(s, x(s))$ in (2.1.13) is a continuous function, and hence $x(t)$ is differentiable with respect to t, and its derivative is equal to $f(t, x(t))$. So the proof of the existence is complete. It remains to show that the solution $x(t)$ on D is unique. Let $y(t)$ be another solution of (2.1.1) with $y(t_0) = x_0$. Then

$$y(t) = x_0 + \int_{t_0}^{t} f(s, y(s))ds.$$

Let

$$N = \sup_{|t - t_0| \leq h} |x(t) - y(t)|.$$

Then by the Lipschitz condition, we have

$$|x(t) - y(t)| \leq K \int_{t_0}^{t} |x(s) - y(s)|ds \tag{2.1.14}$$

or

$$|x(t) - y(t)| \leq KN|t - t_0|.$$

The substitution of this estimate into (2.1.14) yields

$$|x(t) - y(t)| \leq KN|t - t_0|^2/(2!) \text{ for } |t - t_0| \leq h.$$

Repeating this substitution, we obtain

$$|x(t) - y(t)| \leq KN\frac{|t - t_0|^m}{m!}, m = 1, 2, \ldots,$$

for $|t - t_0| \leq h$. Since the right side of this inequality tends to zero as $m \to \infty$, we have that

$$N = \sup_{|t - t_0| \leq h} |x(t) - y(t)| = 0.$$

This completes the proof.   □

We will give another proof of the uniqueness once Gronwall's inequality is introduced. To illustrate the above procedure, we provide the following examples.

Example 2.2. We consider the IVP

$$x'(t) = 2t(1 + x), \ x(0) = 0.$$

Set $x_0(t) = 0$. Then for any $n \geq 1$, we have the recurrent formula

$$x_n(t) = \int_0^t 2s(1 + x_{n-1}(s))ds.$$

For $n = 1$, we have

$$x_1(t) = \int_0^t 2s\,ds = t^2,$$

and for $n = 2$,

$$x_2(t) = \int_0^t 2s(1 + s^2)ds = t^2 + \frac{t^4}{2}.$$

We leave it to the reader to verify that

$$x_n(t) = t^2 + \frac{t^4}{2} + \frac{t^6}{3!} + \ldots, \frac{t^{2n}}{n!} = \sum_{i=0}^n \frac{(t^2)^i}{i!} - 1.$$

Definition 2.1.5. A sequence $\{x_n\}$ of functions in $C([a, b], \mathbb{R})$ converges uniformly to $x \in C([a, b], \mathbb{R})$ if $\lim_{n \to \infty} ||x_n - x|| = 0$.

It was established in Example 2.2 that

$$x_n(t) = \sum_{i=0}^n \frac{(t^2)^i}{i!} - 1$$

and the true solution of the IVP is given by $x(t) = e^{t^2} - 1$ on the interval $[0, 1]$. Thus

$$\lim_{n \to \infty} ||x_n - x|| = \lim_{n \to \infty} \left[\max_{t \in [0,1]} \left| \sum_{i=0}^{n} \frac{(t^2)^i}{i!} - 1 - (e^{t^2} - 1) \right| \right] = \lim_{n \to \infty} \sum_{i=n+1}^{\infty} \frac{1}{i!} = 0.$$

Therefore

$$\lim_{n \to \infty} x_n(t) = \lim_{n \to \infty} \sum_{i=0}^{n} \frac{(t^2)^i}{i!} - 1 = e^{t^2} - 1.$$

Example 2.3. Next, we use Theorem 2.1.2 to find the interval of existence of the unique solution of

$$x'(t) = 1 + x^2, \ x(0) = 1.$$

Let

$$D = \{(t, x) : |t| \leq a, \ |x - 1| \leq b\},$$

where a and b are constants. It is clear that

$$M = \max_{(t,x) \in D} |f(t, x)| = 1 + (1 + b)^2$$

and

$$h = \min\{a, \frac{b}{M}\} = \min\{a, \frac{b}{1 + (1 + b)^2}\}.$$

So we must find the maximum of the function $g(b) = \dfrac{b}{1 + (1 + b)^2}$. Using calculus, we see that the function g has its maximum at $b = \sqrt{2}$ and is given by $g(\sqrt{2}) = \frac{\sqrt{2}}{1+(1+\sqrt{2})^2}$. Thus the interval of existence and uniqueness is

$$|t| \leq \frac{\sqrt{2}}{1 + (1 + \sqrt{2})^2}, \ \text{ or } t \in \left[-\frac{\sqrt{2}}{1 + (1 + \sqrt{2})^2}, \frac{\sqrt{2}}{1 + (1 + \sqrt{2})^2} \right].$$

Now suppose in Example 2.3 we chose $D = \{(t, x) : |t| \leq 10, \ |x - 1| \leq b\}$, that is, $a = 10$. Then the interval of existence found above is much smaller or included in $(-10, 10)$. Recall that the interval of existence is included in D.

The next corollary is an immediate consequence of Theorem 2.1.2 and the discussion leading to (2.1.5).

Corollary 2.1. *Suppose* $D \subset \mathbb{R} \times \mathbb{R}$, $f \in C(D, \mathbb{R})$, *and* $\frac{\partial f}{\partial x}$ *is continuous on* D. *Then for any* $(t_0, x_0) \in D$, *the IVP* (2.1.1) *has a unique solution on an interval containing* t_0 *in its domain.*

As an example, consider

$$x'(t) = tx^{1/2}.$$

Then

$$f(t, x) = tx^{1/2} \quad \text{and} \quad \frac{\partial f}{\partial x} = \frac{t}{2x^{1/2}}$$

are continuous in the upper half-plane defined by $x > 0$. We conclude from Corollary 2.1 that for any point (t_0, x_0), $x_0 > 0$, there is some interval around t_0 on which the given differential equation has a unique solution.

Unlike Theorem 2.1.2, the next Cauchy–Peano theorem only guarantees the existence of a solution and not uniqueness. This is due to the fact that f is not required to be Lipschitz.

Based on Remark 2.1, we may restate Theorem 2.1.2.

Theorem 2.1.3. *If f and $\frac{\partial f}{\partial x}$ are continuous on the rectangle*

$$D^* = \{(t, x) : -a \le t \le a, \ -b \le x \le b\},$$

where a and b are positive constants, then there is a number h in the interval $(0, a]$ such that the IVP

$$x' = f(t, x), \quad x(0) = 0,$$

has a unique solution, denoted by $x(t, 0, 0)$, on the interval $-h \le t \le h$. Furthermore,

$$|x(t)| \le b \text{ for } |t| \le h.$$

Example 2.4. As an application, we consider the IVP

$$x' = t(x + 1)^2, \quad x(0) = 0,$$

on the rectangle

$$D^* = \{(t, x) : -1 \le t \le 1, \ -2 \le x \le 2\}.$$

It readily follows that the IVP has the solution $x(t) = \dfrac{t^2}{2 - t^2}$. We want this solution to be in D^*. This means that $|t| \le 1$ and $\dfrac{t^2}{2 - t^2} \le 2$. This is true if and only if $-\dfrac{2}{\sqrt{3}} \le t \le \dfrac{2}{\sqrt{3}}$, from which it follows that $h = \dfrac{2}{\sqrt{3}}$. If we choose a larger b, then we can arrive at a larger h. However, h must be chosen so that $h < \sqrt{2}$. This is due to the fact that

$$\lim_{t \to \left(\sqrt{2}\right)^-} \frac{t^2}{2 - t^2} = \infty.$$

Theorem 2.1.4. *(Cauchy–Peano existence theorem) Let $f : \mathbb{R} \times \mathbb{R} \to \mathbb{R}$ be continuous in a neighborhood of a point $(t_0, x_0) \in \mathbb{R}^2$. Then there exists $\alpha > 0$ such that the*

IVP (2.1.1) has a solution ϕ on the interval $I := [t_0 - \alpha, t_0 + \alpha]$, that is, there exists a function (not necessarily unique) $\phi = \phi(t)$ defined on I such that

$$\phi'(t) = f(t, \phi(t)), \ t \in I,$$

and $\phi(t_0) = x_0$.

Proof. Since f is continuous in a neighborhood of the point $(t_0, x_0) \in \mathbb{R}^2$, there exists a positive constant a such that the rectangle

$$D = \{(t, x) : |t - t_0| \le a, \ |x - x_0| \le a\}$$

is centered at (t_0, x_0). Moreover, $M = \max_{(t,x) \in D} |f(t, x)|$ exists. Next, we set $\alpha := \dfrac{a}{M}$. By Theorem 2.1.1, we know that $\phi = \phi(t)$ satisfies the IVP (2.1.1) if and only if

$$\phi(t) = x_0 + \int_{t_0}^{t} f(s, \phi(s))ds. \tag{2.1.15}$$

Since f is uniformly continuous on the compact set D, given $\varepsilon > 0$, there exists $\delta = \delta(\varepsilon)$ such that

$$|t - \tilde{t}| < \delta \ \text{ and } \ |x - \tilde{x}| < \delta \ \text{ imply } \ |f(t, x) - f(\tilde{t}, \tilde{x})| < \varepsilon$$

for all $(t, x), (\tilde{t}, \tilde{x}) \in D$. Therefore let $\varepsilon = \varepsilon_n = \dfrac{1}{n}$ and $\delta = \delta_n$, $n = 1, 2, \ldots$. Choose points $t_j^{(n)}$ with $t_0^{(n)} = t_0$ and $t_{k(n)} = t_0 + \alpha$ for $0 \le j \le k(n)$ with $|t_{j+1}^{(n)} - t_j^{(n)}| \le \dfrac{\delta_n}{M}$. We define the approximation ϕ_n on $[x_0, x_0 + \alpha]$ by $\phi_n(t_0) = x_0$ and $\phi_n'(t) = f(t_0, x_0)$, $t_0 \le t \le t_1^{(n)}$. Then it is clear that $x_1^{(n)} = \phi_n(t_1^{(n)})$. Continuing in this manner, we get that

$$\phi_n'(t) = f(t_j^{(n)}, x_j^{(n)}) \ \text{ for } \ t_j^{(n)} \le t \le t_{j+1}^{(n)}. \tag{2.1.16}$$

Note that ϕ_n is piecewise continuous and may have a jump discontinuity in its derivative at the partition points, and hence we define

$$\triangle_n(t) = \begin{cases} \phi_n'(t) - f(t, \phi_n(t)), & t_j^{(n)} < t < t_{j+1}^{(n)}, \\ 0, & t = x_j^{(n)}. \end{cases}$$

As a consequence, we may define

$$\phi_n(t) = x_0 + \int_{t_0}^{t} \phi_n'(s)ds = x_0 + \int_{t_0}^{t} [f(s, \phi_n(s)) + \triangle_n(s)]ds.$$

Next, we try to have an upper bound on $\triangle_n(t)$. Note that from (2.1.16) and the definition of $\triangle_n(t)$ we have that for $t_j^{(n)} < t < t_{j+1}^{(n)}$,

$$|\Delta_n(t)| = \left| f(t_j^{(n)}, x_j^{(n)}) - f(t, \phi_n(t)) \right|, \; t_j^{(n)} < t < t_{j+1}^{(n)}$$

$$\leq |x_j^{(n)} - \phi_n(t)| \leq |x_j^{(n)} - x_{j+1}^{(n)}|$$

$$\leq M|t_j^{(n)} - t_{j+1}^{(n)}|$$

$$\leq M \frac{\delta_n}{M} = \delta_n = \varepsilon_n = \frac{1}{n},$$

where we have used that

$$|s - t_j^{(n)}| \leq |x_{j+1}^{(n)} - x_j^{(n)}| \leq \delta_n.$$

By the Ascoli–Arzelà theorem, there exists a uniformly convergent subsequence of ϕ_n that converges to

$$\phi(t) = \lim_{k \to \infty} \phi_{n_k}(t) \; \text{ for } \; t \in [t_0 - \alpha, t_0 + \alpha],$$

that is,

$$\phi_{n_k}(t) = x_0 + \int_{t_0}^t [f(s, \phi_{n_k}(s)) + \Delta_n(s)] ds,$$

which converges to

$$\phi(t) = x_0 + \int_{t_0}^t [f(s, \phi(s)) ds.$$

This completes the proof. $\qquad\qquad\qquad\qquad\qquad\qquad\square$

Example 2.5. Consider the IVP

$$x'(t) = 2tx^{2/3}, \; x(0) = 0.$$

Then $f(t, x) = 2tx^{2/3}$ is continuous everywhere, whereas $\dfrac{\partial f(t, x)}{\partial x} = \dfrac{4t}{3x^{1/3}}$ is not continuous at the points where $x = 0$. Here $x_0 = 0$, and no matter how small we chose $b > 0$, the interval $[-b, b]$ will always contain 0. Therefore we cannot apply Theorem 2.1.2. Instead, we will use Theorem 2.1.4. Let $a = 1$. Then on the rectangle

$$D = \{(t, x) : |t| \leq 1, \; |x| \leq 1\},$$

we have

$$M = \max_{(t,x) \in D} |f(t, x)| = 2 \; \text{ and } \; \alpha = \frac{1}{2} = 0.5.$$

Therefore by Peano's existence theorem, we have at least one solution $x : [-0.5, 0.5] \to [-1, 1]$. As a matter of fact, the IVP has the two solutions $x(t) = 0$ and $x(t) = \frac{t^6}{27}$, both going through the initial point $(0, 0)$.

Next, we state and prove Gronwall's inequality, which plays an important role throughout this book in proving the boundedness, stability, and uniqueness.

Theorem 2.1.5. *(Gronwall's inequality) Let C be a nonnegative constant, and let u, v be nonnegative continuous functions on $[a, b]$ such that*

$$v(t) \leq C + \int_a^t v(s)u(s)ds, \ a \leq t \leq b. \tag{2.1.17}$$

Then

$$v(t) \leq Ce^{\int_a^t u(s)ds}, \ a \leq t \leq b. \tag{2.1.18}$$

In particular, if $C = 0$, then $v = 0$.

Proof. Let $C > 0$, and let $h(t) = C + \int_a^t v(s)u(s)ds$. Then

$$h'(t) = v(t)u(t) \leq h(t)u(t).$$

So we have the differential inequality

$$h'(t) - h(t)u(t) \leq 0.$$

Multiplying both sides of the above expression by the integrating factor $e^{-\int_a^t u(s)ds}$, we get

$$\left(h(t)e^{-\int_a^t u(s)ds}\right)' \leq 0.$$

Integrating both sides from a to t gives

$$h(t)e^{-\int_a^t u(s)ds} - h(a) \leq 0, \ \text{ or } \ h(t) \leq e^{\int_a^t u(s)ds}.$$

Finally,

$$v(t) \leq h(t) \leq Ce^{\int_a^t u(s)ds}.$$

If $C = 0$, then from (2.1.17) it follows that

$$v(t) \leq \int_a^t v(s)u(s)ds \leq \frac{1}{m} + \int_a^t v(s)u(s)ds, \ a \leq t \leq b$$

for any $m \geq 1$. Then from what we have just proved we arrive at

$$v(t) \leq \frac{1}{m}e^{\int_a^t u(s)ds}, \ a \leq t \leq b.$$

Thus for any fixed $t \in [a, b]$, we can let $m \to \infty$ to conclude that $v(t) \leq 0$, and it follows that $v(t) = 0$ for all $t \in [a, b]$. This completes the proof. $\square$

Now we are in a position to revisit the proof of the uniqueness of the solution in Theorem 2.1.2 by using Gronwall's inequality.

Just for a reminder, we assume that the hypotheses of Theorem 2.1.2 hold. Suppose (2.1.1) has two solutions $x(t)$ and $y(t)$ and set $w(t) = x(t) - y(t)$. Then $w'(t) = x'(t) - y'(t)$. Therefore

$$\int_{t_0}^{t} w'(s)ds = w(t) - w(t_0) = \int_{t_0}^{t} \Big(f(s, x(s)) - f(s, y(s)) \Big) ds \text{ with}$$

$$w(t_0) = x(t_0) - y(t_0) = 0.$$

So we arrive at

$$|w(t)| \leq \int_{t_0}^{t} |f(s, x(s)) - f(s, y(s))| ds$$

$$\leq K \int_{t_0}^{t} |x(s) - y(s)| ds$$

$$= K \int_{t_0}^{t} |w(s)| ds.$$

Applying Gronwall's inequality with $v(t) = |w(t)|$, $C = 0$, and $u(s) = 1$, we get $|w(t)| = 0$ for all t such that $|t - t_0| \leq h$. We conclude that $w(t) = 0$ and hence $x(t) = y(t)$.

The next theorem shows the significance of Theorem 2.1.5.

Theorem 2.1.6. *Suppose $\phi(t)$ is continuous on $[0, \infty)$ and differentiable on $(0, \infty)$. If*

$$\lim_{t \to \infty} \phi(t) = 0$$

and

$$\int_{0}^{\infty} |\phi'| dt < \infty,$$

then all solutions of

$$x'' + (1 + \phi(t))x = 0 \tag{2.1.19}$$

are bounded.

Proof. Multiply both sides of (2.1.19) by x' and integrate the resulting equation from 0 to t to arrive at

$$\frac{(x'(t))^2}{2} + \frac{(x(t))^2}{2} + \int_{0}^{t} \phi(s)x(s)x'(s)ds = c_1,$$

where c_1 is a constant. Integrating by parts the third term from the left and rearranging give

$$x^2(t)\big(1 + \phi(t)\big) = 2c_2 - (x'(t))^2 + \int_0^t \phi'(s)x^2(s)ds$$

$$\leq 2c_2 + \int_0^t \phi'(s)x^2(s)ds,$$

where c_2 is a constant. Since $\lim_{t\to\infty} \phi(t) = 0$, we may choose t_0 large enough so that $1 + \phi(t) \geq \frac{1}{2}$ for all $t \geq t_0$. This translates into

$$x^2(t)(\tfrac{1}{2}) \leq x^2(t)\big(1 + \phi(t)\big) \leq 2c_2 + \int_0^t \phi'(s)x^2(s)ds$$

or

$$x^2(t) \leq 4c_2 + 2\int_0^t |\phi'(s)|x^2(s)ds, \ t \geq t_0.$$

Thus by Theorem 2.1.5 we have that

$$x^2(t) \leq 4c_2 e^{2\int_0^t |\phi'(s)|ds}$$

$$\leq 4c_2 e^{2\int_0^\infty |\phi'(s)|ds} = M < \infty \text{ for some positive constant } M$$

and for all $t \geq 0$. Since $x(t)$ is continuous on $0 \leq t \leq t_0$ and t_0 is arbitrary, it follows that $x(t)$ is bounded. $\qquad\square$

2.2 Existence on Banach spaces

In this brief section we discuss the existence of solutions of the IVP (2.1.1) on Banach spaces using the contraction mapping principle. For reminder, we restate the IVP

$$x' = f(t, x), \ x(t_0) = x_0, \tag{2.2.1}$$

where we assume that $f : D \to \mathbb{R}$ is continuous and D is an open subset of $\mathbb{R} \times \mathbb{R}$. The next theorem is another version of Picard's theorem where f satisfies global Lipschitz condition.

Theorem 2.2.1. *(Picard theorem; global version) Let $f \in C(\mathbb{R} \times \mathbb{R}, \mathbb{R})$, and let f satisfy the global Lipschitz condition (2.1.2), that is, there exists a constant $K > 0$ such that*

$$|f(t, x) - f(t, y)| \leq K|x - y| \text{ for all } (t, x), (t, y) \in D.$$

Let $h = \dfrac{1}{2K}$ and $t_0 \in (-h, h)$. Then there exists a unique $x : (-h, h) \to \mathbb{R}$ satisfying the IVP (2.2.1).

Proof. Define the space

$$\mathbb{B} = C([-h, h], \mathbb{R}).$$

Then $\mathbb{B}$ endowed with the norm

$$||x|| = \max_{t \in [-h,h]} |x(t)|$$

is a Banach space. By Theorem 2.1.1, we have that

$$x(t) = x_0 + \int_{t_0}^{t} f(s, x(s))ds, \quad t \geq t_0. \tag{2.2.2}$$

The problem comes down to finding a fixed point of the operator $\mathcal{P} : \mathbb{B} \to \mathbb{B}$ defined by

$$\mathcal{P}(x)(t) = x_0 + \int_{t_0}^{t} f(s, x(s))ds.$$

By a fixed point we mean that there is an element $x \in \mathbb{B}$ such that $\mathcal{P}(x) = x$, which is equivalent to finding a solution of (2.2.2) and therefore of (2.2.1). Note that $\mathcal{P}$ assigns to any function $x \in \mathbb{B}$ another function, which we denote by $\mathcal{P}x$. Thus for $\mathcal{P}x$ to be defined, we need to evaluate its value at some $t \in [-h, h]$. It is clear that $\mathcal{P}$ is continuous in x. To show that $\mathcal{P}(x)(t)$ is a contraction, we let $x, y \in \mathbb{B}$. Then for $t \in [-h, h]$, we have that

$$\left| \mathcal{P}(x)(t) - \mathcal{P}(y)(t) \right| \leq \int_{t_0}^{t} |f(s, x(s)) - f(s, y(s))| ds$$

$$\leq K \int_{t_0}^{t} |x(s) - y(s)| ds$$

$$\leq h K \|x - y\|$$

$$\leq \frac{1}{2} \|x - y\|.$$

Thus

$$\|\mathcal{P}(x) - \mathcal{P}(y)\| \leq \sup_{t \in [-h,h]} \left| \mathcal{P}(x)(t) - \mathcal{P}(y)(t) \right|$$

$$\leq K \int_{t_0}^{t} |x(s) - y(s)| ds$$

$$\leq \frac{1}{2} \|x - y\|.$$

Thus $\mathcal{P}$ defines a contraction mapping on the Banach space $\mathbb{B}$, and hence it has a unique $x \in \mathbb{B}$ that is a solution of (2.2.1) by Theorem 2.1.1. This completes the proof. $\square$

Theorem 2.2.2. *(Picard Theorem; local version) Let $\mathscr{D} \subset \mathbb{R} \times \mathbb{R}$ be an open set containing $(0, x_0)$, and let $f \in C(\mathscr{D}, \mathbb{R})$ be such that $|f| \leq M$ for a positive constant M. Suppose f satisfies the local Lipschitz condition in x on $\mathscr{D}$ with Lipschitz constant $K > 0$. Choose*

$$h \in (0, \frac{1}{2K}) \ \text{so that} \ [-h, h] \times \{|x - x_0| \leq hM\} \subset \mathscr{D}$$

(note that such a choice of h is possible since the set $\mathscr{D}$ is open). Then there exists a unique $x : (-h, h) \to \mathbb{R}$ satisfying the IVP (2.2.1).

Proof. Let $I = [-h, h]$, where h is as chosen in the hypothesis. Let $\mathbb{B} = \{x \in \mathbb{R} : |y - x_0| \leq hM\}$. Define $\mathscr{P} : C(I, \mathbb{B}) \to C(I, \mathbb{B})$ by

$$\mathscr{P}(x)(t) = x_0 + \int_{t_0}^{t} f(s, x(s))ds.$$

The problem reduces to showing that $\mathscr{P}(x) = x$. First, we show that $\mathscr{P}$ is well-defined. It is clear from the choice of h that if $x \in C(I, \mathbb{B})$, then $f(t, x)$ is well-defined for all $t \in [-h, h]$, and hence $x_0 + \int_{t_0}^{t} f(s, x(s))ds$ is well-defined. Moreover, since $|f| \leq M$, we have that $x_0 + \int_{t_0}^{t} f(s, x(s))ds$ is continuous. Thus $\mathscr{P}(x)(t)$ is a continuous map. Finally,

$$|\mathscr{P}(x)(t) - x_0| \leq \int_{t_0}^{t} |f(s, x(s))|ds \leq hM.$$

Thus $\mathscr{P}(x)(t) \in \mathbb{B}$ for all $t \in [-h, h]$, and hence $\mathscr{P}(x) \in C(I, \mathbb{B})$ assigns to any function $x \in \mathbb{B}$ another function, which we denote by $\mathscr{P}x$. Thus for $\mathscr{P}x$ to be defined, we need to evaluate its value at some $t \in [-h, h]$. It is clear that $\mathscr{P}$ is continuous in x. To show that $\mathscr{P}(x)(t)$ is a contraction, we let $x, y \in C(I, \mathbb{B})$. Then for each $t \in [-h, h]$, we have that

$$\|\mathscr{P}(x) - \mathscr{P}(y)\| \leq \sup_{t \in [-h,h]} |\mathscr{P}(x)(t) - \mathscr{P}(y)(t)|$$

$$\leq K \int_{t_0}^{t} |x(s) - y(s)|ds$$

$$\leq Kt \max_{|s| \leq t} |x(s) - y(s)|$$

$$\leq Kh\|x - y\| = \frac{1}{2}\|x - y\|.$$

Thus $\mathscr{P}$ defines a contraction mapping on the Banach space $C(I, \mathbb{B})$, and hence it has a unique $x : (-h, h) \to \mathbb{R}$ that is a solution of (2.2.1) by Theorem 2.1.1. This completes the proof. $\square$

Consider the differential equation $x'(t) = 1 + x^2$. Theorem 2.2.1 cannot be applied here since $f(t, x) = 1 + x^2$ does not satisfy the global Lipschitz condition on $\mathbb{R}$, whereas Theorem 2.2.2 is applicable. Moreover, Theorem 2.2.2 only requires finding $\mathcal{D} \in \mathbb{R} \times \mathbb{R}$ containing $(0, x_0)$. In addition, in both Theorems 2.2.1 and 2.2.2 the initial time t_0 can be chosen so that $t_0 \neq 0$.

2.3 Existence theorem for linear equations

In this section we eloquently apply Theorem 1.4.4 to show that the linear non-homogeneous differential equation

$$x'(t) = a(t)x(t) + f(t), \tag{2.3.1}$$

where $a, f \in C(\mathbb{R}, \mathbb{R})$, has a unique solution $\phi \in C(\mathbb{R}, \mathbb{R})$ with $\phi(t_0) = x_0$. We have the following theorem.

Theorem 2.3.1. *Let $(t_0, x_0) \in \mathbb{R}^2$, and let $a, f \in C(\mathbb{R}, \mathbb{R})$. Then (2.3.1) has a unique solution $\phi : (-\infty, \infty) \to \mathbb{R}$ with $\phi(t_0) = x_0$.*

Proof. Let $s, M > 0$. Find M such that for a given s, we have $|a(t)| \leq M$ for $t \in [t_0 - s, t_0 + s]$. Let $\mathbb{B} = C([t_0 - s, t_0 + s], \mathbb{R})$. Then the space $\mathbb{B}$ endowed with the supremum norm $\| \cdot \|$ is a Banach space. Theorem 2.1.1 implies that

$$x(t) = x_0 + \int_{t_0}^{t} a(u)x(u)du + \int_{t_0}^{t} f(u)du.$$

For $x \in \mathbb{B}$, define $\mathcal{P} : \mathbb{B} \to \mathbb{B}$ by

$$\mathcal{P}(x)(t) = x_0 + \int_{t_0}^{t} a(u)x(u)du + \int_{t_0}^{t} f(u)ds.$$

Clearly, $\mathcal{P}(x)(t)$ is a continuous map. Let $x, y \in \mathbb{B}$. Then

$$\|\mathcal{P}(x) - \mathcal{P}(y)\| \leq \sup_{t \in [t_0 - s, t_0 + s]} |\mathcal{P}(x)(t) - \mathcal{P}(y)(t)|$$

$$\leq M \int_{t_0}^{t} |x(u) - y(u)|du$$

$$\leq Ms \|x - y\|$$

or

$$\|\mathcal{P}(x) - \mathcal{P}(y)\| \leq Ms \|x - y\|.$$

Now if $Ms = \alpha < 1$, then $\mathcal{P}$ is a contraction, and by Theorem 1.4.3, there is a unique fixed point $\phi \in \mathbb{B}$, that is, $\mathcal{P}(\phi) = \phi$, which is a solution of (2.3.1) on $[t_0 - s, t_0 + s]$. However, we claimed that the solution exists and is unique on $(-\infty, \infty)$. To show this, we use Theorem 1.4.4 by letting s be an arbitrary and find an integer $m > 0$ such that $\mathscr{P}^m$ is a contraction:

$$
\begin{aligned}
|\mathscr{P}^2(x)(t) - \mathscr{P}^2(y)(t)| &= |\mathscr{P}(\mathscr{P}(x))(t) - \mathscr{P}(\mathscr{P}(y))(t)| \\
&\leq \sup_{t \in [t_0 - s,\, t_0 + s]} \left| \int_{t_0}^{t} |a(u)| |\mathscr{P}(x)(u) - \mathscr{P}(y)(u)| du \right| \\
&\leq M^2 \|x - y\| \int_{t_0}^{t} |u - t_0| du| \\
&\leq M^2 \|x - y\| \, |t - t_0|^2 / 2!.
\end{aligned}
$$

Continuing this process, we claim that

$$
|\mathscr{P}^m(x)(t) - \mathscr{P}^m(y)(t)| \leq M^m |t - t_0|^m \|x - y\| / m!
$$

for positive integer m. The claim is already proved for $m = 1, 2$. We assume that it holds for a positive integer k and then show that the result holds for $k + 1$. Indeed,

$$
\begin{aligned}
|\mathscr{P}^{k+1}(x)(t) - \mathscr{P}^{k+1}(y)(t)| &= |\mathscr{P}(\mathscr{P}^k(x))(t) - \mathscr{P}(\mathscr{P}^k(y))(t)| \\
&\leq \sup_{t \in [t_0 - s,\, t_0 + s]} \left| \int_{t_0}^{t} |a(u)| |\mathscr{P}^k(x)(u) - \mathscr{P}^k(y)(u)| du \right| \\
&\leq M^{k+1} \|x - y\| \left| \int_{t_0}^{t} |u - t_0|^k du / k! \right| \\
&\leq M^{k+1} \|x - y\| \, |t - t_0|^{k+1} / (k+1)!.
\end{aligned}
$$

Thus

$$
|\mathscr{P}^m(x)(t) - \mathscr{P}^m(y)(t)| \leq M^m s^m \|x - y\| / m!,
$$

and for large m, we have

$$
\frac{(ms)^m}{m!} < 1,
$$

and the proof is complete by Theorem 1.4.4.   □

2.4 Continuation of solutions

As we mentioned before, the interval of existence and uniqueness guaranteed by Theorem 2.1.1 is, in most cases, shorter than the actual interval of existence and uniqueness. This is misleading, since usually it is not the *maximal interval* of existence. To see this, we consider

$$x'(t) = tx, \ x(0) = 1.$$

Let

$$D = \{(t,x) : |t| \le a, \ |x - 1| \le b\},$$

where a and b are constants. It is clear that

$$h = \min\{a, \frac{b}{M}\} = \min\{a, \frac{b}{a(1+b)}\}.$$

Thus for any positive constants a and b, we will always have $h < 1$, since either $h \le a$ or $h \le \frac{b}{a(1+b)} < 1/a$. On the other hand, the true solution of the IVP is $x(t) = e^{t^2/2}$, which exists for all $t \in \mathbb{R}$.

Consider the IVP

$$x' = f(t,x), \tag{2.4.1}$$

$$x(0) = x_0. \tag{2.4.2}$$

In the following theorems, we only consider $t \ge 0$ ($t \le 0$ is left as an exercise; see Exercise 2.17).

Theorem 2.4.1. *Suppose $D = \mathbb{R} \times \mathbb{R}$, $f \in C(D, \mathbb{R})$, and $\frac{\partial f}{\partial x}$ is continuous in D. If a solution ξ of (2.4.1)–(2.4.2) exists on $[0, \alpha]$ for some positive α, then it can be continued to a solution on some longer interval $[0, \alpha + r]$ for some positive constant r.*

Proof. Without loss of generality, we take $x_0 = 0$. Let

$$R = [\alpha - a, \alpha + a] \times [\xi(\alpha) - a, \xi(\alpha) + a],$$

where a is small enough so that $\alpha - a > 0$. Since f is continuous on the rectangle R containing the point $(\alpha, \xi(\alpha))$, we have, by applying Theorem 2.1.2 to the IVP

$$x' = f(t,x) \tag{2.4.3}$$

$$x(\alpha) = \xi(\alpha), \tag{2.4.4}$$

that a solution ϕ of (2.4.3)–(2.4.4) exists on some interval $[\alpha, \alpha + r]$. We claim that the function $\tilde{\xi}$ given by

$$\tilde{\xi}(t) = \begin{cases} \xi(t) & \text{if } 0 \le t \le \alpha, \\ \phi(t) & \text{if } \alpha < t \le \alpha + r, \end{cases}$$

solves (2.4.3)–(2.4.4) on $[0, \alpha + r]$. To do so, we must show that the left and right derivatives exist at α and are equal. Since $\phi(\alpha) = \xi(\alpha)$, the right derivative is given by

$$\lim_{s \to 0^+} \frac{\tilde{\xi}(\alpha + s) - \tilde{\xi}(\alpha)}{s} = \lim_{s \to 0^+} \frac{\phi(\alpha + s) - \xi(\alpha)}{s}$$

$$= \lim_{s \to 0^+} \frac{\phi(\alpha + s) - \phi(\alpha)}{s}$$

$$= \phi'(\alpha) = f(\alpha, \phi(\alpha))$$

$$= f(\alpha, \xi(\alpha)) = \xi'(\alpha).$$

Similarly, the left derivative is $\xi'(\alpha)$ due to the fact that ξ is a solution on $[0, \alpha]$. This completes the proof. □

Theorem 2.4.2. *The maximal interval of existence is open.*

Proof. Let J be the maximal interval of existence. If α of Theorem 2.4.1 is in J, then the same theorem shows that the interval $(\alpha - r, \alpha + r)$ is also in J. Hence J is open. This completes the proof. □

We have the following theorem.

Theorem 2.4.3. *Suppose $D = \mathbb{R} \times \mathbb{R}$, $f \in C(D, \mathbb{R})$, and $\frac{\partial f}{\partial x}$ is continuous in D. If x is a solution of (2.4.1)–(2.4.2) on some interval J containing $t_0 = 0$, then x is the unique solution on J.*

Proof. Suppose $J = (\alpha, w)$. Without loss of generality, let $x_0 = 0$. Then Theorem 2.2.1 implies that for some positive constant h, x is the unique solution on the closed interval $[-h, h]$. We will prove the uniqueness on the half-open interval $[0, w)$. Suppose there are two distinct solutions z_1 and z_2 of (2.4.1)–(2.4.2) on $[0, w)$. Then we have $z_1 = z_2$ on $[0, h]$, and there exists $t^* \in (h, w)$ such that $z_1(t^*) \neq z_2(t^*)$. Let

$$\xi = \inf\{t \in (0, w) : z_1(t^*) \neq z_2(t^*).$$

It follows that $h < \xi \le t^*$. We claim that $\xi < t^*$. Since $z_1(t^*) \neq z_2(t^*)$ and since z_1 and z_2 are continuous, $z_1 \neq z_2$ on some interval to the left of t^*. This proves the claim. Due to the continuity of z_1 and z_2 and the equality $z_1 = z_2$ on $[0, t^*]$, we must have $z_1(t^*) = z_2(t^*)$. In addition, we know from Theorem 2.4.1 that the IVP

$$x' = f(t, x), \quad x(\xi) = z_1(\xi)$$

has a unique solution on some interval, say $[\xi - \beta, \xi + \beta]$, for some positive constant β. It follows that $z_1 = z_2$ on $[\xi, \xi + \beta]$, which contradicts the definition of ξ. This shows that the solution x is unique on $[0, w)$. Similarly, we may prove the uniqueness on $(\alpha, 0]$. This completes the proof. □

Theorem 2.4.4. *Assume the hypotheses of Theorems 2.4.1 and 2.4.3. If the maximal interval of existence $J = [0, w)$, where $w < \infty$, then the solution $\phi(t)$ of $x' = f(t, x)$, $x(0) = 0$, is unbounded in $[0, w)$.*

Proof. We prove by contradiction by assuming that the solution is bounded on $[0, w)$. Since $f(t, \phi)$ is bounded on $[0, w)$, there is a positive constant M such that $f(t, \phi)| \leq M$. Then by Theorem 2.1.1, we have that

$$\phi(t) = \int_{t_0}^{t} f(s, \phi(s))ds.$$

Let $\{t_j\} \uparrow w$. Then for any j and $k > j$, we have that

$$|\phi(t_k) - \phi(t_j)| = \left| \int_{t_j}^{t_k} f(s, \phi(s))ds \right| \leq M|t_j - t_k|.$$

Now the sequence $\{t_j\}$ is convergent and hence is a Cauchy sequence. Therefore $\{\phi(t_j\}$ must be a Cauchy sequence and so also convergent. Since the sequence $\{t_j\}$ was arbitrary and converging to w, we have that $\lim_{t \to w^-} \phi(t)$ exists. Thus we may extend ϕ to w by letting $\phi(w)$ be the limit. Now consider the IVP

$$x' = f(t, x),$$
$$x(w) = \phi(w).$$

This IVP has a unique solution on $[w - r, w + r]$ for some positive constant r, which is an extension of the solution ϕ a little further. Thus $[0, w]$ is not the maximal positive interval of existence, which is a contradiction. This completes the proof. $\qquad \square$

Consider the IVP

$$x'(t) = x^2, \; x(0) = 1.$$

The function $f(t, x) = x^2 \in C^1(\mathbb{R})$, and according to Theorem 2.1.2, the IVP has a unique solution on an open interval containing 0. Solving the IVP, we obtain the solution $x(t) = \frac{1}{t-1}$. The maximal interval of existence must include $t = 0$. Thus the solution is defined on its maximal interval of existence $(-\infty, 1)$. Furthermore,

$$\lim_{t \to 1^-} x(t) = -\infty.$$

According to Theorem 2.4.4, the IVP of Example 2.4 has the maximal interval of existence $(-\sqrt{2}, \sqrt{2})$. It is the case since

$$\lim_{t \to (\sqrt{2})^-} \frac{t^2}{2 - t^2} = \infty \quad \text{and} \quad \lim_{t \to (\sqrt{2})^+} \frac{t^2}{2 - t^2} = -\infty.$$

2.5 Dependence on initial conditions

One of the main characteristics a mathematical model should enjoy is that a solution should depend continuously on initial conditions. In other words, a small change in

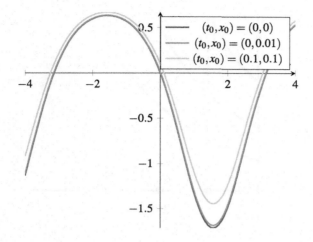

FIGURE 2.2

The three graphs stay close with three different initial conditions $(0, 0)$, $(0, 0.1)$, and $(0.1, 0.1)$.

the initial conditions should result in a small deviation in the solutions. We have seen that a solution $x(t)$ of the IVP

$$x' = f(t, x),\ x(t_0) = x_0 \tag{2.5.1}$$

depends not only on the time t, but also on the initial point (t_0, x_0). Thus, to emphasize such dependence, we use the notation $x(t, t_0, x_0)$ to denote $x(t)$. Consider

$$x'(t) = (x - 1) \cos(t),\ x(t_0) = x_0,$$

which has the solution

$$x(t) = \frac{e^{\sin(t)}}{e^{\sin(t_0)}} (x_0 - 1) + 1.$$

In Fig. 2.2 we plot the graphs of the solution for three different initial conditions, which shows the sensitivity of the solution to the initial data.

Example 2.6. The scalar (IVP)

$$x'(t) = x^2,\ x(0) = x_0,\ t \geq 0$$

has the solution

$$x(t) = \frac{x_0}{1 - x_0 t}$$

with maximal interval $[0, \frac{1}{x_0})$. As the initial value x_0 increases, the maximal interval of existence shrinks as depicted in Fig. 2.3.

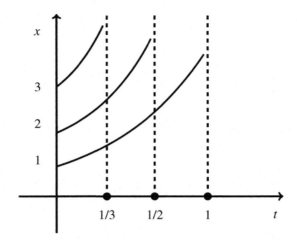

FIGURE 2.3

Dependence on x_0.

Example 2.7. In Fig. 2.4 we display the graph of

$$x' = x\sin(x) + t$$

and show its sensitivity to the initial data.

We have the following theorem.

Theorem 2.5.1. *Let $D \subset \mathbb{R} \times \mathbb{R}$, and suppose f and $\dfrac{\partial f}{\partial x}$ are continuous and bounded in D. Let x and y be two solutions of (2.5.1) passing through the initial conditions (t_0, x_0) and (t_0^*, x_0^*), respectively. Assume that the two solutions exist on a common interval (α, β), where α and β are finite. Then to each $\varepsilon > 0$, there corresponds $\delta > 0$ such that*

$$|t_0 - t_0^*| < \delta, \ |x_0 - x_0^*| < \delta \ \text{imply} \ |x(t) - y(t)| < \varepsilon, \ t, t^* \in (\alpha, \beta).$$

Proof. Since f and $\frac{\partial f}{\partial x}$ are continuous and bounded on D, there are positive constants M and K such that

$$|f(t, x)| \le M, \ |\frac{\partial f}{\partial x}| \le K.$$

By Theorem 2.1.1, we have that for $t, t^* \in (\alpha, \beta)$,

$$x(t, t_0, x_0) = x_0 + \int_{t_0}^{t} f(s, x(s))ds, \ \ y(t^*, t_0^*, x_0^*) = x_0^* + \int_{t_0^*}^{t^*} f(s, y(s))ds.$$

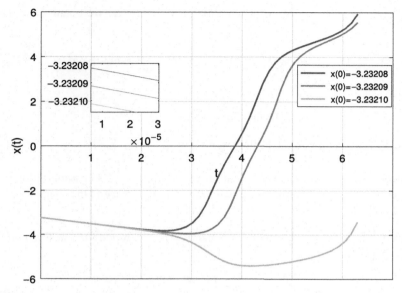

FIGURE 2.4

Sensitivity of solutions of $x' = x \sin(x) + t$ to the initial data.

By noticing that

$$\int_{t_0}^{t} f(s, x(s))ds = \int_{t_0^*}^{t} f(s, x(s))ds + \int_{t_0}^{t_0^*} f(s, x(s))ds,$$

$$\int_{t_0^*}^{t^*} f(s, y(s))ds = \int_{t_0^*}^{t} f(s, y(s))ds + \int_{t}^{t^*} f(s, y(s))ds,$$

we arrive at

$$|x(t, t_0, x_0) - y(t^*, t_0^*, x_0^*)| \le |x_0 - x_0^*| + \left| \int_{t_0^*}^{t} |f(s, x(s)) - f(s, y(s))|ds \right|$$

$$+ \left| \int_{t_0}^{t_0^*} |f(s, x(s))|ds \right| + \left| \int_{t}^{t^*} |f(s, y(s))|ds \right|$$

$$\le |x_0 - x_0^*| + K \left| \int_{t_0^*}^{t} |x(s)) - y(s)|ds \right|$$

$$+ M|t_0^* - t_0| + M|t^* - t|$$

$$\le \delta + K \left| \int_{t_0^*}^{t} |x(s)) - y(s)|ds \right| + 2M\delta.$$

Letting $v(t) = |x(t) - y(t)|$, $C = \delta + 2M\delta$, and $u(s) = K$, we obtain from Theorem 2.1.5 that

$$|x(t, t_0, x_0) - y(t^*, t_0^*, x_0^*)| \le \delta(1 + 2M)e^{K|t - t_0^*|} \le \delta(1 + 2M)e^{K(\alpha - \beta)}$$

or

$$|x(t, t_0, x_0) - y(t^*, t_0^*, x_0^*)| \le \varepsilon$$

for $\delta < \frac{\varepsilon e^{K(\alpha - \beta)}}{1 + 2M}$. This completes the proof. $\square$

Theorem 2.5.1 shows that the solution $x(t, t_0, x_0)$ of (2.5.1) passing through the initial data (t_0, x_0) is a continuous function in (t, t_0, x_0). We end the chapter with the following theorem.

Theorem 2.5.2. *Let f, g be defined in a region $D \subset \mathbb{R} \times \mathbb{R}$ and satisfy the hypotheses of Theorem 2.5.1. Set*

$$M = \sup_{(t,u) \in D} |f(t, u) - g(t, u)|.$$

Let x and y be the solutions of the IVPs

$$x' = f(t, x), \; x(t_0) = x_0 \quad and \quad y' = g(t, y), \; y(t_0) = y_0,$$

respectively. Then the solutions x and y satisfy the estimate

$$|x(t) - y(t)| \le |x_0 - y_0|e^{K|t - t_0|} + \frac{M}{K}\left(e^{K|t - t_0|} - 1\right), \tag{2.5.2}$$

where K is given in the proof of Theorem 2.5.1.

Proof. Using a similar setup as in the proof of Theorem 2.5.1, we get

$$
\begin{aligned}
|x(t) - y(t)| &\le |x_0 - y_0| + \int_{t_0}^t \left|f(s, x(s)) - g(s, y(s))\right| ds \\
&\le |x_0 - y_0| + \int_{t_0}^t \left|f(s, x(s)) - f(s, y(s))\right| ds \\
&\quad + \int_{t_0}^t \left|f(s, x(s)) - g(s, y(s))\right| ds \\
&\le |x_0 - y_0| + \int_{t_0}^t \left[K|x(s) - y(s)| + M\right] ds.
\end{aligned}
$$

Let $U(t) = |x(t) - y(t)|$ and use an improvisation of Exercise 2.9 to obtain the estimate given by (2.5.2). This completes the proof. $\square$

Suppose M in Theorem 2.5.2 is small, that is, $M \le \varepsilon > 0$. If $|t - t_0| \le a$ for positive constant a, then inequality (2.5.2) yields

$$|x(t) - y(t)| \le |x_0 - y_0|e^{Ka} + \frac{\varepsilon}{K}\left(e^{Ka} - 1\right),$$

and thus if $|x_0 - y_0| \le \delta$ for some $\delta > 0$, then the two solutions remain close to each other. In other words, Theorem 2.5.2 says that if two differential equations have their right-hand sides "close together," then their solutions cannot differ very much.

2.6 Exercises

Exercise 2.1. Show that $f(t, x) = \dfrac{x}{x^2 + 1}$ is uniformly Lipschitz on $\mathbb{R}$ and identify the Lipschitz constant K.

Exercise 2.2. Compute the Lipschitz constant K for each of the functions in the regions indicated.

1. $f(t, x) = t^3 + x^4$, $\{(t, x) : |t| \le 2, |x| \le 3\}$.
2. $f(t, x) = t^2 \cos(x) + t \sin(x)$, $\{(t, x) : |t| \le 2, |x| < \infty\}$.
3. $f(t, x) = t|x|$, $\{(t, x) : |t| \le 1, |x| < \infty\}$.
4. $f(t, x) = t^2 \sin(x) \cos(x)$, $\{(t, x) : |t| \le a, |x| < \infty\}$ for a positive constant a.
   Hint: when you perform $|f(t, x) - f(t, y)|$, add and subtract the term $\sin(x) \cos(y)$.

Exercise 2.3. Apply the Picard iterations to the IVPs

1. $x'(t) = tx$, $x(0) = 1$,
2. $x'(t) = 2t(1 + x)$, $x(0) = 2$,

and show that the obtained $\{x_n(t)\}$ of each iterate converges to the true solution of each IVP (the true solution is the solution found by solving the IVP).

Exercise 2.4. Use Theorem 2.1.2 to find the interval of existence of the unique solution of the following differential equations:

1. $x'(t) = x^2$, $x(0) = 2$.
2. $x'(t) = 5 + x^2$, $x(1) = 2$.
3. $x'(t) = x^3$, $x(0) = 2$.
4. $x'(t) = t + x^2$, $x(0) = 0$.
5. $x'(t) = 1 + x^2$, $x(0) = 0$.

Exercise 2.5. Explain why the IVP $x'(t) = tx - \sin(x)$, $x(0) = 2$, has a unique solution in the neighborhood of the point $(0, 2)$.

Exercise 2.6. Let f be scalar and continuous in both arguments. Show that if x is a solution of the second-order differential equations $x''(t) = f(t, x(t))$, $x(t_0) = x_0$, $x'(t_0) = x_1$, then it satisfies

$$x(t) = x_0 + x_1(t - t_0) + \int_{t_0}^{t} (t - s)f(s, x(s))ds. \qquad (2.6.1)$$

Hint: Integrate twice and use the general formula

$$\int_0^t \left[\int_0^s (g(\tau, \phi)\tau)d\tau \right] ds = \int_0^t \left[\int_\tau^t ds \right] (g(\tau, \phi)\tau)d\tau.$$

Exercise 2.7. Explain why neither Theorem 2.1.2 nor Theorem 2.1.4 can be applied to the IVP

$$x'(t) = \frac{x^2}{t}, \quad x(0) = 1.$$

Use separation of variables and find the family of solutions

$$x(t) = \frac{-1}{\text{Ln}(t) + c},$$

for some constant c. Is it possible to find c?

Exercise 2.8. Check for the existence and uniqueness of solutions of the following IVPs and sketch the corresponding rectangle D:

1. $(t - 1)x'(t) = x^2 + t$, $x(0) = 1$.
2. $(t - 1)x'(t) = x^2 + t$, $x(1) = 0$.
3. $x'(t) = t^{1/3}x + x^2$, $x(0) = 3$.
4. $(t^2 - 1)x'(t) = tx$, $x(0) = 1$.
5. $tx^2x'(t) = x^3 - t^3$, $x(1) = 1$.

Exercise 2.9. (*Generalized Gronwall's inequality*) Show that if f, u, v are nonnegative continuous functions on $[a, b]$ such that

$$u(t) \leq f(t) + \int_a^t v(s)u(s)ds, \ a \leq t \leq b,$$

then

$$u(t) \leq f(t) + \int_a^t v(s)f(s)e^{\int_s^t v(r)dr}ds, \ a \leq t \leq b.$$

Exercise 2.10. Find all nonnegative continuous functions $f : [0, 1] \to \mathbb{R}$ such that

$$f(t) \leq \int_0^t f(s)ds, \ 0 \leq t \leq 1.$$

Exercise 2.11. Let K_1, K_2 be positive constants, and let f be a nonnegative continuous function on $[a, b]$ such that

$$f(t) \leq K_1 + K_2 \int_a^t f(s)ds, \ a \leq t \leq b.$$

Show that f satisfies

$$f(t) \leq K_1 e^{K_2(b-a)}.$$

Exercise 2.12. (*Another form of Gronwall's inequality*) Let K_1, K_2 be positive constants, and let C_0 be a nonnegative constant. Suppose f is a nonnegative continuous function on $[0, h]$ such that

$$f(t) \leq C_0 + \int_0^t [K_1 f(s) + K_2]ds, \ 0 \leq t \leq h.$$

Show that f satisfies

$$f(t) \leq C_0 e^{K_1 t} + \frac{K_2}{K_1}\left(e^{K_1 t} - 1\right).$$

Exercise 2.13. Let x be a solution of (2.1.1) and suppose f satisfies the global Lipschitz condition with constant K for all $t \in \mathbb{R}$. Suppose that for all $t \in \mathbb{R}$, $\int_0^t |f(s, 0)|ds \leq M$ for positive constant M. Then $x(t)$ satisfies the inequality

$$|x(t)| \leq (|x_0| + M)e^{K(t-t_0)}.$$

Exercise 2.14. Consider the system

$$x' = -x^3.$$

$$y' = -cy + x^{2\lambda} y \cos^2(y).$$

Use Gronwall's inequality to show that
(a) $y(t)$ is bounded for all $t \geq 0$ whenever $c = 0$ and $\lambda > 1$,
(b) $y(t) \to 0$ as $t \to \infty$ whenever $c > 0$ and $\lambda = 1$.

Exercise 2.15. Consider the scalar differential equation

$$x' = (a + \frac{1}{1+t}b)x, \ t \geq 0, \tag{2.6.2}$$

where a is constant. Show that if $y(t) = e^{-at}x$ and x is a solution of (2.6.2), then

$$|y(t)| \leq |x(0)|e^{c\ln(1+t)} \ \text{ for some constant } c > 0.$$

In addition, show that if $a < 0$, then

$$|x(t)| \to 0 \ \text{ as } t \to \infty.$$

Exercise 2.16. Let $g : [0, \infty) \to [0, \infty)$ be given by $g(x) = \frac{1}{2}e^{-x}$. Use the contraction mapping principle to prove that the graphs of g and the line $y = x$ intersect at a unique point (a, a).

Exercise 2.17. Prove Theorem 2.4.1 for $t \leq 0$.

Exercise 2.18. Consider the IVP

$$x'(t) = 2tx^2, \ x(t_0) = x_0.$$

(a) Use Theorem 2.1.2 to find the largest interval of existence and uniqueness of solutions.
(b) Solve the IVP.
(c) Find the maximal interval of existence when $t_0 = 0$ and $x_0 > 0$ and find the limits at the end points.

Exercise 2.19. Consider the IVP

$$x'(t) = 2(t + 1)(1 + x^2), \ x(0) = 0.$$

(a) Solve the IVP.
(b) Find the maximal interval of existence.

Exercise 2.20. Finish the proof of Theorem 2.5.2 to obtain (2.5.2).

Systems of ordinary differential equations

We briefly discuss how the existence and uniqueness theorems of Chapter 2 are extended to systems. Then we develop the notion of the fundamental matrix as a solution and utilize it to write solutions of non-homogeneous systems. The author made a conscious decision not to dwell too much on the topic of *Jordan canonical forms* since it is a well-developed subject in linear algebra.

3.1 Existence and uniqueness

For motivational purpose, we consider the second-order differential equation

$$x''(t) + a^2 x(t) = 0, \tag{3.1.1}$$

where a is some nonzero constant. Eq. (3.1.1) is commonly used as a model describing a mechanical process that periodically oscillates. For example, it is used to describe oscillations of a pendulum, provided that the angle change is small or of a spring with a small deflection. We use the transformation $x_1(t) = x(t)$, which is the displacement of the oscillating object. (For the pendulum, it is the angle made with the vertical.) Introducing the second transformation, the velocity or angular velocity $x_2(t) = x'(t)$, and differentiating, we get $x_1'(t) = x'(t) = x_2(t)$ and $x_2'(t) = x''(t) = -a^2 x(t) = -a^2 x_1(t)$. We therefore arrive at two first-order differential equations

$$x_1' = x_2,$$

$$x_2' = -a^2 x_1.$$

Thus we consider the general system of ordinary differential equations

$$x_1' = f_1(t, x_1, \ldots, x_n),$$
$$x_2' = f_2(t, x_1, \ldots, x_n),$$
$$\vdots$$
$$x_n' = f_n(t, x_1, \ldots, x_n).$$

Advanced Differential Equations. https://doi.org/10.1016/B978-0-32-399280-0.00009-7

Using the vector notations

$$\mathbf{x} = \begin{pmatrix} x_1 \\ x_2 \\ \vdots \\ x_n \end{pmatrix}$$

and

$$f(t, \mathbf{x}) = \begin{pmatrix} f_1(t, \mathbf{x}) \\ f_2(t, \mathbf{x}) \\ \vdots \\ f_n(t, \mathbf{x}) \end{pmatrix},$$

the above system can be written in the vector form

$$\mathbf{x}' = f(t, \mathbf{x}). \tag{3.1.2}$$

Different norms on $\mathbb{R}^n$ were discussed in Chapter 1. For $\mathbf{x} = (x_1, x_2, \ldots, x_n)$, we consider the norm

$$\|\mathbf{x}\| = \sqrt{\sum_{i=1}^{n} x_i^2}.$$

However, in this section, it is more convenient to use the norm that we denote by $|\mathbf{x}|$ given by

$$|\mathbf{x}| = \sum_{i=1}^{n} |x_i|.$$

It is assumed that $|x_i|$ denotes the absolute value. The two norms are topologically equivalent in the sense that the open sets in $\mathbb{R}^n$ are the same under either norm. A set S in $\mathbb{R}^n$ is said to be open if for each $\mathbf{x}^* \in S$, there is a $\delta(\varepsilon) > 0$ such that $\mathbf{x} \in S$ whenever $|\mathbf{x} - \mathbf{x}^*| < \delta(\varepsilon)$. It can be easily shown that $\|\mathbf{x}\| \le |\mathbf{x}| \le \sqrt{n}\|\mathbf{x}\|$ for all $\mathbf{x} \in \mathbb{R}^n$. This shows the two norms are equivalent. As a direct result of this, if ϕ^i is a sequence of vector-valued functions such that for some function ϕ, $\|\phi^i - \phi\| \to 0$, $i \to \infty$, then $|\phi^i - \phi| \to 0$, $i \to \infty$, and the converse holds too.

Next, we extend some definitions and theorems from Chapter 1 to systems.

Definition 3.1.1. The function $f : D \subset \mathbb{R} \times \mathbb{R}^n \to \mathbb{R}^n$ is said to satisfy the *global* Lipschitz condition in $\mathbf{x}$ if there exists a positive constant K (called the Lipschitz constant) such that

$$|f(t, \mathbf{x}) - f(t, \mathbf{y})| \le K|\mathbf{x} - \mathbf{y}| \text{ for } (t, \mathbf{x}), (t, \mathbf{y}) \in D. \tag{3.1.3}$$

If for every subset $Q \subset D$, there is a positive constant K such that (3.1.3) holds, then the function f satisfies a *local* Lipschitz condition in $\mathbf{x}$. Now we discuss how to determine the Lipschitz constant K. Let $R \subset D$ be such that

$$R = \{(t, \mathbf{x}) : |t - t_0| \leq a, \ |\mathbf{x} - \mathbf{x_0}| \leq b\}$$

and assume that f and all its partial derivatives $\frac{\partial f_i}{\partial x_j}$ are continuous in R. Then the Lipschitz constant K can be defined as

$$K = \sum_{i=1}^{N} \sum_{j=1}^{N} \max_{(t,\mathbf{x}) \in R} \left| \frac{\partial f_i}{\partial x_j} \right|. \tag{3.1.4}$$

Throughout this section, we assume $f : D \to \mathbb{R}^n$ is continuous, where $D \subset \mathbb{R} \times \mathbb{R}^n$ is open.

Definition 3.1.2. Let $(t_0, \mathbf{x_0}) \in D$. We say that $x \in \mathbb{R}^n$ is a solution of the IVP

$$\mathbf{x}' = f(t, \mathbf{x}), \ \mathbf{x}(t_0) = \mathbf{x_0}, \tag{3.1.5}$$

where $\mathbf{x_0} = (x_{01}, x_{02}, \ldots, x_{0n})^T$, on an interval I if $t_0 \in I$, $\mathbf{x} : I \to \mathbb{R}^n$ is differentiable, $(t, \mathbf{x(t)}) \in D$ for $t \in I$, $\mathbf{x}' = f(t, \mathbf{x})$ for $t \in I$, and $\mathbf{x}(t_0) = \mathbf{x_0}$.

In preparation for the next theorem we observe that the IVP (3.1.5) given in Definition 3.1.2 is equivalent to the integral relation

$$\mathbf{x}(t) = \mathbf{x_0} + \int_{t_0}^{t} f(s, \mathbf{x}(s)) ds. \tag{3.1.6}$$

We have the following definition.

Definition 3.1.3. We say $\mathbf{x} : I \to \mathbb{R}^n$ is a solution of the integral equation given by (3.1.6) on an interval I if $t_0 \in I$, $\mathbf{x}$ is continuous on I, $(t, \mathbf{x(t)}) \in D$ for $t \in I$, and (3.1.6) is satisfied for $t \in I$.

The proof of the next theorem is identical to that of Theorem 2.1.1.

Theorem 3.1.1. *Let D be an open subset of $\mathbb{R}^{n+1}$, and let $(t_0, x_0) \in D$. Then $\mathbf{x}$ is a solution of (3.1.5) on an interval I if and only if $\mathbf{x}$ satisfies the integral equation given by (3.1.6) on I.*

Theorem 3.1.2. *(Picard's local existence and uniqueness) Let $D \subset \mathbb{R} \times \mathbb{R}^n$ be defined by*

$$D = \{(t, \mathbf{x}) : |t - t_0| \leq a, \ |\mathbf{x} - \mathbf{x_0}| \leq b\},$$

where a and b are positive constants. Assume that $f \in C(D, \mathbb{R}^n)$ satisfies the Lipschitz condition (3.1.3) with Lipschitz constant K given by (3.1.4). Let

$$M = \max_{(t,\mathbf{x}) \in D} |f(t, \mathbf{x})| \tag{3.1.7}$$

and

$$h = \min\{a, \frac{b}{M}\}. \tag{3.1.8}$$

Then the IVP

$$\mathbf{x}' = f(t, \mathbf{x}), \ \mathbf{x}(t_0) = \mathbf{x_0}, \tag{3.1.9}$$

where $\mathbf{x_0} = (x_{01}, x_{02}, \ldots, x_{0n})^T$, *has a unique solution, denoted by* $\mathbf{x}(t, t_0, \mathbf{x_0})$, *on the interval* $|t - t_0| \leq h$ *and passing through* $(t_0, \mathbf{x_0})$ *with* $|\mathbf{x} - \mathbf{x_0}| \leq b$.

Proof. We only give a sketch of the proof since it is identical to that of Theorem 2.1.2. Using Theorem 2.1.1 we have

$$\mathbf{x}(t) = \mathbf{x_0} + \int_{t_0}^t f(s, \mathbf{x}(s))ds.$$

In other words, if $\mathbf{x}(t)$ is a solution to the above integral, then it is to be understood that

$$\mathbf{x}_1(t) = x_{01} + \int_{t_0}^t f_1(s, x_1(s), \ldots, x_n(s))ds,$$

$$\mathbf{x}_2(t) = x_{02} + \int_{t_0}^t f_2(s, x_1(s), \ldots, x_n(s))ds,$$

$$\vdots$$

$$\mathbf{x}_n(t) = x_{0n} + \int_{t_0}^t f_n(s, x_1(s), \ldots, x_n(s))ds.$$

As for Picard's iterations, we define

$$\mathbf{x}_{k+1,1}(t) = x_{01} + \int_{t_0}^t f_1(s, x_{k,1}(s), \ldots, x_{k,n}(s))ds,$$

$$\mathbf{x}_{k+1,2}(t) = x_{02} + \int_{t_0}^t f_2(s, x_{k,1}(s), \ldots, x_{k,n}(s))ds,$$

$$\vdots$$

$$\mathbf{x}_{k+1,n}(t) = x_{0n} + \int_{t_0}^t f_n(s, x_{k,1}(s), \ldots, x_{k,n}(s))ds,$$

or

$$\mathbf{x}_{k+1}(t) = \mathbf{x_0} + \int_{t_0}^t f(s, \mathbf{x}_k(s))ds, \ k = 0, 1, \ldots, n, \tag{3.1.10}$$

where

$$\mathbf{x_0} = (x_{01}, x_{02}, \ldots, x_{0n})^T,$$

and

$$\mathbf{x}_k(t) = (\mathbf{x}_{k,1}(t), \mathbf{x}_{k,2}(t), \ldots, \mathbf{x}_{k,n}(t))^T.$$

Now use (3.1.10) and Theorem 2.1.2 to finish the proof.   □

To illustrate Picard's iteration, we present the following example.

Example 3.1. Consider the linear 2×2 system

$$x_1'(t) = 2x_1 - 3x_2,$$
$$x_2'(t) = x_1 - 2x_2,$$

with the initial conditions given by the vector

$$\mathbf{x}(0) = \mathbf{x_0} = \begin{pmatrix} x_{01} \\ x_{02} \end{pmatrix} = \begin{pmatrix} 1 \\ -1 \end{pmatrix}.$$

Setting $k = 0$, the first iteration is

$$\mathbf{x}_{1,1}(t) = 1 + \int_0^t (2 - 3)ds = 1 - t,$$

$$\mathbf{x}_{1,2}(t) = -1 + \int_0^t (1 - 2)ds = -1 - t.$$

Next, setting $k = 1$ gives

$$\mathbf{x}_{2,1}(t) = 1 + \int_0^t [2(1 - s) - 3(-1 - s)]ds = 1 + 5t + t^2/2,$$

$$\mathbf{x}_{2,2}(t) = -1 + \int_0^t [(1 - s) - 2(-1 - s)]ds = -1 + 3t + t^2/2.$$

Using the method of the next section, we can show that the true solution is

$$\mathbf{x} = \begin{pmatrix} 3e^t - 2e^{-t} \\ e^t - 2e^{-t} \end{pmatrix}.$$

Remark 3.1. It is essential to know that Theorems 2.1.4, 2.2.1, 2.2.2, 2.4.1, 2.4.2, 2.4.3, and 2.5.1 can be easily extended to the IVP (3.1.9). The instructor may assign them as exercises.

Next, we consider the nth-order differential equation

$$x^{(n)}(t) = g\left(t, x, x', \ldots, x^{(n-1)}\right)$$

and use the transformation

$$x_i(t) = x^{(i-1)}(t), \quad i = 1, 2, \ldots, n$$

to arrive at the following system:

$$x_1' = x_2,$$
$$x_2' = x_3,$$
$$\vdots$$
$$x_n' = g(t, x_1, x_2, \ldots, x_n).$$

The right side of this system defines a vector function $f(t, \mathbf{x})$. To fully describe the solution, we impose the initial conditions

$$x^{(i-1)}(t_0) = x_{0i} \quad \text{for } i = 1, 2, \ldots, n,$$

where x_{0i} for $i = 0, 1, \ldots, n-1$ are given constants. Thus our existence and uniqueness results mean that we can impose initial conditions on the function $\mathbf{x}$ and its first $n-1$ derivatives.

Example 3.2. Consider the system

$$x_1' = 1,$$
$$x_2' = \frac{x_3}{x_1^2},$$
$$x_3' = -\frac{x_2}{x_1^2}$$

with initial vector

$$\begin{pmatrix} x_1(\frac{1}{\pi}) \\ x_2(\frac{1}{\pi}) \\ x_3(\frac{1}{\pi}) \end{pmatrix} = \begin{pmatrix} \frac{1}{\pi} \\ 0 \\ -1 \end{pmatrix}.$$

Then the function

$$f(t, x) = \begin{pmatrix} 1 \\ \frac{x_3}{x_1^2} \\ -\frac{x_2}{x_1^2} \end{pmatrix}$$

is continuous with continuous partial derivatives on the set $D = \{(x_1, x_2, x_3) \in \mathbb{R}^3 : x_1 > 0\}$, and hence D is the region of existence of solutions.

To find the maximal interval, we try to solve the system. From $x_1' = 1$ we get $x_1(t) = t$, where we have applied the given initial condition. Substituting $x_1 = t$ into the second and third components of the system, and solving for t^2, we arrive at

$$\frac{x_3}{x_2'} = -\frac{x_2}{x_3'},$$

from which we have

$$\frac{d}{dt}(x_2^2 + x_3^2) = 0, \text{ or } x_2^2 + x_3^2 = C.$$

Applying the initial conditions, we get $C = 1$, that is,

$$x_2^2 + x_3^2 = 1.$$

It is clear that $x_2(t) = \sin(\frac{1}{t})$ and $x_3(t) = \cos(\frac{1}{t})$ satisfy $x_2^2 + x_3^2 = 1$ and the initial conditions. Thus the solution is given by

$$\mathbf{x}(t) = \begin{pmatrix} t \\ \sin(\frac{1}{t}) \\ \cos(\frac{1}{t}) \end{pmatrix},$$

and hence the maximal interval of existence is $(0, \infty)$. Note that $\lim\limits_{t \to 0^+} \mathbf{x}(t)$ does not exist.

Theorem 3.1.3. *(Global existence) Let β be a positive constant, and let $D = \mathbb{R}^n$. Let $t_0 \geq 0$, and let $J = [t_0, \beta]$ be an interval in $\mathbb{R}$. Suppose $f : J \times D \to \mathbb{R}^n$ is continuous and satisfies the global Lipschitz condition on $J \times D$ with Lipschitz constant $k > 0$. Assume the existence of two positive continuous functions $M(t)$ and $N(t)$ such that*

$$|f(t, \mathbf{x})| \leq M(t) + N(t)|\mathbf{x}| \text{ on } J \times D. \tag{3.1.11}$$

Then the nonlinear initial value problem

$$\mathbf{x}' = f(t, \mathbf{x}), \ \mathbf{x}(t_0) = \mathbf{x_0} \tag{3.1.12}$$

has a unique solution on the entire interval J.

Proof. First, we show a solution $\mathbf{x} : J \to \mathbb{R}^n$ exists on $[t_0, \beta]$. Due to the continuity of the functions $M(t)$ and $N(t)$, there are two positive constants M_1 and N_1 such that $M(s) \leq M_1$ and $N(s) \leq N_1$ on $[t_0, \beta]$. Thus, for $t_0 \leq t \leq \beta$, we have that

$$|\mathbf{x}(t)| = \left| \mathbf{x_0} + \int_{t_0}^t f(s, \mathbf{x}(s))ds \right|$$

$$\leq |\mathbf{x_0}| + \int_{t_0}^t \left(M_1 + N_1|\mathbf{x}(s)| \right)ds$$

$$\leq |\mathbf{x_0}| + M_1(\beta_1 - t_0) + \int_{t_0}^t N_1|\mathbf{x}(s)|ds.$$

Applying Theorem 2.1.5, we arrive at the inequality

$$|\mathbf{x}(t)| \leq \left[|\mathbf{x_0}| + M_1(\beta_1 - t_0) \right]e^{N_1(\beta_1 - t_0)}.$$

This shows $|\mathbf{x}(t)|$ is bounded on $[t_0, \beta]$. Now we show uniqueness. Suppose (3.1.12) has two solutions $\mathbf{x}(t)$ and $\mathbf{y}(t)$ and set $\mathbf{w}(t) = \mathbf{x}(t) - \mathbf{y}(t)$. Then $\mathbf{w}'(t) = \mathbf{x}'(t) - \mathbf{y}'(t)$. Therefore

$$\int_{t_0}^{t} \mathbf{w}'(s)ds = \mathbf{w}(t) - \mathbf{w}(t_0) = \int_{t_0}^{t} \Big(f(s, \mathbf{x}(s)) - f(s, \mathbf{y}(s)) \Big) ds \ \ \text{with}$$

$$\mathbf{w}(t_0) = \mathbf{x}(t_0) - \mathbf{y}(t_0) = 0.$$

So we arrive at

$$|\mathbf{w}(t)| \leq \int_{t_0}^{t} |f(s, \mathbf{x}(s)) - f(s, \mathbf{y}(s))|ds$$

$$\leq k \int_{t_0}^{t} |\mathbf{x}(s) - \mathbf{y}(s)|ds$$

$$= k \int_{t_0}^{t} |\mathbf{w}(s)|ds.$$

It follows from Gronwall's inequality that $|\mathbf{w}(t)| = 0$ for all $t \in [t_0, \beta]$ and the solution is unique on $[t_0, \beta]$. This completes the proof. $\square$

3.2 $x' = A(t)x$

Consider the linear system of ordinary differential equations of the form

$$x_1' = a_{11}(t)x_1(t) + a_{12}(t)x_2(t) + \dots, a_{1n}(t)x_n(t) + g_1(t),$$
$$x_2' = a_{21}(t)x_1(t) + a_{22}(t)x_2(t) + \dots, a_{2n}(t)x_n(t) + g_2(t),$$

$$\dots$$

$$x_n' = a_{n1}(t)x_1(t) + a_{n2}(t)x_2(t) + \dots, a_{nn}(t)x_n(t) + g_n(t),$$

where the functions a_{ij}, g_i, $1 \leq i \leq n$, $1 \leq j \leq n$, are continuous real-valued functions on an interval I. Using vector and matrix notations, the above system is equivalent to the vector equation

$$x'(t) = A(t)x(t) + g(t), \tag{3.2.1}$$

where

$$x := \begin{pmatrix} x_1 \\ x_2 \\ \vdots \\ x_n \end{pmatrix}, \ \ x' := \begin{pmatrix} x_1' \\ x_2' \\ \vdots \\ x_n' \end{pmatrix},$$

and

$$A(t) := \begin{pmatrix} a_{11} & a_{12} & \cdots & a_{1n} \\ a_{21} & a_{22} & \cdots & a_{2n} \\ \vdots & \vdots & \ddots & \vdots \\ a_{n1} & a_{n2} & \cdots & a_{nn} \end{pmatrix}, \quad g(t) := \begin{pmatrix} g_1(t) \\ g_2(t) \\ \vdots \\ g_n(t) \end{pmatrix}$$

for $t \in I$. Note that the matrix functions A and g are continuous on an interval I if and only if all their entries are continuous on I.

Definition 3.2.1. We say the $n \times 1$ vector y is a solution of (3.2.1) on I if y is continuously differentiable on I and

$$y'(t) = A(t)y(t) + g(t)$$

for all $t \in I$.

We denote the space of n-tuple continuously differentiable functions $x : I \to \mathbb{R}^n$ by $C^1(I, \mathbb{R}^n)$. Next, we define matrix norm.

Definition 3.2.2. A norm of a square matrix A is a non-negative real number denoted by $\|A\|$. There are several different ways of defining a matrix norm, but they all share the following properties:

1. $\|A\| \geq 0$ for any square matrix.
2. $\|A\| = 0$ if and only if $A = 0$ (zero matrix).
3. $\|kA\| = |k|\|A\|$ for any constant k.
4. $\|A + B\| \leq \|A\| + \|B\|$ for square matrices A and B.
5. $\|AB\| \leq \|A\|\|B\|$.

Example 3.3. Let A be a matrix with entries a_{ij}, $1 \leq i \leq n$, $1 \leq j \leq n$, that is, $A = (a_{ij})$. We define the following possible norms on the matrix A.

1. The 1-norm:

$$\|A\|_1 = \max_{1 \leq j \leq n} \left(\sum_{i=1}^{n} |a_{ij}| \right)$$

(the maximum absolute column sum).
2. The infinity norm:

$$\|A\|_\infty = \max_{1 \leq i \leq n} \left(\sum_{j=1}^{n} |a_{ij}| \right)$$

(the maximum of the sums of the absolute values along each row).
3. The Euclidean norm:

$$\|A\|_E = \sqrt{\sum_{i=1}^{n} \sum_{j=1}^{n} (a_{ij})^2}$$

(the square root of the sum of squares of all entries).

Note that if I is the identity matrix, then $||I||_1 = ||I||_\infty = 1$.

Next, we state and prove our first existence and uniqueness result regarding (3.2.1).

Theorem 3.2.1. *Suppose $A(t)$ and $g(t)$ are continuous on some interval $a \le t \le b$. Then (3.2.1) has a unique solution $\phi(t)$ on the interval $a \le t \le b$ with $\phi(t_0) = \eta$ and $a \le t_0 \le b$.*

Proof. Let $\| \cdot \|$ be a suitable matrix norm. For the existence, it suffices to show that solutions remain bounded for all $a \le t \le b$. We know from Theorem 3.1.1 that ϕ given by

$$\phi(t) = \eta + \int_{t_0}^{t} [A(s)\phi(s) + g(s)]ds$$

is a solution of (3.2.1) with $\phi(t_0) = \eta$. Let

$$K_1 = |\eta|, \quad K_2 = \max_{t \in [a,b]} |g(t)|(b-a), \quad \text{and} \quad K_3 = \max_{t \in [a,b]} ||A(t)||.$$

Then

$$|\phi(t)| \le |\eta| + \left| \int_{t_0}^{t} [A(s)\phi(s) + g(s)]ds \right|$$

$$\le K_1 + \int_{t_0}^{t} ||A(s)|| |\phi(s)|ds + \int_{t_0}^{t} |g(s)|ds$$

$$\le K_1 + \max_{t \in [a,b]} ||A(t)|| \int_{t_0}^{t} |\phi(s)|ds + \max_{t \in [a,b]} |g(t)| \int_{t_0}^{t} ds$$

$$\le K_1 + K_3 \int_{t_0}^{t} |\phi(s)|ds + \max_{t \in [a,b]} |g(t)|(b-a)$$

$$= \left(K_1 + K_2 \right) + K_3 \int_{t_0}^{t} |\phi(s)|ds. \tag{3.2.2}$$

Applying Theorem 2.1.5 (Gronwall's inequality) we arrive at the inequality

$$|\phi(t)| \le \left(K_1 + K_2 \right) e^{K_3(t-t_0)}$$

$$\le \left(K_1 + K_2 \right) e^{K_3(b-a)}. \tag{3.2.3}$$

From (3.2.3) we see that solutions are bounded for all $t \in [a, b]$ and hence exist. This completes the proof of the existence. For the uniqueness, we assume the existence of two solutions ϕ_1 and ϕ_2 of (3.2.1) passing through (t_0, η). Then imitating the proof of the existence, we arrive at

$$
\begin{aligned}
|\phi_1(t) - \phi_2(t)| &\leq \left| \int_{t_0}^{t} A(s)[\phi_1(s) - \phi_2(s)]ds \right| \\
&\leq \int_{t_0}^{t} ||A(s)|| |\phi_1(s) - \phi_2(s)|ds \\
&\leq \max_{t \in [a,b]} ||A(t)|| \int_{t_0}^{t} |\phi_1(s) - \phi_2(s)|ds \\
&\leq K_3 \int_{t_0}^{t} |\phi_1(s) - \phi_2(s)|ds.
\end{aligned}
$$

By Theorem 2.1.5 (Gronwall's inequality) it follows that

$$
|\phi_1(t) - \phi_2(t)| = 0 \ \text{ for all } \ t \in [a, b],
$$

and hence $\phi_1(t) = \phi_2(t)$ for all $t \in [a, b]$. The proof is complete. $\qquad\square$

As a consequence of Theorem 3.2.1, we have the following corollary.

Corollary 3.1. *If $A(t)$ and $g(t)$ are continuous on $\mathbb{R}$, then the unique solution $\phi(t)$ of (3.2.1) is defined for all $t \in \mathbb{R}$.*

In the next discussion, we try to estimate the error between two solutions. Let ϕ_1 and ϕ_2 be two solutions of (3.2.1) with $\phi_1(t_0) = \eta_1$ and $\phi_2(t_0) = \eta_2$, respectively. Then

$$
\begin{aligned}
|\phi_1(t) - \phi_2(t)| &\leq |\phi_1(t_0) - \phi_2(t_0)| + \left| \int_{t_0}^{t} A(s)[\phi_1(s) - \phi_2(s)]ds \right| \\
&\leq |\eta_1 - \eta_2| + \int_{t_0}^{t} ||A(s)|| |\phi_1(s) - \phi_2(s)|ds.
\end{aligned}
$$

Using Gronwall's inequality, we arrive at

$$
|\phi_1(t) - \phi_2(t)| \leq |\eta_1 - \eta_2| e^{\int_{t_0}^{t} ||A(s)||ds}, \tag{3.2.4}
$$

which is an estimate of the error between the two solutions at two different initial conditions.

Example 3.4. Consider the differential system

$$
\begin{aligned}
x_1' &= -x_2 \sin(t) + 4, \\
x_2' &= -x_1 + 2tx_2 - x_3 + e^t, \\
x_3' &= 3x_1 \cos(t) + x_2 + \frac{1}{t}x_3 - 5t^2.
\end{aligned}
$$

Consider two solutions $\phi_1(t)$ and $\phi_2(t)$ on the interval $(1, 3)$ with initial vectors

$$\phi_1(2) = \begin{pmatrix} 7 \\ 3 \\ -2 \end{pmatrix}$$

and

$$\phi_2(2) = \begin{pmatrix} 6.7 \\ 3.2 \\ -1.9 \end{pmatrix}.$$

Next, we compute the norm of the matrix

$$A(t) = \begin{pmatrix} 0 & -\sin(t) & 0 \\ -1 & 2t & -1 \\ 3\cos(t) & 1 & 1/t \end{pmatrix}.$$

Using the 1-norm, we get

$$||A(t)|| = \max\{1 + 3|\cos(t)|, |\sin(t)| + 2t + 1, 1 + \frac{1}{t}\}$$

or

$$||A(t)|| \leq \max\{4, 2 + 2t, 1 + \frac{1}{t}\} = 2 + 2t \leq 8, \ 1 < t < 3.$$

Using the estimate formula given by (3.2.4), we get

$$|\phi_1(t) - \phi_2(t)| \leq (0.3 + 0.2 + 0.1)e^{\int_2^t ||A(s)||ds}$$
$$\leq 0.6e^{\int_2^t 8ds}$$
$$\leq 0.6e^{8(t-2)} < 0.6e^{8(3-2)} \approx 1790.$$

Example 3.5. (Pendulum with friction) Fig. 3.1 shows a simple pendulum with mass-less string of length L and a bob of mass m acting under the external influence of gravity. Let g be the acceleration of gravity, m be the mass of the bob, L be the length of the rod, and b be the coefficient of friction (damping), which all are positive constants. The period is the time it takes for the bob to swing from its farthest right position to its left farthest position and back. Let $s(t)$ be the distance along the arc from the lowest point of the bob at time t, with displacement to the right considered positive. Let $\theta(t)$ be the corresponding angle with respect to the vertical. The gravitational force is directed downward and has the magnitude mg. The force acting in the tangential direction is $-mg\sin(\theta)$. Then the motion is modeled by the second-order differential equation

$$\frac{d^2s}{dt^2} = -g\sin(\theta).$$

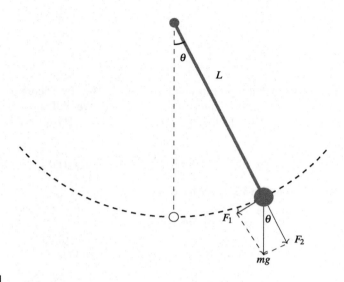

FIGURE 3.1

Pendulum with friction.

Now $s = L\theta$ as L is the radius of the circle, and so

$$\frac{d^2\theta}{dt^2} = -\frac{g}{L}\sin(\theta).$$

We add damping to the model and assume that the damping force is proportional to the velocity. As θ is our independent variable, the damping is represented as the force proportional to the angular velocity, say, $-b\frac{d\theta}{dt}$. The negative sign indicates that the damping force is in the opposite direction of the motion. Then the model takes the form

$$mL\theta'' + mg\sin(\theta) + bL\theta' = 0. \tag{3.2.5}$$

Assume the initial data

$$\theta(t_0) = \theta_0, \quad \theta'(t_0) = \omega_0. \tag{3.2.6}$$

To apply Theorem 2.4.4, we put (3.2.5) into a system. Thus we let $x_1(t) = \theta(t)$ and $x_2(t) = \theta'(t)$ and obtain the nonlinear two-dimensional system

$$x_1' = x_2,$$
$$x_2' = -\frac{g}{L}\sin(x_1) - \frac{b}{m}x_2, \tag{3.2.7}$$

with corresponding initial conditions

$$x_1(t_0) = \theta_0, \quad x_2(t_0) = \omega_0. \tag{3.2.8}$$

Let

$$f(t, x) = \begin{pmatrix} x_2 \\ -\frac{g}{L} \sin(x_1) - \frac{b}{m} x_2 \end{pmatrix}.$$

Clearly, f satisfies the global Lipschitz condition on $\mathbb{R}^2$. Thus by Theorems 2.4.3 and 2.4.4, the solution exists on some interval $[t_0, \alpha)$, $\alpha > 0$. We will show that $\alpha = \infty$ by appealing to Theorem 2.4.4. Define the *energy function* $E : [t_0, \alpha) \times \mathbb{R}^2 \to [0, \infty)$ by

$$E(t) := E(t, x_1, x_2) = \frac{1}{2} x_2^2(t) + \frac{g}{L} \left(1 - \cos(x_1(t)) \right). \tag{3.2.9}$$

Then along the solutions of (3.2.7) we have that

$$\frac{d}{dt} E(t) = x_2 \, x_2' + \frac{g}{L} \sin(x_1) \, x_1'$$

$$= x_2 \left(-\frac{g}{L} \sin(x_1) - \frac{b}{m} x_2 \right) + \frac{g}{L} \sin(x_1) x_2$$

$$= -\frac{b}{m} x_2^2 \leq 0.$$

Let $t_0 \leq t < \alpha$ and integrate the above inequality from t_0 to t to get

$$E(t) - E(t_0) \leq 0$$

or

$$E(t) \leq \frac{1}{2} \omega_0^2 + \frac{g}{L} \left(1 - \cos(\theta_0) \right).$$

We have from (3.2.9) that $\frac{1}{2} x_2^2(t) \leq E(t)$, so that

$$|x_2(t)| \leq \sqrt{ 2 \left[\frac{1}{2} \omega_0^2 + \frac{g}{L} \left(1 - \cos(\theta_0) \right) \right] } := M.$$

Using this in the first equation of (3.2.7), after integrating, we find

$$|x_1(t)| \leq |\theta_0| + M(t - t_0) \text{ for } t_0 \leq t < \alpha.$$

Now suppose for contradiction that $\alpha < \infty$. Then on $[t_0, \alpha)$,

$$|x_1(t)| \leq |\theta_0| + M(\alpha - t_0) \text{ and } |x_2(t)| \leq M.$$

Thus

$$|x| := \sqrt{x_1^2 + x_2^2} \leq \sqrt{M} + |\theta_0| + M(\alpha - t_0),$$

which is bounded on $[t_0, \alpha)$. This is a contradiction to Theorem 2.4.4, and hence $\alpha = \infty$. So the maximal interval of existence is $[t_0, \infty)$. It is worth noting that the *energy function* will be referred to as a *Lyapunov function*, which we will study in more detail in Chapter 7.

In Example 3.5 the energy function had to be chosen just right in order to yield the right results. Generally speaking, for any second-order differential equation of the form

$$x'' + f(x)x' + g(x) = 0 \tag{3.2.10}$$

and under more conditions on g, the energy function may take the form

$$E(t) := E(t, x_1, x_2) = \frac{1}{2}x_2^2(t) + \int_0^{x_1} g(s)ds \tag{3.2.11}$$

by transforming (3.2.10) into a system using the transformations

$$x_i(t) = x^{(i-1)}(t), \quad i = 1, 2.$$

On the other hand, if we do not transform (3.2.10) into a system, then we may use the energy function

$$E(t) := E(t, x) = \frac{1}{2}(x')^2 + \int_0^{x(t)} g(s)ds, \tag{3.2.12}$$

where $\frac{1}{2}(x')^2$ corresponds to the *kinetic energy*, and $\int_0^{x(t)} g(s)ds$ represents the *potential energy*.

3.2.1 Fundamental matrix

Solving the non-homogeneous equation

$$x'(t) = A(t)x(t) + g(t) \tag{3.2.13}$$

requires solving the homogeneous system

$$x'(t) = A(t)x(t), \tag{3.2.14}$$

where $A(t)$ is an $n \times n$ matrix of coefficients $a_{ij}(t)$, which are assumed to be continuous on an interval I. Recall that a solution $x(t)$ of (3.2.14) is an n-tuple of C^1 functions $x_i : I \to \mathbb{R}$. We adopt the notation

$$x(t) = \begin{pmatrix} x_1(t) \\ x_2(t) \\ \vdots \\ x_n(t) \end{pmatrix}.$$

The solution x maybe considered as a C^1 vector-valued function $x : I \to \mathbb{R}^n$. The space of such functions is denoted by $C^1(I, \mathbb{R}^n)$. If $\mathscr{S}$ is the *solution space* of (3.2.14), then $\mathscr{S} \subset C^1(I, \mathbb{R}^n)$.

Definition 3.2.3. We say that a set $\{f_1, f_2, \ldots, f_n\}$ of $n \times 1$ vector functions defined on an interval I is *linearly dependent* on I if there exist constants $c_1, c_2, \ldots, c_n$, not all zero, such that

$$c_1 f_1 + c_2 f_2 + \ldots + c_n f_n = 0$$

for every $t \in I$. If the set of functions is not *linearly dependent* on the interval I, it is said to be linearly independent.

We have the following important theorem regarding the solution space $\mathscr{S}$.

Theorem 3.2.2. *The solution space $\mathscr{S}$ is a linear vector space.*

Proof. Let $x \in C^1(I, \mathbb{R}^n)$ be a solution of (3.2.14). It is clear that $x = 0$ is a solution of (3.2.14). If $x(t_*) = 0$ for some $t_* \in I$, then $x(t) = 0$ for all $t \in I$. This is true due to the uniqueness of the solution. Thus the zero solution is in $\mathscr{S}$. Next, we show that $\mathscr{S}$ is closed under addition and scalar multiplication. Let $\phi_1(t)$ and $\phi_2(t)$ be two arbitrary functions in $\mathscr{S}$, and let c_1 c_2 be constants in $\mathbb{R}$. Set $y(t) = c_1\phi_1(t) + c_2\phi_2(t)$. Then

$$\begin{aligned}
y'(t) &= (c_1\phi_1(t) + c_2\phi_2(t))' \\
&= c_1\phi_1'(t) + c_2\phi_2'(t) \\
&= c_1 A(t)\phi_1(t) + c_2 A(t)\phi_2(t) \\
&= A(t)\Big(c_1\phi_1(t) + c_2\phi_2(t)\Big) \\
&= A(t)y(t).
\end{aligned}$$

Next, we show that the set of solutions of (3.2.14) is linearly independent. Let

$$\{\phi_1(t), \phi_2(t), \ldots, \phi_n(t)\}$$

be a set of solutions of (3.2.14). We show that if $\{\phi_1(t), \phi_2(t), \ldots, \phi_n(t)\}$ is linearly independent at some $t_* \in I$, then the set is linearly independent for all $t \in I$, that is,

$$c_1\phi_1(t) + c_2\phi_2(t) + \ldots + c_n\phi_n(t) = 0$$

only in the case $c_1 = c_2 = \ldots = c_n = 0$. Suppose the contrary, that is, there are c_i, $i = 1, 2, \ldots, n$, not all zero such that

$$c_1\phi_1(t_1) + c_2\phi_2(t_1) + \ldots + c_n\phi_n(t_1) = 0$$

for some $t_1 \in I$, $t_1 \neq t_*$. Now we use the same constants c_i to define a function for all $t \in I$,

$$x(t) = c_1\phi_1(t) + c_2\phi_2(t) + \ldots + c_n\phi_n(t). \tag{3.2.15}$$

Then by the above relation we must have $x(t_1) = 0$, since it is a solution of (3.2.14). Thus by uniqueness $x(t) = 0$ for all $t \in I$. This is a contradiction to the fact that

$$c_1\phi_1(t_*) + c_2\phi_2(t_*) + \ldots + c_n\phi_n(t_*) = 0$$

only in the case $c_1 = c_2 = \ldots = c_n = 0$. This completes the proof of the linear independence. Left to show that $\mathscr{S}$ is an n-dimensional subspace of the vector space $C^1(I, \mathbb{R}^n)$. Let $e_1, e_2, \ldots, e_n$ be the standard basis of $\mathbb{R}^n$. Then by Corollary 3.1 there exists a set $\{\phi_1(t), \phi_2(t), \ldots, \phi_n(t)\}$ of n solutions that form a set of solutions of (3.2.14) satisfying

$$\phi_1(t_0) = e_1, \phi_2(t_0) = e_2, \ldots, \phi_n(t_0) = e_n. \tag{3.2.16}$$

This set of solutions is linearly independent for all $t \in I$ and hence forms a linearly independent set in the solution space $\mathscr{S}$. Let $\phi(t)$ be an arbitrary solution of (3.2.14) and define

$$\phi_0 = \phi(t_0). \tag{3.2.17}$$

Now since $\phi_0 \in \mathbb{R}^n$, there are constants $c_i, i = 1, 2, \ldots, n$, not all zero, such that

$$\phi_0 = c_1 e_1 + c_2 e_2 + \ldots + c_n e_n. \tag{3.2.18}$$

Using the same constants of (3.2.18), we may define a new function $\psi : I \to \mathbb{R}^n$,

$$\psi(t) = c_1 \phi_1(t) + c_2 \phi_2(t) + \ldots + c_n \phi_n(t). \tag{3.2.19}$$

Since each $\phi_i, i = 1, 2, \ldots, n$, is a solution of (3.2.14), so is $\psi(t)$. By (3.2.16) and (3.2.18), we have that $\psi(t_0) = \phi_0$. This and (3.2.17) imply that ψ and ϕ have the same initial condition, and hence by the uniqueness of the solution we have that

$$\psi(t) = \phi(t) \text{ for all } t \in \mathbb{R}.$$

From (3.2.19) it follows that the arbitrary solution may be written as

$$\psi(t) = c_1 \phi_1(t) + c_2 \phi_2(t) + \cdots + c_n \phi_n(t),$$

which is a linear combination of the solutions $\phi_i, i = 1, 2, \ldots, n$. Therefore the set of solutions

$$\{\phi_1(t), \phi_2(t), \ldots, \phi_n(t)\}$$

is a basis for the solution space $\mathscr{S}$. This implies that the dimension of $\mathscr{S}$ is n. This completes the proof. $\qquad\square$

Definition 3.2.4. (*Fundamental set of solutions*) A set of n solutions of the linear differential system (3.2.14), all defined on the same open interval I, is called a fundamental set of solutions on I if the solutions are linearly independent functions on I.

We have the following corollary.

Corollary 3.2. *Let $A(t)$ be an $n \times n$ matrix of continuous coefficients $a_{ij}(t)$ on an interval I. If $\{\phi_1(t), \phi_2(t), \ldots, \phi_n(t)\}$ form a fundamental set of solutions on I, then* *the general solution of* (3.2.14) *is given by*

$$x(t) = c_1\phi_1(t) + c_2\phi_2(t) + \cdots + c_n\phi_n(t)$$

with constants $c_i, i = 1, 2, \ldots, n$.

The following definition is needed for the next theorem.

Definition 3.2.5. Let f and g be two continuous functions on $\mathbb{R}$. Then we write

$$f(x) = O\big(g(x)\big)$$

if there exists a positive real constant M and a real number x^* such

$$|f(x)| \leq Mg(x) \quad \text{for all} \quad x \geq x^*.$$

Theorem 3.2.3. *If* $\{\phi_1(t), \phi_2(t), \ldots, \phi_n(t)\}$ *form a fundamental set of solutions of* (3.2.14) *on I and X is the matrix function with columns* $\{\phi_1(t), \phi_2(t), \ldots, \phi_n(t)\}$, *then*

$$\det X(t) = \det X(t_0)e^{\int_{t_0}^t \operatorname{tr}(A(s))ds} \quad \text{for } t \in I. \tag{3.2.20}$$

Proof. Let the Wronskian $w(t) := \det X(t)$. Let $t_0 \in I$ and expand $X(t)$ around t_0:

$$\begin{aligned} X(t) &= X(t_0) + (t - t_0)X'(t_0) + O\big((t - t_0)^2\big) \\ &= X(t_0) + (t - t_0)A(t_0)X(t_0) + O\big((t - t_0)^2\big) \\ &= \big[I + (t - t_0)A(t_0)\big]X(t_0) + O\big((t - t_0)^2\big). \end{aligned}$$

As a consequence, we have

$$\det X(t) = \det\Big(\big[I + (t - t_0)A(t_0)\big]\Big)\det X(t_0)$$

or

$$W(t) = \det\big[I + (t - t_0)A(t_0)\big]W(t_0).$$

By the Terence Tao matrix identity, which says

$$\det(C + hB) = \det(C) + h\operatorname{tr}\big(\operatorname{adj}(C)B\big) + O(h^2),$$

we have that

$$\det(I + (t - t_0)A(t_0))W(t_0) = W(t_0)\big[1 + (t - t_0)\operatorname{tr}\big(A(t_0)\big)\big]$$

or

$$W(t) = W(t_0)\big[1 + (t - t_0)\operatorname{tr}\big(A(t_0)\big)\big]. \tag{3.2.21}$$

Next, we expand $W(t)$ in Taylor series around t_0:

$$W(t) = W(t_0) + (t - t_0)W'(t_0) + O\big((t - t_0)^2\big).$$

Substituting the value of $W(t)$ given by (3.2.21) into the above equality gives

$$W'(t_0) = W(t_0)\text{tr}\big(A(t_0)\big).$$

Since t_0 is arbitrary, for any time $t \in \mathbb{R}$, we have the equation

$$W'(t) = W(t)\text{tr}\big(A(t)\big),$$

which has the solution

$$W(t) = W(t_0)e^{\int_{t_0}^{t} \text{tr}(A(s)ds)} \quad \text{for } t \in I.$$

This completes the proof. □

Definition 3.2.6. (*Fundamental matrix*) A matrix solution Φ is called a *fundamental matrix solution* (or, shortly, fundamental matrix) of (3.2.14) on an interval I if its columns form a fundamental set of solutions. If, in addition, $\Phi(t_0) = I$, then a fundamental matrix solution is called the *principal fundamental matrix solution* (or, shortly, principal matrix).

Theorem 3.2.4. *The matrix Φ is a fundamental matrix of $x' = Ax$ at t_0 if and only if*

(a) Φ *is a solution of $x' = Ax$*
  and
(b) $\det\Phi(t_0) \neq 0$.

Proof. We know that $\Phi' = A\Phi$. Let

$$\Phi = \big[\phi_1(t), \phi_2(t), \ldots, \phi_n(t)\big],$$

where the columns of Φ consist of $\phi_1(t), \phi_2(t), \ldots, \phi_n(t)$. Then

$$\Phi' = \big[\phi_1'(t), \phi_2'(t), \ldots, \phi_n'(t)\big],$$

and

$$A\Phi = \big[A\phi_1(t), A\phi_2(t), \ldots, A\phi_n(t)\big].$$

Since two matrices are equal if and only if their corresponding columns are the same,

$$\Phi'(t) = A\Phi(t)$$

if and only if

$$x'_j = Ax_j$$

or if and only if each column of Φ is a solution.

Condition (b) is equivalent to the linear independence of

$$\phi_1(t_0), \phi_2(t_0), \ldots, \phi_n(t_0).$$

This completes the proof. $\qquad\qquad\qquad\qquad\qquad\qquad\qquad\qquad\square$

We claim that the matrix

$$\Phi(t) = \begin{pmatrix} e^{-t} & te^{-t} \\ -e^{-t} & (1-t)e^{-t} \end{pmatrix}$$

is the fundamental matrix for the system

$$x'(t) = \begin{pmatrix} 0 & 1 \\ -1 & -2 \end{pmatrix} \begin{pmatrix} x_1 \\ x_2 \end{pmatrix}.$$

To see this, we have

$$\Phi'(t) = \begin{pmatrix} -e^{-t} & e^{-t} - te^{-t} \\ e^{-t} & -e^{-t} - (1-t)e^{-t} \end{pmatrix} = \begin{pmatrix} -e^{-t} & (1-t)e^{-t} \\ e^{-t} & -2e^{-t} + te^{-t} \end{pmatrix}.$$

On the other hand,

$$\begin{pmatrix} 0 & 1 \\ -1 & -2 \end{pmatrix} \Phi = \begin{pmatrix} 0 & 1 \\ -1 & -2 \end{pmatrix} \begin{pmatrix} e^{-t} & te^{-t} \\ -e^{-t} & (1-t)e^{-t} \end{pmatrix} = \begin{pmatrix} -e^{-t} & (1-t)e^{-t} \\ e^{-t} & te^{-t} - 2e^{-t} \end{pmatrix}.$$

This proves the claim. For the next two theorems, we suppress the argument t.

Theorem 3.2.5. *If Φ is a fundamental matrix of (3.2.14), then ΦC, where C is an arbitrary $n \times n$ nonsingular constant matrix, is a fundamental matrix of (3.2.14). Conversely, if Ψ is another fundamental matrix to (3.2.14), then there exists a constant nonsingular $n \times n$ matrix C such that $\Psi = \Phi C$ for all $t \in I$.*

Proof. Let Φ be a fundamental matrix of (3.2.14) on I. Then

$$(\Phi C)' = \Phi' C = (A(t)\Phi)C = A(t)(\Phi C).$$

So ΦC solves (3.2.14). Next,

$$\det(\Phi C) = \det\Phi \det C \neq 0,$$

and by Theorem 3.2.4, ΦC is a fundamental matrix of (3.2.14). Conversely, let Φ and Ψ be two fundamental matrix solutions of (3.2.14). Since $\Phi\Phi^{-1} = I$, we have that

$\Phi'\Phi^{-1} + \Phi(\Phi^{-1})' = 0$ or $(\Phi^{-1})' = -\Phi^{-1}\Phi'\Phi^{-1}$. Now we differentiate $\Phi^{-1}\Psi$ and obtain

$$\begin{aligned}
(\Phi^{-1}\Psi)' &= (\Phi^{-1})'\Psi + \Phi^{-1}\Psi' \\
&= -\Phi^{-1}\Phi'\Phi^{-1}\Psi + \Phi^{-1}\Psi' \\
&= \left(-\Phi^{-1}A(t)\Phi\Phi^{-1} + \Phi^{-1}A(t)\right)\Psi \\
&= \left(-\Phi^{-1}A(t) + \Phi^{-1}A(t)\right)\Psi \\
&= 0.
\end{aligned}$$

Therefore integrating the above expression gives $\Phi^{-1}\Psi = C$ for some $n \times n$ constant matrix C, or $\Psi = \Phi C$. Furthermore, as Ψ and Φ are fundamental matrix solutions, we have $\det\Phi \neq 0$ and $\det\Psi \neq 0$, and hence $\det C \neq 0$. This completes the proof. $\qquad\square$

We remark that it is not true in general that $C\Phi$ is a fundamental matrix of (3.2.14).

Theorem 3.2.6. *If Φ is a fundamental matrix of (3.2.14) on an interval I, then Φc solves (3.2.14) on I for every $n \times 1$ constant vector c.*

Proof. Let Φ be a fundamental matrix of (3.2.14) on I. Then

$$(\Phi c)' = \Phi'c = (A(t)\Phi)c = A(t)(\Phi c).$$

So Φc solves (3.2.14). $\qquad\square$

Theorem 3.2.7. *If $\Phi(t)$ is a fundamental matrix of (3.2.14) on an interval I, then $\Phi(t)c$ with $c = \Phi^{-1}(t_0)x_0$ is a solution of (3.2.14) with $x(t_0) = x_0$.*

Proof. First, $\Phi^{-1}(t_0)$ exists by the definition of a fundamental matrix. Hence by Theorem 3.2.6 $x(t) = \Phi(t)\Phi^{-1}(t_0)x_0$ solves (3.2.14). Moreover, $x(t_0) = \Phi(t_0) \times \Phi^{-1}(t_0)x_0 = Ix_0 = x_0$. This completes the proof. $\qquad\square$

3.2.2 $x' = Ax$

Now we consider homogeneous systems with constant matrices. In particular, we look at

$$x'(t) = Ax(t), \qquad (3.2.22)$$

where A is an $n \times n$ constant matrix. The fundamental solution of (3.2.22) can be constructed from a knowledge of the eigenvalues and eigenvectors of A. We begin with the following definition.

Definition 3.2.7. Let A be an $n \times n$ constant matrix, shortly, a "matrix." A number λ is said to be an *eigenvalue* of A if there exists a nonzero vector v such that

$$Av = \lambda v. \qquad (3.2.23)$$

The solution vector v is said to be an *eigenvector* corresponding to the eigenvalue λ. We may refer to λ and v as an *eigenpair*.

Now consider a solution x of (3.2.22) of the form

$$x(t) = v e^{\lambda t}$$

for $v \in \mathbb{R}^n$ and $\lambda \in \mathbb{R}$. Then we have

$$\lambda v e^{\lambda t} = A v e^{\lambda t},$$

or

$$(A - \lambda I)v = 0 \text{ if and only if } \det(A - \lambda I) = 0.$$

Thus we have the following straightforward result.

Theorem 3.2.8. *If λ_0, v_0 is an eigenpair of A, then*

$$x(t) = e^{\lambda_0 t} v_0 = v_0 e^{\lambda_0 t}$$

is a solution of $x' = Ax$.

Proof. Let $x(t) = e^{\lambda_0 t} v_0$. Then

$$x'(t) = \lambda_0 e^{\lambda_0 t} v_0 = \lambda_0 v_0 e^{\lambda_0 t} = A v_0 e^{\lambda_0 t} = A e^{\lambda_0 t} v_0 = Ax,$$

as desired. This completes the proof. $\qquad\qquad\qquad\qquad\qquad\qquad\qquad\square$

Theorem 3.2.9. *(Independent eigenvectors) Let $v_1, v_2, \ldots, v_p$ be the corresponding eigenvectors to the distinct eigenvalues $\lambda_1, \lambda_2, \ldots, \lambda_p$ of a matrix A. Then $v_1, v_2, \ldots, v_p$ are linearly independent.*

Proof. Suppose $v_1, v_2, \ldots, v_j$ are linearly independent for positive integer j. If $j < p$, then v_{j+1} can be written as a linear combination of the vectors $v_1, v_2, \ldots, v_j$, that is, there are constants $c_1, c_2, \ldots, c_j$ such that

$$v_{j+1} = c_1 v_1 + c_2 v_2 + \cdots + c_j v_j.$$

Multiply from the left by the matrix A and apply the fact that $A v_i = \lambda_i v_i$ for $i = 1, 2, \ldots j$ to arrive at

$$\begin{aligned}
A v_{j+1} &= \lambda_{j+1} v_{j+1} \\
&= \lambda_{j+1}\left(c_1 v_1 + c_2 v_2 + \ldots + c_j v_j\right) \\
&= c_1 \lambda_{j+1} v_1 + c_2 \lambda_{j+1} v_2 + \ldots + c_j \lambda_{j+1} v_j.
\end{aligned}$$

On the other hand,

$$Av_{j+1} = A\big(c_1 v_1 + c_2 v_2 + \ldots + c_j v_j\big)$$
$$= c_1 A v_1 + c_2 A v_2 + \ldots + c_j A v_j$$
$$= c_1 \lambda_1 v_1 + c_2 \lambda_2 v_2 + \ldots + c_j \lambda_j v_j.$$

Subtracting the two equations gives

$$c_1(\lambda_{j+1} - \lambda_1)v_1 + c_2(\lambda_{j+1} - \lambda_2)v_2 + \ldots + c_j(\lambda_{j+1} - \lambda_j)v_j = 0.$$

Since $v_1, v_2, \ldots, v_j$ are linearly independent, we must have that

$$c_1(\lambda_{j+1} - \lambda_1) = 0, \; c_2(\lambda_{j+1} - \lambda_2) = 0, \; \ldots, \; c_j(\lambda_{j+1} - \lambda_j) = 0.$$

Since

$$\lambda_{j+1} - \lambda_j \neq 0, \;\; \text{for all} \;\; j = 1, 2, \ldots, n,$$

this could only be possible if

$$c_1 = c_2, \ldots, c_j = 0.$$

This implies that

$$v_{j+1} = 0 \;\; \text{(zero vector)},$$

which is a contradiction, since v_{j+1} is the eigenvector corresponding to λ_{j+1}. This completes the proof. $\qquad\square$

As a result of Definition 3.2.4, Corollary 3.2, and Theorem 3.2.9, we have the following theorem concerning systems with constant matrices such as (3.2.22).

Theorem 3.2.10. *(Distinct eigenvalues) Let $\lambda_1, \lambda_2, \ldots, \lambda_n$ be n distinct real eigenvalues of the matrix A of (3.2.22), and let $K_1, K_2, \ldots, K_n$ be the corresponding eigenvectors. Then the general solution of (3.2.22) on the interval $I = (-\infty, \infty)$ is given by*

$$x(t) = c_1 K_1 e^{\lambda_1 t} + c_2 K_2 e^{\lambda_2 t}, \ldots, c_n K_n e^{\lambda_n t}$$

for constants $c_i, i = 1, 2, \ldots, n$.

Remark 3.2. If we let

$$\phi_i = K_i e^{\lambda_i t}, \;\; i = 1, 2, \ldots, n,$$

then the fundamental matrix Φ of (3.2.22) is formed by taking its columns to be

$$\Phi = \big[\phi_1(t), \phi_2(t), \ldots, \phi_n(t)\big].$$

Example 3.6. (*Distinct eigenvalues*) Consider the linear system

$$x' = \begin{pmatrix} -4 & 1 & 1 \\ 1 & 5 & -1 \\ 0 & 1 & -3 \end{pmatrix} \begin{pmatrix} x_1 \\ x_2 \\ x_3 \end{pmatrix}.$$

Then the eigenpairs are given by

$$\lambda_1 = -3, K_1 = \begin{pmatrix} 1 \\ 0 \\ 1 \end{pmatrix}, \quad \lambda_2 = -4, K_2 = \begin{pmatrix} 10 \\ -1 \\ 1 \end{pmatrix}, \quad \lambda_3 = 5, K_3 = \begin{pmatrix} 1 \\ 8 \\ 1 \end{pmatrix},$$

and hence the general solution can be written as

$$x(t) = c_1 \begin{pmatrix} 1 \\ 0 \\ 1 \end{pmatrix} e^{-3t} + c_2 \begin{pmatrix} 10 \\ -1 \\ 1 \end{pmatrix} e^{-4t} + c_3 \begin{pmatrix} 1 \\ 8 \\ 1 \end{pmatrix} e^{5t}.$$

Moreover, the fundamental matrix $\Phi(t)$ is given by

$$\Phi(t) = \begin{pmatrix} e^{-3t} & 10e^{-4t} & e^{5t} \\ 0 & -e^{-4t} & 8e^{5t} \\ e^{-3t} & e^{-4t} & e^{5t} \end{pmatrix}. \tag{3.2.24}$$

Now suppose we impose that every solution satisfies the initial condition

$$x(t_0) := x_0 = \begin{pmatrix} x_{01} \\ x_{02} \\ x_{03} \end{pmatrix}.$$

Then by Theorem 3.2.7, the unique solution is

$$x(t) = \Phi(t)\Phi^{-1}(t_0)x_0,$$

where $\Phi(t)$ is given by (3.2.24).

Next, we discuss repeated roots. For example, the system

$$x' = \begin{pmatrix} 3 & -18 \\ 2 & -19 \end{pmatrix} \begin{pmatrix} x_1 \\ x_2 \end{pmatrix} \tag{3.2.25}$$

has the repeated eigenvalue $\lambda_1 = \lambda_2 = -3$. If $K_1 = \begin{pmatrix} k_1 \\ k_2 \end{pmatrix}$ is the corresponding eigenvector, then we have the two equations $6k_1 - 18k_2 = 0$, $2k_1 - 6k_2 = 0$, which are both equivalent to $k_1 = 3k_2$. As a consequence, we have the single eigenvector $K_1 = \begin{pmatrix} 3 \\ 1 \end{pmatrix}$, and so one solution is given by

$$\phi_1 = \begin{pmatrix} 3 \\ 1 \end{pmatrix} e^{-3t}.$$

Since we are interested in finding the general solution, we need to examine the question of finding another solution.

In general, if m is a positive integer and $(\lambda - \lambda_1)^m$ is a factor of the characteristic equation $\det(A - \lambda I) = 0$, whereas $(\lambda - \lambda_1)^{m+1}$ is not a factor, then λ_1 is said to be an eigenvalue of multiplicity m. We discuss two such scenarios:

(a) For some $n \times n$ matrix A, it may be possible to find m linearly independent eigenvectors $K_1, K_2, \ldots, K_n$ corresponding to an eigenvalue λ_1 of multiplicity $m \le n$. In this case the general solution of the system contains the linear combination

$$c_1 K_1 e^{\lambda_1 t} + c_2 K_2 e^{\lambda_2 t} + \ldots + c_n K_n e^{\lambda_n t}.$$

(b) If there is one eigenvector corresponding to an eigenvalue of multiplicity m, then we can find m linearly independent solutions of the form

$$\phi_1 = K_{11} e^{\lambda_1 t},$$
$$\phi_2 = K_{21} t e^{\lambda_1 t} + K_{22} e^{\lambda_1 t},$$
$$\vdots$$
$$\phi_m = K_{m1} \frac{t^{m-1}}{(m-1)!} e^{\lambda_1 t} + K_{m2} \frac{t^{m-2}}{(m-2)!} e^{\lambda_1 t} + K_{mm} e^{\lambda_1 t},$$

where K_{ij} are column vectors, known as *generalized eigenvectors*. For an illustration of case (b), we suppose λ_1 is an eigenvalue of multiplicity two with only one corresponding eigenvector. To find the second eigenvector, in general, we assume a second solution of

$$x' = Ax$$

of the form

$$\phi_2(t) = K t e^{\lambda_1 t} + P e^{\lambda_1 t}, \tag{3.2.26}$$

where

$$P = \begin{pmatrix} p_1 \\ p_2 \\ \vdots \\ p_n \end{pmatrix} \quad \text{and} \quad K = \begin{pmatrix} k_1 \\ k_2 \\ \vdots \\ k_n \end{pmatrix}$$

are to be found. Differentiate $\phi_2(t)$ and substitute back into $x' = Ax$ to get

$$(AK - \lambda_1 K) t e^{\lambda_1 t} + (AP - \lambda_1 P - K) e^{\lambda_1 t} = 0.$$

Since this equation must hold for all t, we have

$$(A - \lambda_1 I)K = 0 \tag{3.2.27}$$

and

$$(A - \lambda_1 I)P = K. \tag{3.2.28}$$

Eq. (3.2.27) implies that K must be an eigenvector of A associated with the eigenvalue λ_1, and hence we have one solution $\phi_1(t) = Ke^{\lambda_1 t}$. To find the second solution given by (3.2.26), we must solve for the vector P in (3.2.28). Now we go back to finding a second solution for (3.2.24). We already know K given by K_1. Let $P = \begin{pmatrix} p_1 \\ p_2 \end{pmatrix}$. Then from (3.2.28) we have $(A + 3I)P = K$, which implies that $6p_1 - 18p_2 = 3$ or $2p_1 - 6p_2 = 1$. Since these two equations are equivalent, we may choose $p_1 = 1$ and find $p_2 = 1/6$. However, for simplicity, we choose $p_1 = 1/2$, so that $p_2 = 0$. Using (3.2.26), we find that

$$\phi_2(t) = \begin{pmatrix} 3 \\ 1 \end{pmatrix} te^{-3t} + \begin{pmatrix} 1/2 \\ 0 \end{pmatrix} e^{-3t}.$$

Thus the general solution is

$$x(t) = c_1\phi_1(t) + c_2\phi_2(t) = c_1 \begin{pmatrix} 3 \\ 1 \end{pmatrix} e^{-3t} + c_2 \left(\begin{pmatrix} 3 \\ 1 \end{pmatrix} te^{-3t} + \begin{pmatrix} 1/2 \\ 0 \end{pmatrix} e^{-3t} \right).$$

Moreover, the fundamental matrix $\Phi(t)$ is given by

$$\Phi(t) = \begin{pmatrix} 3e^{-3t} & 3te^{-3t} + \frac{1}{2}e^{-3t} \\ e^{-3t} & te^{-3t} \end{pmatrix}.$$

3.2.3 Exponential matrix e^{At}

In the previous section we looked at the concept of a fundamental matrix $\Phi(t)$ for the linear system $x'(t) = A(t)x(t)$ and applied it only to constant matrices. In this section we study in detail such constant matrices; namely, we consider

$$x'(t) = Ax(t), \quad x(t_0) = x_0, \tag{3.2.29}$$

where A is an $n \times n$ constant matrix, and $x \in \mathbb{R}^n$. Note that

$$\|Ax - Ay\| \le K\|x - y\|,$$

where $K > 0$ is constant, and $\| \cdot \|$ is the matrix norm of A defined by

$$\|A\| = \sup_{x \in \mathbb{R}^n - \{0\}} \frac{|Ax|}{|x|}.$$

Then by Picard's theorem (Theorem 3.1.2) there exists a solution on the interval $|t - t_0| < h$ for $x_0 : [-h, h] \to \mathbb{R}^n$. We may define Picard's iteration by

$$x_{n+1}(t) = x_0 + \int_{t_0}^{t} Ax_n(s)ds, \quad x_0(t) = x_0, \tag{3.2.30}$$

with the understanding that

$$x_i = (\phi_1^i, \phi_2^i, \ldots, \phi_n^i)^T, \ i = 1, 2, \ldots, n,$$

where $\phi_j^i \in \mathbb{R}^n$, $j = 1, 2, \ldots, n$. Setting $n = 0$ in (3.2.30), we get

$$x_1 = x_0 + \int_{t_0}^t Ax_0 ds = x_0 + (t - t_0)Ax_0.$$

Similarly, for $n = 1$, we have

$$x_2 = x_0 + \int_{t_0}^t Ax_1(s)ds = x_0 + \int_{t_0}^t A(s - t_0)Ax_0 ds$$

$$= x_0 + (t - t_0)Ax_0 + \frac{(t - t_0)^2}{2}A^2 x_0.$$

Using an induction argument, we arrive at

$$x_n = x_0 + (t - t_0)Ax_0 + \frac{(t - t_0)^2}{2}A^2 x_0 + \ldots + \frac{(t - t_0)^n}{n!}A^n x_0$$

$$= \sum_{k=0}^n \frac{(t - t_0)^k}{k!}A^k x_0.$$

Letting $n \to \infty$, we have

$$x(t) = e^{A(t - t_0)}x_0, \tag{3.2.31}$$

where

$$e^{A(t - t_0)} = \sum_{n=0}^\infty \frac{(t - t_0)^n}{n!}A^n. \tag{3.2.32}$$

It readily follows from the ratio test that the infinite series in (3.2.32) converges absolutely for all $t - t_0 \in \mathbb{R}$. Thus $e^{A(t - t_0)}$ exists for all times. Note that the infinite series given by (3.2.32) can be differentiated term by term, and hence

$$\frac{d}{dt}e^{A(t - t_0)} = A\sum_{n=0}^\infty n\frac{(t - t_0)^n}{n!}A^n$$

$$= \sum_{n=1}^\infty \frac{(t - t_0)^{n-1}}{(n - 1)!}A^n$$

$$= A\sum_{n=1}^\infty \frac{(t - t_0)^{n-1}}{(n - 1)!}A^{n-1}$$

$$= A \sum_{k=0}^{\infty} \frac{(t-t_0)^k}{k!} A^k$$

$$= A e^{A(t-t_0)}.$$

This shows that if

$$x(t) = e^{A(t-t_0)} x_0,$$

then $x(t)$ satisfies $x' = Ax$. Moreover,

$$x(t) = e^{A(t-t_0)} x_0 = \left[I + (t-t_0)A + \frac{(t-t_0)^2}{2} A^2 + \ldots + \frac{(t-t_0)^n}{n!} A^n + \ldots \right] x_0,$$

and hence

$$x(t_0) = I x_0 = x_0.$$

We have the following definition.

Definition 3.2.8. Let A be an $n \times n$ constant matrix. Then we define the *exponential matrix* function by $e^{A(t-t_0)}$, which is the solution of $x' = Ax$, and $e^{A(0)} = I$ (identity matrix).

Thus we have already proved that

$$x(t) = e^{A(t-t_0)} x_0 \tag{3.2.33}$$

is the unique solution of

$$x'(t) = Ax(t), \quad x(t_0) = x_0,$$

for all $t \in \mathbb{R}$.

Example 3.7. Consider the system

$$x' = \begin{pmatrix} 3 & 0 \\ 0 & -1 \end{pmatrix} \begin{pmatrix} x_1 \\ x_2 \end{pmatrix}, \quad x(0) = \begin{pmatrix} 2 \\ -2 \end{pmatrix}.$$

Then the matrix A is diagonal and given by

$$A = \begin{pmatrix} 3 & 0 \\ 0 & -1 \end{pmatrix}.$$

Since A is a diagonal matrix, by induction we can easily show that

$$A^n = \begin{pmatrix} 3^n & 0 \\ 0 & (-1)^n \end{pmatrix}, \quad n = 1, 2, \ldots.$$

Hence

$$I + tA + \frac{(tA)^2}{2} + \ldots + \frac{(tA)^n}{n!}$$

$$= \begin{pmatrix} 1 + \frac{3t}{1!} + \frac{(3t)^2}{2!} + \ldots + \frac{(3t)^n}{n!} & 0 \\ 0 & 1 + \frac{(-t)}{1!} + \frac{(-t)^2}{2!} + \ldots + \frac{(-t)^n}{n!} \end{pmatrix}.$$

We deduce from the exponential function properties that the exponential matrix e^{At} is given by

$$e^{At} = \begin{pmatrix} e^{3t} & 0 \\ 0 & e^{-t} \end{pmatrix},$$

and the solution is

$$x(t) = e^{At} x_0 = \begin{pmatrix} e^{3t} & 0 \\ 0 & e^{-t} \end{pmatrix} \begin{pmatrix} 2 \\ -2 \end{pmatrix}.$$

In the next theorem we list some properties of the exponential matrix. For simplicity, we start at the initial time zero, that is, we take $t_0 = 0$. We are concerned with

$$x' = Ax, \tag{3.2.34}$$

where A is an $n \times n$ constant matrix, and $x \in \mathbb{R}^n$.

Theorem 3.2.11. *Let A and B be $n \times n$ constant matrices. Then:*

(1) $\frac{d}{dt} e^{At} = A e^{At}$, $t \in \mathbb{R}$;

(2) $\det(e^{At}) \neq 0$, $t \in \mathbb{R}$, and e^{At} is a fundamental matrix solution for (3.2.34);

(3) $e^{At} = I + At + \frac{1}{2}(At)^2 + \frac{1}{6}(At)^3 + \cdots + \frac{1}{k!}(At)^k + \cdots$;

(4) $e^{At} e^{As} = e^{A(t+s)}$, $t, s \in \mathbb{R}$;

(5) $\left(e^{At}\right)^{-1} = e^{-At}$;

(6) *If $AB = BA$, then $e^{At} B = B e^{At}$;*

(7) *If $AB = BA$, then $e^{At} e^{Bt} = e^{(A+Bt)}$, $t \in \mathbb{R}$;*

(8) *If P is a nonsingular matrix, then $e^{PBP^{-1}} = P e^B P^{-1}$.*

Proof. The proof of (1) is already done. For (2), we already know that e^{At} solves (3.2.34). Moreover, at $t = 0$, $e^{At} = I$, and hence $\det(e^{A0}) = \det(I) \neq 0$. By the uniqueness of solution this must hold for all $t \in \mathbb{R}$. Now since e^{At} solves (3.2.34), we have that e^{At} is a fundamental matrix for (3.2.34). The proof of (3) is already done. For (4), we let $L(t) = e^{At}e^{As} - e^{A(t+s)}$, where $s \in \mathbb{R}$ is fixed, and $t \in \mathbb{R}$. Then

$$L'(t) = Ae^{At}e^{As} - Ae^{A(t+s)} = A\left(e^{At}e^{As} - e^{A(t+s)}\right)$$

or $L'(t) = AL(t)$, from which we conclude that $L(t)$ solves (3.2.34). On the other hand, $L(0) = e^{As} - e^{As} = 0$, and hence by the uniqueness of solutions we get that $L(t) = 0$ for all $t \in \mathbb{R}$. This completes the proof of (4).

By the proof of (4), we have that

$$e^{At}e^{-At} = e^{At-At} = e^{A(t+(-t))} = I.$$

This implies that

$$\left(e^{At}\right)^{-1} = e^{-At}, \ t \in \mathbb{R}. \qquad \square$$

Remark 3.3. In the case where A is an $n \times n$ constant matrix, the exponential matrix is equivalent to

$$e^{A(t-t_0)} = \Phi(t)\Phi^{-1}(t_0),$$

which is the principal matrix.

Next, we discuss an important topic that relies on the *Jordan canonical form*, which helps us find the exponential matrix e^{At}. The next theorem is well known in linear algebra, and you may consult Brauer and Nohel [17] for its proof.

Theorem 3.2.12. *(Jordan canonical form) For every $n \times n$ matrix A, we can construct a matrix C such that*

$$C^{-1}AC = \begin{pmatrix} J_1 & 0 & \cdots & 0 \\ 0 & J_2 & \ddots & \vdots \\ \vdots & \ddots & \ddots & \vdots \\ 0 & \cdots & 0 & j_k \end{pmatrix}, \qquad (3.2.35)$$

where

$$J_k = \lambda_k I + N_k$$

with

$$N_k = \begin{pmatrix} 0 & 1 & 0 & 0 & \cdots & 0 & 0 \\ 0 & 0 & 1 & 0 & \cdots & 0 & 0 \\ \cdot & \cdot & \cdot & \cdot & \cdots & \cdot & \cdot \\ 0 & 0 & 0 & 0 & \cdots & 0 & 1 \\ 0 & 0 & 0 & 0 & \cdots & 0 & 0 \end{pmatrix}.$$

Remark 3.4. (The simplest case) If A has n real and distinct eigenvalues $\lambda_1, \lambda_2,$ $\ldots, \lambda_n$, then the matrix $C^{-1}AC$ will be a diagonal matrix with entries λ_i, that is,

$$C^{-1}AC = \begin{pmatrix} \lambda_1 & 0 & \cdots & 0 \\ 0 & \lambda_2 & \ddots & \vdots \\ \vdots & \ddots & \ddots & \vdots \\ 0 & \cdots & 0 & \lambda_n \end{pmatrix}.$$

In this case the matrix C is formed from the corresponding eigenvectors of the matrix A.

If (3.2.35) holds, then let

$$B = C^{-1}AC,$$

and hence

$$e^{tA} = e^{tCBC^{-1}} = Ce^{tB}C^{-1}.$$

In Section 3.2.1, we encountered different scenarios on the type and nature of the eigenvalues when computing the eigenvectors. The next result, called *Putzer algorithm*, allows us to find the fundamental matrix or, in particular, the exponential matrix e^{At} when the matrix is constant, regardless of the order or multiplicities of the eigenvalues. Putzer's algorithm depends on the next theorem, known as the Cayley–Hamilton theorem.

Theorem 3.2.13. *(Cayley–Hamilton) Every $n \times n$ constant matrix satisfies its characteristic equation*

$$p_A(\lambda) := \det(A - \lambda I) = 0.$$

In other words, if

$$p_A(\lambda) = (-1)^n \left(\lambda^n + c_1 \lambda^{n-1} + \cdots + c_{n-1} \lambda + c_n \right),$$

then

$$p_A(A) = (-1)^n \left(A^n + c_1 A^{n-1} + \cdots + c_{n-1} A + c_n I \right) = 0.$$

Theorem 3.2.14. *(Putzer algorithm) Suppose A is an $n \times n$ constant matrix with eigenvalues $\lambda_1, \lambda_2, \ldots, \lambda_n$, which may come in any order or multiplicities. Suppose $\mu_1(t), \mu_2(t), \ldots, \mu_n(t)$ satisfy the corresponding initial value problems*

$$\mu_1' = \lambda_1 \mu_1, \ \mu_1(0) = 1; \ \mu_j' = \lambda_j \mu_j + \mu_{j-1}, \ \mu_j(0) = 0, \ j = 2, 3, \ldots, n.$$

$$(3.2.36)$$

Let the matrices $P_0, P_1, \ldots, P_n$ be defined recursively by

$$P_0 = I, \quad P_j = \prod_{k=1}^{j}(A - \lambda_k I), \quad j = 1, 2, \ldots, n. \qquad (3.2.37)$$

Then

$$e^{At} = \sum_{j=0}^{n-1} \mu_{j+1}(t) P_j. \qquad (3.2.38)$$

Proof. Let

$$\Phi(t) = \sum_{j=0}^{n-1} \mu_{j+1}(t) P_j.$$

Then

$$\begin{aligned}
\Phi(0) &= \sum_{j=0}^{n-1} \mu_{j+1}(0) P_j \\
&= \mu_1(0) P_0(0) + \mu_2(0) P_1(0) + \ldots + \mu_n(0) P_{n-1}(0) \\
&= \mu_1(0) P_0(0) + 0 + \ldots + 0 \quad \text{(by (3.2.36))} \\
&= \mu_1(0) I = I \quad \text{(by (3.2.37))}.
\end{aligned}$$

Next, we differentiate Φ with respect to t:

$$\begin{aligned}
\Phi'(t) &= \sum_{j=0}^{n-1} \mu'_{j+1}(t) P_j \\
&= \lambda_1 \mu_1 P_0 + \sum_{j=1}^{n-1} \mu'_{j+1}(t) P_j, \quad \text{(by (3.2.36))} \\
&= \lambda_1 \mu_1 P_0 + \sum_{j=1}^{n-1} (\lambda_{j+1} \mu_{j+1} + \mu_j) P_j, \quad \text{(by (3.2.36))}. \qquad (3.2.39)
\end{aligned}$$

On the other hand,

$$A\Phi = \sum_{j=0}^{n-1} \mu_{j+1}(t) A P_j.$$

From (3.2.37) we have

$$P_j = \prod_{k=1}^{j}(A - \lambda_k I) = (A - \lambda_j I) \prod_{k=1}^{j-1}(A - \lambda_k I) = (A - \lambda_j I) P_{j-1}.$$

This implies that

$$AP_j = P_{j+1} + \lambda_{j+1}P_j, \quad j = 1, 2, \ldots n - 1.$$

Substituting AP_j into $A\Phi$ in the above expression yields

$$A\Phi = \sum_{j=0}^{n-1} \mu_{j+1}(t)[P_{j+1} + \lambda_{j+1}I P_j],$$

$$A\Phi = \lambda_1\mu_1(t)P_0 + (\lambda_2\mu_2(t) + \mu_1(t))P_1 + \ldots$$
$$+ (\lambda_n\mu_n(t) + \mu_{n-1}(t))P_{n-1} + \mu_n(t)P_n.$$

By the Cayley–Hamilton theorem $P_n = 0$, and hence the last expression of $A\Phi$ is the same as the right side of (3.2.39) so that we have $\Phi'(t) = A\Phi(t)$ with $\Phi(0) = I$. Thus expression (3.2.38) holds by the uniqueness of the solution. This completes the proof. $\qquad\square$

Example 3.8. Consider the linear system

$$x' = \begin{pmatrix} 2 & 2 & -1 \\ 2 & -1 & 2 \\ -1 & 2 & 2 \end{pmatrix} \begin{pmatrix} x_1 \\ x_2 \\ x_3 \end{pmatrix}$$

satisfying the initial condition

$$x(0) := x_0 = \begin{pmatrix} 1 \\ -1 \\ 2 \end{pmatrix}.$$

The matrix

$$A = \begin{pmatrix} 2 & 2 & -1 \\ 2 & -1 & 2 \\ -1 & 2 & 2 \end{pmatrix}$$

has the eigenvalues $3, 3, -3$. We may order them arbitrarily. Let, for example, $\lambda_1 = 3$, $\lambda_2 = 3$, and $\lambda_3 = -3$. Using (3.2.37) with $j = 1, 2$, we see that

$$P_0 = I, \quad P_1 = A - 3I, \quad P_2 = (A - 3I)^2.$$

Similarly, by (3.2.36)

$$\mu_1' = 3\mu_1, \ \mu_1(0) = 1;$$
$$\mu_2' = 3\mu_2 + \mu_1(t), \ \mu_2(0) = 0;$$
$$\mu_3' = -3\mu_3 + \mu_2(t), \ \mu_3(0) = 0.$$

Solving the above initial value problems gives

$$\mu_1(t) = e^{3t}, \quad \mu_2(t) = te^{3t}, \quad \text{and} \quad \mu_3(t) = \frac{te^{3t}}{6} - \frac{e^{3t}}{36} + \frac{e^{-3t}}{36}.$$

By using (3.2.38) with $n = 3$, we obtain

$$e^{At} = \sum_{j=0}^{2} \mu_{j+1}(t) P_j = \mu_1 P_0 + \mu_2 P_1 + \mu_3 P_2,$$

and the unique solution is given by

$$x(t) = e^{At} x_0.$$

Example 3.9. Assume we have three masses m_1, m_2, and m_3 that are connected to each other by four springs with spring constants k_1, k_2, k_3, and k_4, respectively. As seen in Fig. 3.2, the first and last springs are attached to two fixed walls, and the masses slide along a frictionless horizontal surface. Displacement of the masses from their equilibrium positions is determined by the coordinates x, y, and z, each positive to the right. The motion of the masses depends strongly on the initial conditions, that is, by hand, any of the masses is displaced by pulling forward or backward, and we let them go. Next, we attempt to find the equations of motion. The springs operate according to Hooke's law: Force = spring constant multiplied by elongation. The springs restore after compression and extension. In addition, the first three springs are elongated by x, $y - x$, $z - y$, respectively. Then using Newton's second law, Force = sum of the Hooke forces, we arrive at the system of second-order differential equations

$$m_1 x''(t) = -k_1 x(t) + k_2 [y(t) - x(t)],$$
$$m_2 y''(t) = -k_2 [y(t) - x(t)] + k_3 [z(t) - y(t)],$$
$$m_3 z''(t) = -k_3 [z(t) - y(t)] - k_4 z(t). \tag{3.2.40}$$

System (3.2.40) can be written as a second-order vector–matrix system

$$Mx''(t) = Ax(t),$$

where the *displacement* $\mathbf{x}$, *mass matrix* M, and *stiffness matrix* A are given by

$$\mathbf{x} = \begin{pmatrix} x \\ y \\ z \end{pmatrix}, \quad M = \begin{pmatrix} m_1 & 0 & 0 \\ 0 & m_2 & 0 \\ 0 & 0 & m_3 \end{pmatrix}, \quad A = \begin{pmatrix} -k_1 - k_2 & k_2 & 0 \\ k_2 & -k_2 - k_3 & k_3 \\ 0 & k_3 & -k_3 - k_4 \end{pmatrix}.$$

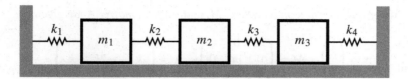

FIGURE 3.2

Three masses connected by four springs.

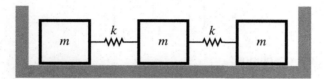

FIGURE 3.3

Three identical boxcars connected by identical springs.

Set $m_1 = m_2 = m_3 = 1$, $k_1 = 2$, $k_2 = 1$, $k_3 = 1$, and $k_4 = 2$ in the second-order vector–matrix system

$$M\mathbf{x}''(t) = A\mathbf{x}(t).$$

Then using Putzer's algorithm, we find the solution

$$\begin{pmatrix} x \\ y \\ z \end{pmatrix} = \left(a_1 \cos(t) + b_1 \sin(t) \right) \begin{pmatrix} 1 \\ 2 \\ 1 \end{pmatrix}$$

$$+ \left(a_2 \cos(\sqrt{3}t) + b_2 \sin(\sqrt{3}t) \right) \begin{pmatrix} 1 \\ 0 \\ -1 \end{pmatrix}$$

$$+ \left(a_3 \cos(2t) + b_3 \sin(2t) \right) \begin{pmatrix} 1 \\ -1 \\ 1 \end{pmatrix},$$

where a_1, a_2, b_1, b_2, a_3, and b_3 are arbitrary constants.

Example 3.10. (Boxcars) A special case of mass-spring system is *boxcars*. In Fig. 3.3 we assume three identical boxcars on a level track connected by two identical springs. Excepts for the springs on fixed ends, this problem is the same as in Example 3.9. With this in mind, we have that $k_1 = k_4 = 0$ and $k_2 = k_3 = k$. In addition, $m_1 = m_2 = m_3 = m$. Consequently, we arrive at the system

$$M\mathbf{x}''(t) = A\mathbf{x}(t),$$

where the *displacement* **x**, *mass matrix M*, and *stiffness matrix A* are given by

$$\mathbf{x} = \begin{pmatrix} x \\ y \\ z \end{pmatrix}, \quad M = \begin{pmatrix} m & 0 & 0 \\ 0 & m & 0 \\ 0 & 0 & m \end{pmatrix}, \quad A = \begin{pmatrix} -k & k & 0 \\ k & -2k & k \\ 0 & k & -k \end{pmatrix}.$$

For illustration, we take $\frac{k}{m} = 1$. Then we have the second-order system

$$\mathbf{x}'' = \begin{pmatrix} -1 & 1 & 0 \\ 1 & -2 & 1 \\ 0 & 1 & -1 \end{pmatrix} \mathbf{x},$$

which has the vector solution

$$\begin{pmatrix} x \\ y \\ z \end{pmatrix} = (a_1 + b_1 t) \begin{pmatrix} 1 \\ 1 \\ 1 \end{pmatrix}$$

$$+ (a_2 \cos(t) + b_2 \sin(t)) \begin{pmatrix} 1 \\ 0 \\ -1 \end{pmatrix}$$

$$+ (a_3 \cos(\sqrt{3}t) + b_3 \sin(\sqrt{3}t)) \begin{pmatrix} 1 \\ -2 \\ 1 \end{pmatrix},$$

where a_1, a_2, a_3, b_1, b_2, and b_3 are arbitrary constants.

3.3 $x' = A(t)x + g(t)$

Now we consider the non-homogeneous system

$$x' = A(t)x + g(t), \tag{3.3.1}$$

where $A(t)$ is an $n \times n$ matrix, and $g(t)$ is an $n \times 1$ vector, both continuous on some interval I. It was proven in Theorem 3.2.1 that system (3.3.1) has a unique solution on I if an initial condition is specified for $t_0 \in I$. We are interested in finding the *general solution* of the non-homogeneous system (3.3.1). With the non-homogeneous system we associate the complementary homogeneous system

$$x' = A(t)x. \tag{3.3.2}$$

Knowing the solution of (3.3.2) is enough to find the general solution of (3.3.1), and such a solution will be given by the *variation of parameters formula*.

Definition 3.3.1. Let x_h be the homogeneous solution of (3.3.2), and let x_p be the particular solution of (3.3.1) on an interval I. Then the general solution $x : I \to \mathbb{R}^n$ of (3.3.1) on the interval I is given by

$$x(t) = x_h(t) + x_p(t).$$

We have seen that if Φ is the fundamental matrix of the homogeneous system (3.3.2), then $x_h = \Phi(t)k$ for some constant $n \times 1$ vector k.

Theorem 3.3.1. *Let $A(t)$ be an $n \times n$ matrix, and let $g(t)$ be an $n \times 1$ vector, both continuous on some interval I. Suppose Φ is the fundamental matrix of the homogeneous system (3.3.2). Then $x(t)$ is a solution of*

$$x' = A(t)x + g(t), \quad x(t_0) = x_0, \ t \geq t_0, \tag{3.3.3}$$

on I if and only if x satisfies

$$x(t) = \Phi(t)\Phi^{-1}(t_0)x_0 + \int_{t_0}^{t} \Phi(t)\Phi^{-1}(s)g(s)ds, \tag{3.3.4}$$

where $t_0 \in I$ and $x_0 \in \mathbb{R}^n$.

Proof. First, note that the uniqueness of the solution of (3.3.3) follows from Theorem 3.2.1. Let Φ be the fundamental matrix of the homogeneous system (3.3.2). Suppose x satisfies (3.3.4). Taking the derivative in t in (3.3.4) gives

$$x'(t) = \Phi'(t)\Phi^{-1}(t_0)x_0 + \Phi'(t)\int_{t_0}^{t} \Phi^{-1}(s)g(s)ds + \Phi(t)\Phi^{-1}(t)g(t)$$

$$= A(t)\Phi(t)\Phi^{-1}(t_0)x_0 + A(t)\Phi(t)\int_{t_0}^{t} \Phi^{-1}(s)g(s)ds + g(t)$$

$$= A(t)\left[\Phi(t)\Phi^{-1}(t_0)x_0 + \Phi(t)\int_{t_0}^{t} \Phi^{-1}(s)g(s)ds\right] + g(t)$$

$$= A(t)x(t) + g(t).$$

Finally, replacing t by t_0 in (3.3.4) gives $x(t_0) = Ix_0 = x_0$. Now let $x(t)$ be the unique solution of (3.3.3), and let us show that it satisfies (3.3.4). As we have mentioned earlier, $x_h = \Phi(t)k$ for some constant $n \times 1$ vector k. To derive the particular solution x_p, we look for a particular solution of the form

$$x_p(t) = \Phi(t)u(t)$$

for some continuous function u on I. Differentiating x_p and plugging back into (3.3.3), we have

$$x_p' = \Phi'(t)u(t) + \Phi(t)u'(t)$$

or

$$\Phi'(t)u(t) + \Phi(t)u'(t) = A(t)\Phi(t)u(t) + g(t).$$

Since

$$\Phi'(t) = A(t)\Phi(t),$$

the above expression reduces to

$$\Phi(t)u'(t) = g(t).$$

Multiplying the above expression by $\Phi^{-1}(t)$ from the left, we get $u'(t) = \Phi^{-1}(t)g(t)$ or $u(t) = \int_{t_0}^{t} \Phi^{-1}(s)g(s)ds$. Finally, the equality $x_p(t) = \Phi(t)u(t)$ gives

$$x_p(t) = \Phi(t)\int_{t_0}^{t} \Phi^{-1}(s)g(s)ds.$$

It follows that the general solution is

$$x(t) = x_h(t) + x_p(t) = \Phi(t)k + \int_{t_0}^{t} \Phi(t)\Phi^{-1}(s)g(s)ds. \tag{3.3.5}$$

Using the initial condition in (3.3.5), we get

$$x_0 = \Phi(t_0)k + 0$$

or

$$k = \Phi^{-1}(t_0)x_0.$$

A substitution of k into (3.3.5) gives (3.3.4). This completes the proof. $\qquad\square$

Eq. (3.3.4) is called the *variation of parameters formula*.

Remark 3.5. In the case where A is an $n \times n$ constant matrix the exponential matrix is equivalent to

$$e^{A(t-t_0)} = \Phi(t)\Phi^{-1}(t_0),$$

which is the principal matrix. In this case, (3.3.4) reduces to or takes the form

$$x(t) = e^{A(t-t_0)}x_0 + \int_{t_0}^{t} e^{A(t-s)}g(s)ds. \tag{3.3.6}$$

Example 3.11. Solve

$$x'(t) = \begin{pmatrix} 4 & 2 \\ 3 & -1 \end{pmatrix} x + \begin{pmatrix} -15te^{-2t} \\ -4te^{-2t} \end{pmatrix}, \quad x(0) = \begin{pmatrix} 7 \\ 3 \end{pmatrix}.$$

Using Putzer's algorithm, we find that

$$e^{At} = \frac{1}{7} \begin{pmatrix} e^{-2t} + 6e^{5t} & -2e^{-2t} + 2e^{5t} \\ -3e^{-2t} + 3e^{5t} & 6e^{-2t} + e^{5t} \end{pmatrix}.$$

Note that $e^{A0} = I$. Using formula (3.3.6), we have

$$x(t) = \frac{1}{7} \begin{pmatrix} e^{-2t} + 6e^{5t} & -2e^{-2t} + 2e^{5t} \\ -3e^{-2t} + 3e^{5t} & 6e^{-2t} + e^{5t} \end{pmatrix} \begin{pmatrix} 7 \\ 3 \end{pmatrix}$$

$$+ \int_{t_0}^{t} \frac{1}{7} \begin{pmatrix} e^{-2(t-s)} + 6e^{5(t-s)} & -2e^{-2(t-s)} + 2e^{5(t-s)} \\ -3e^{-2(t-s)} + 3e^{5(t-s)} & 6e^{-2(t-s)} + e^{5(t-s)} \end{pmatrix} \begin{pmatrix} -15se^{-2s} \\ -4se^{-2s} \end{pmatrix} ds$$

$$= \frac{1}{14} \begin{pmatrix} (6 + 28t - 7t^2)e^{-2t} + 92e^{5t} \\ (-4 + 14t + 21t^2)e^{-2t} + 46e^{5t} \end{pmatrix}.$$

3.4 Discussion

When the matrix A is not constant and we wish to compute the fundamental matrix of the system

$$x' = A(t)x, \tag{3.4.1}$$

where $A(t)$ is an $n \times n$ matrix with continuous components on some interval I, then the Putzer algorithm will not help much. Suppose we are able to find two matrices Λ and S such that

$$A = S\Lambda S^{-1}. \tag{3.4.2}$$

Then we claim that also

$$e^{At} = Se^{\Lambda t}S^{-1} \tag{3.4.3}$$

for all t. To prove the claim, we use (3.2.32). As a consequence of (3.4.2), we have that $At = S(\Lambda t)S^{-1}$, and hence

$$(At)^k = \left(S(\Lambda t)S^{-1} \right)\left(S(\Lambda t)S^{-1} \right) \ldots \left(S(\Lambda t)S^{-1} \right).$$

Using (3.2.32) with $t_0 = 0$, we get

$$e^{At} = I + At + \frac{1}{2}(At)^2 + \frac{1}{6}(At)^3 + \ldots + \frac{1}{k!}(At)^k + \ldots$$

$$= I + S(\Lambda t)S^{-1} + \frac{1}{2}S(\Lambda t)^2 S^{-1} + \frac{1}{6}S(\Lambda t)^3 S^{-1} + \ldots + \frac{1}{k!}S(\Lambda t)^k S^{-1} + \ldots$$

$$= S\left[I + \Lambda t + \frac{1}{2}(\Lambda t)^2 + \frac{1}{6}(\Lambda t)^3 + \ldots + \frac{1}{k!}(\Lambda t)^k + \ldots\right]S^{-1}$$
$$= Se^{\Lambda t}S^{-1},$$

as claimed. Thus we state the general theorem.

Theorem 3.4.1. *For every $n \times n$ matrix A with entries in complex numbers, we can find an invertible matrix S and an upper triangular matrix Λ such that (3.4.2) holds.*

Another point of discussion is that if we integrate (3.4.1) from t_0 to t, then we get $x(t) = e^{\int_{t_0}^t A(s)ds}$. Now let

$$\Phi(t) = e^{\int_{t_0}^t A(s)ds}. \tag{3.4.4}$$

For $\Phi(t)$ to be fundamental matrix solution, we must have

$$A(t)\left(\int_{t_0}^t A(s)ds\right) = \left(\int_{t_0}^t A(s)ds\right)A(t). \tag{3.4.5}$$

Let us see why. Let $J = \int_{t_0}^t A(s)ds$. Then

$$e^J = I + J + \frac{1}{2!}J^2 + \frac{1}{3!}J^3 + \ldots + \frac{1}{k!}J^k + \cdots,$$

and

$$\frac{d}{dt}e^{\int_{t_0}^t A(s)ds} = \frac{d}{dt}\left(I + \int_{t_0}^t A(s)ds + \frac{1}{2!}\left(\int_{t_0}^t A(s)ds\right)^2\right.$$
$$\left. + \ldots + \frac{1}{(k-1)!}\left(\int_{t_0}^t A(s)ds\right)^{k-1}A(t) + \frac{1}{k!}\left(\int_{t_0}^t A(s)ds\right)^k + \ldots\right)$$
$$= A(t) + \int_{t_0}^t A(s)ds\, A(t) + \frac{1}{2!}\left(\int_{t_0}^t A(s)ds\right)^2 A(t)$$
$$+ \ldots + \frac{1}{(k-1)!}\left(\int_{t_0}^t A(s)ds\right)^{k-1}A(t) + \frac{1}{k!}\left(\int_{t_0}^t A(s)ds\right)^k A(t) + \ldots$$
$$= \left[I + \int_{t_0}^t A(s)ds + \frac{1}{2!}\left(\int_{t_0}^t A(s)ds\right)^2 + \ldots\right.$$
$$\left. + \frac{1}{(k-1)!}\left(\int_{t_0}^t A(s)ds\right)^{k-1}A(t) + \frac{1}{k!}\left(\int_{t_0}^t A(s)ds\right)^k + \ldots\right]A(t)$$
$$\neq A(t)\left[I + \int_{t_0}^t A(s)ds + \frac{1}{2!}\left(\int_{t_0}^t A(s)ds\right)^2\right.$$
$$\left. + \ldots + \frac{1}{(k-1)!}\left(\int_{t_0}^t A(s)ds\right)^{k-1}A(t) + \frac{1}{k!}\left(\int_{t_0}^t A(s)ds\right)^k + \ldots\right]$$
$$= A(t)\Phi(t).$$

Thus if (3.4.5) holds, then $\Phi'(t) = A(t)\Phi(t)$. We have the following application of the above discussion.

Example 3.12. Find the fundamental matrix $\Phi(t)$ for the system

$$x' = \begin{pmatrix} 1 & t \\ t & 1 \end{pmatrix} \begin{pmatrix} x_1 \\ x_2 \end{pmatrix}.$$

Let

$$A(t) = \begin{pmatrix} 1 & t \\ t & 1 \end{pmatrix}$$

and

$$\Phi(t) = e^{\int_0^t A(s)ds}.$$

Now we verify condition (3.4.5):

$$A(t) \int_{t_0}^{t} A(s)ds = \begin{pmatrix} 1 & t \\ t & 1 \end{pmatrix} \begin{pmatrix} t & t^2/2 \\ t^2/2 & t \end{pmatrix} = \begin{pmatrix} t+t^3/2 & 3t^2/2 \\ 3t^2/2 & t+t^3/2 \end{pmatrix}.$$

On the other hand,

$$\left(\int_{t_0}^{t} A(s)ds \right) A(t) = \begin{pmatrix} t & t^2/2 \\ t^2/2 & t \end{pmatrix} \begin{pmatrix} 1 & t \\ t & 1 \end{pmatrix} = \begin{pmatrix} t+t^3/2 & 3t^2/2 \\ 3t^2/2 & t+t^3/2 \end{pmatrix}.$$

The matrix $\int A(t)dt$ has the eigenvalues

$$\lambda_1 = t + \frac{t^2}{2} \quad \text{and} \quad \lambda_2 = t - \frac{t^2}{2}.$$

Luckily, the corresponding eigenvectors are independent of t. The eigenpairs are

$$\lambda_1, K_1 = \begin{pmatrix} 1 \\ 1 \end{pmatrix}; \ \lambda_2, K_2 = \begin{pmatrix} -1 \\ 1 \end{pmatrix}.$$

Let

$$H = \begin{pmatrix} 1 & -1 \\ 1 & 1 \end{pmatrix}.$$

Then the inverse of H is

$$H^{-1} = \frac{1}{2} \begin{pmatrix} 1 & 1 \\ -1 & 1 \end{pmatrix}.$$

Next, we try to diagonalize $\int_0^t A(s)ds$ by the transformation

$$J = H^{-1} \left(\int_0^t A(s)ds \right) H := \begin{pmatrix} t + \frac{t^2}{2} & 0 \\ 0 & t - \frac{t^2}{2} \end{pmatrix}.$$

As consequence of (3.4.3), we get the exponential matrix of J,

$$e^J = \begin{pmatrix} e^{t+\frac{t^2}{2}} & 0 \\ 0 & e^{t-\frac{t^2}{2}} \end{pmatrix}.$$

We can now calculate the fundamental matrix $\Phi(t)$:

$$\Phi(t) = e^{\int_{t_0}^{t} A(s)ds} = He^J H^{-1}$$

$$= \frac{e^t}{2} \begin{pmatrix} e^{\frac{t^2}{2}} + e^{-\frac{t^2}{2}} & e^{\frac{t^2}{2}} - e^{-\frac{t^2}{2}} \\ e^{\frac{t^2}{2}} - e^{-\frac{t^2}{2}} & e^{\frac{t^2}{2}} + e^{-\frac{t^2}{2}} \end{pmatrix}.$$

3.5 Exercises

Exercise 3.1. Find the region of existence of the following system and then find the maximal interval of existence:

$$x_1' = -\frac{x_3}{x_2^2},$$

$$x_2' = 1,$$

$$x_3' = \frac{x_1}{x_2^2},$$

with initial vector

$$\begin{pmatrix} x_1(\frac{1}{\pi}) \\ x_2(\frac{1}{\pi}) \\ x_3(\frac{1}{\pi}) \end{pmatrix} = \begin{pmatrix} -1 \\ \frac{1}{\pi} \\ 0 \end{pmatrix}.$$

Exercise 3.2. Show that $||A||_1$ is a norm, that is, verify all the items in Definition 3.2.2.

Exercise 3.3. Show that if A is an $n \times n$ invertible matrix, then

$$||A|| > 0, \quad \text{and} \quad ||A^{-1}|| \geq \frac{1}{||A||}.$$

Exercise 3.4. Suppose an $n \times n$ matrix $A(t)$ and an $n \times 1$ vector $g(t)$ are continuous on $-\infty < t < \infty$, $\int_{-\infty}^{\infty} |A(t)|dt < \infty$, and $\int_{-\infty}^{\infty} |g(t)|dt < \infty$. Show that the solution $\phi(t)$ of $y'(t) = A(t)y(t) + g(t)$, exists for $-\infty < t < \infty$. In addition, compute the upper bound for $|\phi(t)|$ on $-\infty < t < \infty$.

Exercise 3.5. In Example 3.4 keep the norm of the matrix A as a function of t and obtain a slightly better estimate of the error.

Exercise 3.6. Prove that every solution of

$$x'' - x' + t^2 x = t^4$$

exists for all $t \in \mathbb{R}$.

Exercise 3.7. Use the method of *energy function* to show that the solutions

$$x'' + \cos^2(x)x' - \sin^2(x) = 0$$

exist for all times, that is, the maximal interval of existence is $[0, \infty)$.

Exercise 3.8. Obtain two iterations of the 3×3 linear system

$$x_1'(t) = 5x_1 - x_2,$$
$$x_2'(t) = -5x_2 + 9x_3,$$
$$x_3'(t) = -5x_1 - x_2,$$

with the initial condition

$$\mathbf{x}(0) = \mathbf{x_0} = \begin{pmatrix} x_{01} \\ x_{02} \\ x_{03} \end{pmatrix} = \begin{pmatrix} 1 \\ -1 \\ 0 \end{pmatrix}.$$

Exercise 3.9. Use Theorem 3.2.10 to solve, for all $t \geq t_0$, the linear system

$$x' = \begin{pmatrix} 1 & -1 & 2 \\ -1 & 1 & 0 \\ -1 & 0 & 1 \end{pmatrix} \begin{pmatrix} x_1 \\ x_2 \\ x_3 \end{pmatrix}$$

with initial condition

$$x(t_0) := x_0 = \begin{pmatrix} 1 \\ 0 \\ 2 \end{pmatrix}.$$

Exercise 3.10. Use Theorem 3.2.10 to solve the linear system with the initial condition

$$x' = \begin{pmatrix} 1 & -12 & -14 \\ 1 & 2 & -3 \\ 1 & 1 & -2 \end{pmatrix} \begin{pmatrix} x_1 \\ x_2 \\ x_3 \end{pmatrix}, \quad x(0) = x_0 := \begin{pmatrix} 4 \\ 6 \\ -7 \end{pmatrix}.$$

Exercise 3.11. (a) Let

$$x'(t) = Ax(t),$$

where A is a constant $n \times n$ matrix. Suppose A has only one eigenvector K associated with an eigenvalue λ_1 of multiplicity three. We can find a second solution of the form (3.2.26), where the eigenvector P satisfies (3.2.28). Show that if we assume a third solution $\phi_3(t)$ of the form

$$\phi_3(t) = K\frac{t^2}{2}e^{\lambda_1 t} + Pte^{\lambda_1 t} + Qe^{\lambda_1 t},$$

then the vector Q is given by the equation

$$(A - \lambda_1 I)Q = P.$$

(b) Find the fundamental matrix for the linear system

$$x' = \begin{pmatrix} 1 & 0 & 0 \\ 2 & 2 & -1 \\ 0 & 1 & 0 \end{pmatrix} \begin{pmatrix} x_1 \\ x_2 \\ x_3 \end{pmatrix}.$$

Exercise 3.12. (a) Let

$$x'(t) = Ax(t),$$

where A is an $n \times n$ constant matrix with real entries. Let K be an eigenvector corresponding to the complex eigenvalue $\lambda_1 = \alpha + i\beta$ (with real α and β). Show that the corresponding two linearly independent solutions $\phi_1(t)$ and $\phi_2(t)$ are given by

$$\phi_1(t) = \Big(\mathrm{Re}(K)\cos(\beta t) - \mathrm{Im}(K)\sin(\beta t)\Big)e^{\alpha t}$$

and

$$\phi_2(t) = \Big(\mathrm{Im}(K)\cos(\beta t) + \mathrm{Re}(K)\sin(\beta t)\Big)e^{\alpha t}.$$

For example, if $K = \begin{pmatrix} 1 \\ 2i \end{pmatrix}$, then

$$K = \begin{pmatrix} 1 \\ 0 \end{pmatrix} + \begin{pmatrix} 0 \\ 2i \end{pmatrix} = \mathrm{Re}(K) + \mathrm{Im}(K).$$

(b) Find the fundamental matrix for the linear system

$$x' = \begin{pmatrix} 1 & 2 \\ -1/2 & 1 \end{pmatrix} \begin{pmatrix} x_1 \\ x_2 \end{pmatrix}.$$

Exercise 3.13. Let Φ be a fundamental matrix of $x' = Ax$, and let C be a nonsingular matrix with $AC = CA$. Prove that now $C\Phi$ is a fundamental matrix of $x' = Ax$. (See Theorem 3.2.5.)

Exercise 3.14. Prove parts (5), (6), and (7) of Theorem 3.2.11.

Exercise 3.15. Finish the calculation in Example 3.8 and show that

$$e^{At} = \cosh(3t)I + \frac{1}{3}\sinh(3t)A.$$

Exercise 3.16. Use Theorem 3.2.14 to compute e^{At} for

(a)

$$A = \begin{pmatrix} 2 & 0 & 0 \\ 1 & 2 & 0 \\ 1 & 0 & 3 \end{pmatrix},$$

(b)

$$A = \begin{pmatrix} 5 & -4 & 0 \\ 1 & 0 & 2 \\ 0 & 2 & 5 \end{pmatrix}.$$

Exercise 3.17. Use Theorem 3.2.14 to compute e^{At} for

$$x' = \begin{pmatrix} -1 & -2 & 2 \\ 3 & 5 & -3 \\ 3 & 4 & -2 \end{pmatrix} \begin{pmatrix} x_1 \\ x_2 \\ x_3 \end{pmatrix}$$

satisfying the initial condition

$$x(0) = x_0 := \begin{pmatrix} 1 \\ -1 \\ 1 \end{pmatrix}.$$

Exercise 3.18. Consider two non-homogeneous systems $x' = Ax + g_1(t)$ and $x' = Ax + g_2(t)$ with solutions z_1 and z_2, respectively. Let $g(t) = g_1(t) + g_2(t)$. Show that $z_1 + z_2$ solves $x' = Ax + g(t)$. Here g_1, g_2 are $n \times 1$ vector functions that are continuous on $\mathbb{R}$.

Exercise 3.19. Use the variation of parameters formula to solve the given system subject to the indicated initial condition.

(a)

$$x'(t) = \begin{pmatrix} 1 & 1 & 0 \\ 1 & 1 & 0 \\ 0 & 0 & 3 \end{pmatrix} x + \begin{pmatrix} e^t \\ e^{2t} \\ te^{3t} \end{pmatrix}, \quad x(0) = \begin{pmatrix} 0 \\ 1 \\ 1 \end{pmatrix},$$

(b)

$$x'(t) = \begin{pmatrix} 3 & -1 & -1 \\ 1 & 1 & -1 \\ 1 & -1 & 1 \end{pmatrix} x + \begin{pmatrix} 0 \\ t \\ 2e^t \end{pmatrix}, \quad x(0) = \begin{pmatrix} 1 \\ 2 \\ 0 \end{pmatrix}.$$

Exercise 3.20. Imitate the proof that leads to the variation of parameters formula (1.5.2) for non-homogeneous scalar differential equations to derive (3.3.6) for the non-homogeneous system

$$x'(t) = Ax(t) + g(t), \quad x(t_0) = x_0,$$

where A is an $n \times n$ constant matrix, and g is continuous on some interval I.

Exercise 3.21. Find the fundamental matrix $\Phi(t)$ for the system

$$x' = \begin{pmatrix} -\frac{1}{t} & \frac{2}{t^2} \\ -1 & \frac{3}{t} \end{pmatrix} \begin{pmatrix} x_1 \\ x_2 \end{pmatrix}.$$

Exercise 3.22. Assume the fundamental matrix $\Phi(t)$ of a certain 2×2 differential system is given by

$$\Phi(t) = \begin{pmatrix} 2 & t \\ t & t^2 \end{pmatrix}.$$

Find the corresponding system of differential equations.

Exercise 3.23. Write the system

$$x_1' = -x_1 + x_2 e^{2t}, \quad x_2' = -x_2$$

in the form of

$$x'(t) = A(t)x(t).$$

Compute the eigenvalues of $A(t)$ and then find the solution of the system. Are solutions bounded or unbounded?

Stability of linear systems

We look at the stability of linear systems via the variation of parameters. The chapter also includes a nice section on Floquet theory with its application to Mathieu's equation.

4.1 Definitions and examples

Consider the scalar differential equation

$$x' = ax, \ t \geq 0, \tag{4.1.1}$$

for constant a. Then (4.1.1) has a solution $x(t) = Ce^{at}$ for every constant C. Note that $x(t) = 0$ is a solution of (4.1.1) since it satisfies both sides of the equation. It is clear that solutions go to infinity for $a > 0$ and decay to zero for $a < 0$ as $t \to \infty$. Thus the behavior of the solutions depends on the sign of the constant a. It seems that the solution $x = 0$ is an attractor asymptotically stable in the case $a < 0$, which we will discuss in detail in this chapter.

Notice that if $a = 0$, then solutions are constants. Thus if we start near the zero solution $x = 0$, then we stay near the zero solution, and hence the zero solution is stable but not asymptotically stable.

In the above example, we were able to find the solution. However, this is not the case for the nonlinear differential equation

$$x'(t) = \frac{x(x-1)(x+2)}{1 + \sin^2(x)}, \ t \geq 0, \ x(0) = x_0. \tag{4.1.2}$$

Eq. (4.1.2) is nonlinear, and the true solution cannot be explicitly found. On the other hand, we notice that $x = -2, 0, 1$ are all *constant solutions* or *equilibrium solutions* of (4.1.2). However, we can use calculus to analyze those constant solutions. Note that the right side of (4.1.2) is continuous on $\mathbb{R}$, and hence a solution exists for all $t \geq 0$. Moreover, any solution passing through (t_0, x_0) is unique. For the next discussion, we refer the reader to Fig. 4.1.

(a) For $x < -2$, we have from (4.1.2) that $x' < 0$, and hence solutions grow negatively unbounded; in particular, they pull away from the constant solution -2.

Advanced Differential Equations. https://doi.org/10.1016/B978-0-32-399280-0.00010-3

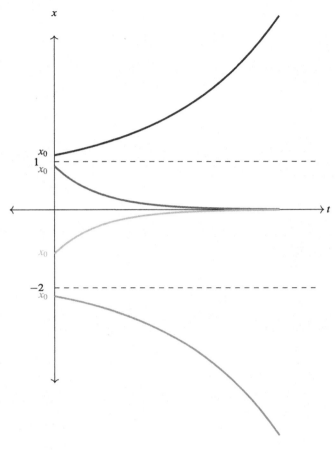

FIGURE 4.1

Diagram for the constant solutions.

(b) For $-2 < x < 0$, we have from (4.1.2) that $x' > 0$, and hence solutions increase toward the constant solution $x = 0$ and away from the constant solution $x = -2$.

(c) For $0 < x < 1$, we have from (4.1.2) that $x' < 0$, and hence solutions decrease toward the constant solution $x = 0$ and away from the constant solution $x = 1$. Note that in (b) and (c) solutions cannot cross the constant solution $x = 0$ due to the uniqueness. Finally,

(d) For $x > 1$, we have from (4.1.2) that $x' > 0$, and hence solutions unboundedly increase and away from the constant solution $x = 1$.

Thus by the above discussion and Fig. 4.1, we see that the constant solution $x = 0$ is an attractor (asymptotically stable), and the other constant solutions $x = -2, 1$ are repellers (unstable), which we discuss later.

Consider the system of ordinary differential equations

$$x' = f(t, x), \tag{4.1.3}$$

where $f \in C\big([0, \infty) \times D, \mathbb{R}^n\big)$, and $D \subset \mathbb{R}^n$ is open. We say that a vector $x^* \in \mathbb{R}^n$ is an *equilibrium*, or *constant solution*, or *equilibrium solution* of (4.1.3) if

$$f(t, x^*) = 0, \ \ 0 \le t < w,$$

for positive constant w. In practice, it is easier to talk about a *zero solution*, or $x = 0$. To do so, we translate every nonzero equilibrium point to zero by the change of variables

$$\tilde{x}(t) = x(t) - x^*,$$

and from (4.1.3) we have that

$$\tilde{x}'(t) = (x(t) - x^*)' = x'(t) = f(t, \tilde{x}(t) + x^*).$$

In this case, $\tilde{x}(t) \equiv 0$ is an equilibrium point of (4.1.3), which we may call the *zero solution* of (4.1.3). Thus, in most cases, we may require $f(t, 0) = 0, 0 \le t < w$, when we talk about the zero solution.

Example 4.1. (*Lotka–Volterra predator–prey model*) Let $x(t)$ and $y(t)$ be the number of preys and predators at time t, respectively. To keep the model simple, we make the following assumptions:
• The predator species are dependent on a single prey species as its only food supply;
• The prey species have an unlimited food supply; and
• There is no threat to the prey other than the specific predator.
In the absence of predation, the prey population would grow at a natural rate:

$$\frac{dx}{dt} = ax(t), \ a > 0.$$

On the other hand, in the absence of prey, the predator population would decline at a natural rate:

$$\frac{dy}{dt} = -cy(t), \ c > 0.$$

The effects of predator eating prey is an interaction rate of decline $(-bxy, \ b > 0)$ in the prey population x and an interaction rate of growth $(dxy, \ d > 0)$ of predator population y. Hence we obtain the predator–prey model

$$\frac{dx}{dt} = ax - bxy,$$
$$\frac{dy}{dt} = -cy + dxy. \tag{4.1.4}$$

The Lotka–Volterra model consists of a system of linked differential equations that cannot be separated from each other and that cannot be solved in closed form. Thus we search for the equilibrium solutions by setting the right sides of both equations in (4.1.4) equal to zero, that is,

$$x(a - by) = 0, \quad y(-c + dx) = 0.$$

From the first equation we obtain $x = 0$, $y = a/b$, and from the second equation we get $y = 0$, $x = c/d$. Now if $x = 0$, then from the second equation we have that $-cy = 0$, and hence $y = 0$. Thus we have the equilibrium solution

$$x_1^* = (0, 0).$$

Similarly, if $y = a/b$, then $-\dfrac{c}{b}a + dx\dfrac{a}{b} = 0$, which implies that $x = c/d$. So the other equilibrium solution is

$$x_2^* = (c/d, a/b).$$

The first equilibrium solution x_1^* effectively represents the extinction of both species. If both populations are at 0, then they will continue to be so indefinitely. The second equilibrium solution x_2^* represents a fixed point at which both populations sustain their current non-zero numbers and do so indefinitely. The levels of the population at which this equilibrium is achieved depend on the chosen values of the parameters a, b, c, and d. We further discuss the stability of both equilibrium solutions x_1^* and x_2^* and hope that x_1^* is unstable. (See Fig. 4.2.)

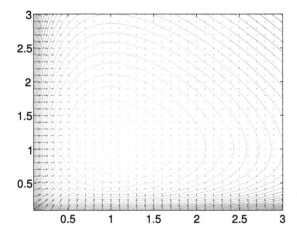

FIGURE 4.2

Prey–predator dynamics as described by the level curves. The arrows describe the velocity and direction of solutions for $a = b = c = d = 1$.

Next, we consider (4.1.3) with the initial condition $x(t_0) = x_0$ with the assumption that $f(t, 0) = 0$.

Definition 4.1.1. The zero solution $x = 0$ of (4.1.3):

(a) is *stable* (S) if for all $\varepsilon > 0$ and $t_0 \geq 0$, there is $\delta = \delta(t_0, \varepsilon) > 0$ such that $|x(t_0)| < \delta$ implies $|x(t, t_0, x_0)| < \varepsilon$,

(b) is *uniformly stable* (US) if δ is independent of t_0,

(c) is *unstable* if it is not stable,

(d) is *asymptotically stable* (AS) if it is stable and $\lim\limits_{t \to \infty} |x(t, t_0, x_0)| = 0$.

(e) is *uniformly asymptotically stable* (UAS) if it is US and there exists $\gamma > 0$ such that for each $\mu > 0$, there exists $T = T(\mu) > 0$ such that $|x(t_0)| < \gamma$, $t \geq t_0 + T$, implies $|x(t, t_0, x_0)| < \mu$.

There are various stability definitions that can be found in many of the referenced books. As far as this book is concerned, the stability definitions that we have listed are the most important ones. We furnish a few examples to illustrate the above definitions.

Example 4.2. In this example we show that the zero solution is US and AS but not UAS. We consider the initial value problem given by

$$(1 - t)x'(t) = x, \; x(t_0) = x_0, \; t \geq t_0 > 1.$$

For simplicity, we may write our problem as

$$x' = -\frac{x}{t - 1}, \; x(t_0) = x_0, \; t \geq t_0 > 1. \tag{4.1.5}$$

Separating the variables in (4.1.5), integrating from t_0 to t, and then applying the initial data, we arrive at the solution

$$x(t, t_0, x_0) = \frac{t_0 - 1}{t - 1} x_0, \; t \geq t_0 > 1. \tag{4.1.6}$$

For any $\varepsilon > 0$, set $\delta = \varepsilon$, so that for $|x_0| < \delta$ and $t \geq t_0 > 1$, we have that

$$|x(t, t_0, x_0)| = \left|\frac{t_0 - 1}{t - 1} x_0\right| = \left|\frac{t_0 - 1}{t - 1}\right| |x_0| \leq |x_0| < \delta = \varepsilon.$$

This shows that the zero solution is US. Moreover,

$$\lim_{t \to \infty} |x(t, t_0, x_0)| = \lim_{t \to \infty} \left|\frac{t_0 - 1}{t - 1}\right| |x_0| = 0,$$

and so the zero solution is AS. On the other hand, if we evaluate the solution $x(t, t_0, x_0) = \dfrac{t_0 - 1}{t - 1} x_0$ at $t = t_0 + T$, then we get $x(t_0 + T, t_0, x_0) = \dfrac{t_0 - 1}{t_0 + T - 1} x_0 \to$ 1 as $t_0 \to \infty$. This implies that $|x(t, t_0, x_0)| \not< \varepsilon$ for $t \geq t_0 + T$. We conclude that the zero solution is not UAS. So this example shows that US plus AS do not necessarily imply UAS.

Example 4.3. The differential equation

$$x' = 0, \ x(t_0) = x_0,$$

has the constant solution $x(t) = x_0$, and hence the zero solution is US but not AS. On the other hand,

$$x' = x, \ x(t_0) = x_0,$$

has the solution $x(t) = x_0 e^{t-t_0}$, and its zero solution is unstable.

Example 4.4. The differential equation

$$x' = -x, \ x(t_0) = x_0,$$

has the solution

$$x(t, t_0, x_0) = x_0 e^{-(t-t_0)}, \ t \geq t_0 \geq 0,$$

and its zero solution is US and UAS. To see this, for any $\varepsilon > 0$, we set $\delta = \varepsilon$. Then for $|x_0| < \delta$, we have that $x(t, t_0, x_0)| = |x_0 e^{-(t-t_0)}| \leq |x_0| \leq \varepsilon = \delta$. So the zero solution is US. For the UAS, we must show that there exists $\gamma > 0$ such that for each $\mu > 0$, there exists $T = T(\mu) > 0$ such that $|x(t_0)| < \gamma, t \geq t_0 + T$, implies $|x(t, t_0, x_0)| < \mu$. Set $\delta = 1$ of the uniform stability, and let $\gamma = \delta = 1$. Let $T(\mu) > -\text{Ln}(\mu)$. Then for $t \geq t_0 + T(\mu)$,

$$|x(t, t_0, x_0)| = |x_0 e^{-(t-t_0)}| \leq e^{-T(\mu)} \leq e^{\text{Ln}(\mu)} = \mu.$$

This shows the zero solution is UAS.

4.2 $x' = A(t)x$

Now we are in a position to discuss the stability of the zero solution of the linear system

$$x'(t) = A(t)x(t), \ x(t_0) = x_0, \tag{4.2.1}$$

where $A(t)$ is an $n \times n$ matrix with continuous entries on $[0, \infty)$, using the concept of the fundamental matrix. We begin with the following theorem.

Theorem 4.2.1. *Let* $\Phi(t)$ *be the fundamental matrix of* (4.2.1). *Then the zero solution of* (4.2.1) *is*

(a) *stable if and only if there exists a positive constant M such that*

$$|\Phi(t)| \leq M, \ t \geq 0;$$

(b) *asymptotically stable if and only if*

$$|\Phi(t)| \to 0 \ as \ t \to \infty.$$

Proof. (a) ($\Leftarrow$) Let $\Phi(t)$ be the fundamental matrix of (4.2.1). Then the unique solution x of (4.2.1) is given by $x(t) = \Phi(t)\Phi^{-1}(t_0)x_0$. Thus for any $\varepsilon > 0$ such that $|x_0| < \varepsilon$, set $\delta = \frac{\varepsilon}{|\Phi^{-1}(t_0)|M}$. Then

$$|x(t)| = |\Phi(t)\Phi^{-1}(t_0)x_0| \le |\Phi(t)||\Phi^{-1}(t_0)||x_0| \le M|\Phi^{-1}(t_0)|\delta = \varepsilon.$$

($\Rightarrow$) Set $\varepsilon = 1$ from the stability proof. Then

$$|x(t)| = |\Phi(t)\Phi^{-1}(t_0)x_0| < 1 \text{ for } t \ge t_0 \text{ if } |x_0| < \delta(1, t_0),$$

which implies that

$$|\Phi(t)\Phi^{-1}(t_0)| < \frac{1}{\delta(1, t_0)}.$$

Therefore

$$|\Phi(t)| = |\Phi(t)\Phi^{-1}(t_0)\Phi(t_0)| \le |\Phi(t)\Phi^{-1}(t_0)||\Phi(t_0)| \le \frac{1}{\delta(1, t_0)}|\Phi(t_0)| := M.$$

This completes the proof of (a).

Next, we prove (b). We already know that the zero solution is stable. Now

$$|x(t)| = |\Phi(t)\Phi^{-1}(t_0)x_0| \to 0 \text{ as } t \to \infty$$

if and only if

$$|\Phi(t)| \to 0 \text{ as } t \to \infty.$$

This completes the proof. $\qquad\qquad\square$

Theorem 4.2.2. *Let $\Phi(t)$ be the fundamental matrix of (4.2.1). Then the zero solution of (4.2.1) is uniformly stable if and only if there exists a positive constant M such that*

$$|\Phi(t)\Phi^{-1}(s)| \le M, \ t \ge s \ge 0. \tag{4.2.2}$$

Proof. ($\Leftarrow$) Let $\Phi(t)$ be the fundamental matrix of (4.2.1). Then the unique solution x of (4.2.1) is given by $x(t) = \Phi(t)\Phi^{-1}(s)x(s)$. Since (4.2.2) holds for all s such that $t \ge s \ge 0$, we have that

$$|x(t)| = |\Phi(t)\Phi^{-1}(s)x(s)| \le |\Phi(t)\Phi^{-1}(s)||x_0|.$$

Thus for any $\varepsilon > 0$, choosing $\delta = \frac{\varepsilon}{M}$, we get that for $|x_0| < \delta$,

$$|x(t)| \le |\Phi(t)||\Phi^{-1}(s)||x_0| \le M\delta = \varepsilon.$$

($\Rightarrow$) Set $\varepsilon = 1$ from the proof of uniform stability. Then

$$|x(t)| = |\Phi(t)\Phi^{-1}(s)x(s)| < 1 \text{ for } t \ge s \text{ if } |x(s)| < \delta(1),$$

which implies that

$$|\Phi(t)\Phi^{-1}(s)| \leq \frac{1}{\delta(1)} := M.$$

This completes the proof. □

The next theorem provides necessary and sufficient conditions for the UAS.

Theorem 4.2.3. *Let $\Phi(t)$ be the fundamental matrix of (4.2.1). Then the zero solution of (4.2.1) is uniformly asymptotically stable if and only if there exist positive constants M and β such that*

$$|\Phi(t)\Phi^{-1}(s)| \leq Me^{-\beta(t-s)}, \quad t \geq s \geq 0. \tag{4.2.3}$$

Proof. ($\Leftarrow$) Let $\Phi(t)$ be the fundamental matrix of (4.2.1). Then the unique solution x of (4.2.1) is given by $x(t) = \Phi(t)\Phi^{-1}(s)x(s)$. It is evident from (4.2.3) that the zero solution is US. For all $t \geq s \geq 0$, we have that

$$|x(t)| = |\Phi(t)\Phi^{-1}(s)x(s)| \leq |\Phi(t)\Phi^{-1}(s)||x_0|.$$

As a result, we may set $\gamma = 1$. Now we take $T(\mu) = -\frac{1}{\beta}\text{Ln}(\frac{\mu}{M})$, $0 < \mu < M$. Then for $|x_0| < 1$ and $t \geq t_0 + T(\mu)$, we have

$$|x(t)| = |\Phi(t)\Phi^{-1}(t_0)x_0| < Me^{-\beta(t-t_0)} = Me^{-\beta T(\mu)} \leq \mu.$$

($\Rightarrow$) Since the zero solution is UAS, there exists $\gamma > 0$ such that for each $\mu > 0$, there exists $T = T(\mu) > 0$ such that $|x(t_0)| < \gamma$, $t \geq t_0 + T$, implies $|x(t, t_0, x_0)| < \mu$. Let $\alpha = \frac{\mu}{\gamma}$. Then $|x_0| \leq \gamma$ and $t \geq t_0 + T(\mu)$ yield

$$|\Phi(t)\Phi^{-1}(t_0)x_0| \leq \mu. \tag{4.2.4}$$

Therefore

$$|\Phi(t + T(\mu))\Phi^{-1}(t)| \leq \alpha < 1, \quad t \geq 0. \tag{4.2.5}$$

For $t \geq t_0$, there is a positive integer k such that

$$t_0 + kT(\mu) \leq t \leq t_0 + (k+1)T(\mu).$$

Now let $t_n = t_0 + (k+1)T(\mu)$, $n = 1, 2, \ldots, k$. Then by (4.2.4) and (4.2.5)

$$|\Phi(t)\Phi^{-1}(t_0)| = \left|\Phi(t)\Phi^{-1}(t_k)\Phi(t_k)\Phi^{-1}(t_{k-1})\ldots\Phi(t_1)\Phi^{-1}(t_0)\right|$$

$$\leq \left|\Phi(t)\Phi^{-1}(t_k)\right|\prod_{n=0}^{k-1}\Phi(t_{n+1})\Phi^{-1}(t_n)$$

$$\leq M\alpha^k.$$

Let $\beta = -\frac{1}{T(\gamma)} \mathrm{Ln}(\alpha) > 0$. Then

$$\begin{aligned}
|\Phi(t)\Phi^{-1}(t_0)| &\leq \frac{M}{\alpha}\alpha^{k+1} \\
&= \frac{M}{\alpha}e^{-(k+1)\beta T(\gamma)} \\
&\leq \frac{M}{\alpha}e^{-\beta(t-t_0)} \\
&:= Le^{-\beta(t-t_0)}.
\end{aligned}$$

This completes the proof. $\square$

Example 4.5. The nonlinear system

$$x'(t) = \begin{pmatrix} 0 & 1 \\ \frac{-2}{t^2} & \frac{2}{t} \end{pmatrix}\begin{pmatrix} x_1 \\ x_2 \end{pmatrix}, \; t > 0,$$

has the fundamental matrix

$$\Phi(t) = \begin{pmatrix} t^2 & t \\ 2t & 1 \end{pmatrix}, \; t > 0.$$

By Theorem 4.2.1, its zero solution is unstable, since there is no constant M such that $\|\Phi(t)\| \leq M$.

Example 4.6. The system

$$x'(t) = \begin{pmatrix} 0 & 1 \\ -2 & -2 \end{pmatrix}\begin{pmatrix} x_1 \\ x_2 \end{pmatrix}, \; t > 0,$$

has the fundamental matrix

$$\Phi(t) = e^{-t}\begin{pmatrix} \cos(t) & \sin(t) \\ -(\cos(t) + \sin(t)) & \cos(t) - \sin(t) \end{pmatrix}, \; t > 0.$$

After some calculations, we can verify that for some positive constant K,

$$\|\Phi(t)\Phi^{-1}(s)\| \leq Ke^{-(t-s)}, \; t \geq s \geq 0,$$

and by Theorem 4.2.3 the zero solution is UAS.

4.3 Floquet theory

This section is concerned with homogeneous systems of the form

$$x'(t) = A(t)x(t), \tag{4.3.1}$$

where the matrix $A(t)$ is periodic with period T (constant), that is,

$$A(t + T) = A(t) \text{ for all } t \in \mathbb{R}.$$

For the rest of this section, we assume that $A(t)$ is a periodic matrix with period T. Our aim is to find an expression for the fundamental matrix in this case. Unlike the case of a constant coefficient matrix, the fundamental matrix cannot be expressed as $\Phi(t) = (\phi_1, \ldots, \phi_n)$ where the ϕ_i, $i = 1, 2, \ldots, n$, satisfy

$$\phi_i(t) = k_i e^{\lambda_i t},$$

and (λ_i, K_i), $i = 1, 2, \ldots n$, is eigenpairs for the constant matrix A. To reinforce this notion, we offer the following example.

Example 4.7. Consider the nonlinear system

$$x'(t) = \begin{pmatrix} \sin(t) & 0 \\ 0 & 2 \end{pmatrix} \begin{pmatrix} x_1 \\ x_2 \end{pmatrix}, \ t \in \mathbb{R}.$$

Then the corresponding $A(t)$ matrix is periodic with period $T = 2\pi$. Moreover, $A(t)$ has the eigenpairs

$$\left(\sin(t), \begin{pmatrix} 1 \\ 0 \end{pmatrix} \right), \left(2, \begin{pmatrix} 0 \\ 1 \end{pmatrix} \right).$$

Without thinking and not paying attention to the fact that A is not a constant matrix, using the obtained eigenpairs, we form the fundamental matrix

$$\Phi(t) = \begin{pmatrix} e^{\sin(t)} & 0 \\ 0 & e^{2t} \end{pmatrix}.$$

It is clear that the first column $\phi_1(t) = \begin{pmatrix} e^{\sin(t)} \\ 0 \end{pmatrix}$ is not a vector solution of the system since

$$\begin{pmatrix} \sin(t) & 0 \\ 0 & 2 \end{pmatrix} \begin{pmatrix} e^{\sin(t)} \\ 0 \end{pmatrix} = \begin{pmatrix} \sin(t)e^{\sin(t)} \\ 0 \end{pmatrix} \neq \phi_1'(t).$$

In other words,

$$A(t)\phi_1(t) \neq \phi_1'(t).$$

Hence we cannot use the eigenpairs to obtain the fundamental matrix for the periodic differential system.

Example 4.7 is a living proof of the need to develop a new approach to find the fundamental matrix of systems in the form (4.3.1). To motivate the subject, we consider the scalar differential equation

$$x'(t) = a(t)x(t), \tag{4.3.2}$$

where $a : \mathbb{R} \to \mathbb{R}$ is continuous and periodic with minimal period T. Then every solution $\phi(t)$ of (4.3.2) has the form

$$\phi(t) = c e^{\int_0^t a(s)ds}. \tag{4.3.3}$$

Set the constant $c = 1$, and for a constant λ, we observe that

$$\int_t^{t+T} a(s)ds = \lambda T, \quad \text{a constant,}$$

for all $t \in \mathbb{R}$, since $a(t + T) = a(t)$. Using (4.3.3) with $c = 1$, we have

$$\phi(t + T) = e^{\int_0^{t+T} a(s)ds} = e^{\int_0^t a(s)ds} e^{\int_t^{t+T} a(s)ds} = \phi(t)e^{\lambda t}. \tag{4.3.4}$$

The number λ in (4.3.4) has a profound effect on the behavior of the solution ϕ. At a glance, if ϕ is periodic with period T, that is, $\phi(t + T) = \phi(t)$, then (4.3.4) implies that

$$e^{\lambda t} = 1.$$

This holds if and only if $\lambda = \frac{2\pi i}{T}$. The number λ is called a *characteristic exponent* for (4.3.2), and $e^{\lambda t}$ is called the corresponding *characteristic multiplier*. More information about the form of the solutions can be revealed by letting the new function

$$p(t) = \phi(t)e^{-\lambda t}.$$

Then the function p is periodic with period T since

$$p(t + T) = \phi(t + T)e^{-\lambda(t+T)} = \phi(t)e^{\lambda T} e^{-\lambda(t+T)} = \phi(t)e^{-\lambda t} = p(t).$$

Thus any solution of (4.3.2) can be written in the form

$$\phi(t) = p(t)e^{\lambda t}. \tag{4.3.5}$$

Eq. (4.3.5) is very significant in the sense that it allows us to study the behavior of all solutions on $-\infty < t < +\infty$. As noted above, ϕ is periodic with period T if and only if $\lambda = \frac{2n\pi i}{T}$. Similarly, $\phi(t) \to 0$ as $t \to \infty$ and becomes unbounded as $t \to -\infty$ if and only if Re $\lambda < 0$.

Now we go back to the periodic system (4.3.1).

Theorem 4.3.1. *Let $A(t + T) = A(t)$ for all $t \in \mathbb{R}$. If $\Phi(t)$ is a fundamental matrix of (4.3.1), then so is $\Phi(t + T)$, and there exists a nonsingular constant matrix B such that*

(a) $\Phi(t + T) = \Phi(t)B$ *for all $t \in \mathbb{R}$;*

(b) $\det(B) = e^{\int_0^t \operatorname{tr}(A(s))ds}$.

Proof. (*a*) Set $Z(t) = \Phi(t + T)$. Then

$$Z'(t) = \Phi'(t + T) = A(t + T)\Phi(t + T) = A(t)\Phi(t + T) = Z(t)\Phi(t + T).$$

Moreover, $\det\big(Z(t)\big) = \det\big(\Phi(t + T)\big) \neq 0$, since $\Phi(t)$ is a fundamental matrix. This proves that $\Phi(t + T)$ is also a fundamental matrix. The second part of (*a*) follows from Theorem 3.2.5. This completes the proof of (*a*). For the proof of (*b*), we let $W(t) = \det\big(\Phi(t)\big)$. From Theorem 3.2.3 we have that

$$W(t) = W(t_0)e^{\int_{t_0}^{t} \mathrm{tr}\big(A(s)\big)ds},$$

and hence

$$
\begin{aligned}
W(t + T) &= W(t_0)e^{\int_{t_0}^{t+T} \mathrm{tr}(A(s))ds} \\
&= W(t_0)e^{\left(\int_{t_0}^{t} \mathrm{tr}\big(A(s)\big)ds + \int_{t}^{t+T} \mathrm{tr}\big(A(s)\big)ds\right)} \\
&= W(t_0)e^{\int_{t_0}^{t} \mathrm{tr}\big(A(s)\big)ds} \cdot e^{\int_{t}^{t+T} \mathrm{tr}\big(A(s)\big)ds} \\
&= W(t)e^{\int_{0}^{T} \mathrm{tr}\big(A(s)\big)ds}.
\end{aligned}
\tag{4.3.6}
$$

In addition,

$$\Phi(t + T) = \Phi(t)B$$

or

$$\det\big(\Phi(t + T)\big) = \det\big(\Phi(t)\big)\det\big(B\big).$$

Thus

$$W(t + T) = W(t)\det(B). \tag{4.3.7}$$

A comparison of (4.3.6) and (4.3.7) gives

$$\det(B) = e^{\int_{0}^{T} \mathrm{tr}\big(A(s)\big)ds}.$$

$\square$

Remark 4.1. Note that since the matrix B is constant, we can uniquely determine it by evaluating both sides of (*a*) at $t = 0$:

$$B = \Phi^{-1}(0)\Phi(T).$$

We have the following definition of Floquet multipliers.

Definition 4.3.1. Let $\Phi(t)$ be the fundamental matrix for the Floquet system (4.3.1). Then the eigenvalues $\rho_1, \rho_2, \ldots, \rho_n$ of

$$B := \Phi^{-1}(0)\Phi(T)$$

are called the *Floquet multipliers* of the Floquet system (4.3.1).

The *Floquet exponents* are $\mu_1, \mu_2, \ldots, \mu_n$ satisfying

$$\rho_1 = e^{\mu_1 T}, \rho_2 = e^{\mu_2 T}, \ldots, \rho_n = e^{\mu_n T}. \tag{4.3.8}$$

Note that μ_j, $j = 1, 2, \ldots, n$, may be complex.

Remark 4.2. **(i)** If $\Phi(0) = I$, then from item (b) of Theorem 4.3.1 it follows that

$$\det(B) = \prod_{j=1}^{n} \rho_j = e^{\int_0^T \operatorname{tr}\left(A(s)\right)ds}.$$

(ii) Since the trace is the sum of the eigenvalues, we also have

$$\operatorname{tr}(B) = \sum_{j=1}^{n} \rho_j.$$

(iii) The exponential exponents are not unique since if $\rho_j = e^{\mu_j T}$, then $\rho_j = e^{(\mu_j + \frac{2\pi i}{T})T}$.

(iv) The Floquet multipliers ρ_j are intrinsic parameters of system (4.3.1) and do not depend on the choice of the fundamental matrix $\Phi(t)$.

Proof (of (iv)*).* Suppose $\Psi(t)$ is another fundamental matrix. Then by Theorem 4.3.1, there is a nonsingular matrix B such that

$$\Psi(t + T) = \Psi(t)B.$$

Now by Theorem 3.2.5

$$\Psi(t) = \Phi(t)C,$$

where C is a nonsingular matrix. Thus

$$\Psi(t + T) = \Phi(t + T)C$$

or

$$\Psi(t)B = (\Phi(t)B)C.$$

This gives

$$\Phi(t)CB = \Phi(t)BC, \quad \text{or } CB = BC.$$

As a consequence,

$$CBC^{-1} = B,$$

so the eigenvalues of B and C are the same. $\qquad\square$

Theorem 4.3.2. *Let ρ be the Floquet multiplier of* (4.3.1) *with the corresponding Floquet exponent μ. Then there exists a solution $x(t)$ of* (4.3.1) *such that*

(a) $x(t+T) = \rho x(t)$;
(b) *There exists a continuously differentiable periodic $n \times 1$ vector function p with period T such that $x(t) = e^{\mu t} p(t)$.*

Proof. Let ρ be an eigenvalue of the matrix B with corresponding eigenvector b. Let $\Phi(t)$ be a fundamental matrix of the Floquet system (4.3.1) such that $x(t) = \Phi(t)b$. Then

$$
\begin{aligned}
x(t+T) &= \Phi(t+T)b \\
&= \Phi(t)Bb \\
&= \rho\Phi(t)b \\
&= \rho x(t).
\end{aligned}
$$

This proves (a). As for (b), we take $p(t) = x(t)e^{-\mu t}$. Then

$$
\begin{aligned}
p(t+T) &= x(t+T)e^{-\mu(t+T)} \\
&= \rho x(t)e^{-\mu(t+T)} \\
&= \rho e^{-\mu T} x(t)e^{-\mu t} \\
&= x(t)e^{-\mu t} \\
&= p(t),
\end{aligned}
$$

where we have used $\rho = e^{\mu T}$. This shows that $p(t)$ is periodic with period T. As a result, we have that $x(t) = e^{\mu t} p(t)$. This completes the proof of (b) and the theorem. $\qquad\square$

Remark 4.3. Suppose a Floquet multiplier ρ is complex. Since ρ is an eigenvalue of the matrix B, then so is its conjugate $\bar{\rho}$. Then we have two solutions, x and $\bar{x}$. Let the Floquet exponent $\mu = \zeta + i\eta$, and consider the periodic complex vector $p(t) = q(t) + ir(t)$. We may write the corresponding two solutions as

$$
x(t) = e^{\mu t} p(t) \ \text{ and } \ x(t) = e^{\bar{\mu} t} \bar{p}(t).
$$

Explicitly, we have

$$
\begin{aligned}
x(t) &= e^{(\zeta + i\eta)t}(q(t) + ir(t)) \\
&= e^{\zeta t}\Big[\big(q(t)\cos(\eta t) - r(t)\sin(\eta t)\big) + i\big(r(t)\cos(\eta t) + q(t)\sin(\eta t)\big)\Big]
\end{aligned}
$$

and

$$
\begin{aligned}
\bar{x}(t) &= e^{(\zeta - i\eta)t}(q(t) - ir(t)) \\
&= e^{\zeta t}\Big[\big(q(t)\cos(\eta t) - r(t)\sin(\eta t)\big) - i\big(r(t)\cos(\eta t) + q(t)\sin(\eta t)\big)\Big].
\end{aligned}
$$

As a consequence, we may write the linearly independent real solutions

$$x_R = \text{Re}\big[e^{\mu t} p(t)\big] = e^{\zeta t}\Big[q(t)\cos(\eta t) - r(t)\sin(\eta t)\Big], \tag{4.3.9}$$

$$x_I = \text{Im}\big[e^{\mu t} p(t)\big] = e^{\zeta t}\Big[q(t)\sin(\eta t) + r(t)\cos(\eta t)\Big]. \tag{4.3.10}$$

Lemma 4.1. *(Logarithmic matrix) If B is an $n \times n$ matrix, then there exists an $n \times n$ (complex) matrix C such that $B = e^C$.*

Proof. Let λ_j, $j = 1, 2, \ldots, m$, with $m \leq n$ denote the eigenvalues of B. We already know that

$$e^{P^{-1}CP} = P^{-1}e^C P.$$

Therefore we may assume that P is in Jordan canonical form, that is,

$$P = diag\big(J_1, J_2, \ldots, J_n\big)$$

with

$$J_j = \lambda_j + N_j, \ \lambda_j \neq 0,$$

where J_j and N_j are defined in Theorem 3.2.12. Suppose that

$$B = diag\big(B_1, B_2, \ldots, B_n\big),$$

where

$$B_j = \begin{pmatrix} \lambda_j & 1 & & \\ & \lambda_j & \ddots & \\ & & & 1 \\ & & & \lambda_j \end{pmatrix}, \ j = 1, 2, \ldots m.$$

Let $l_j \times l_j$ be the size of $B_j = \lambda_j I_j + N_j$, where $N^{l_j} = 0$ (zero matrix) for all $k \geq m$. Thus $\sum_{k=1}^{m} l_k = n$, and so we have constructed a matrix B of size $n \times n$. Thus if we can prove that for every B_j, there exists a matrix C_j such that $B_j = e^{C_j}$, then $B = e^C$. Define

$$C_j := (\text{Ln}(\Lambda_j)I_j + L_j,$$

where

$$L_j := -\sum_{k=1}^{m} \frac{(-N_j)^k}{k\lambda_j^k} \equiv \text{Ln}(I_j + \frac{N_j}{\lambda_j}).$$

Recall that at $x = 0$,

$$\text{Ln}(1 + x) = \sum_{k=1}^{\infty} \frac{(-x)^k}{k}, \ |x| < 1,$$

which implies that

$$1 + x = e^{\text{Ln}(1+x)} = 1 + (x - x^2/2 + x^3/3 + \ldots) + \frac{1}{2!}(x - x^2/2$$
$$+ x^3/3 + \ldots)^2 + \frac{1}{3!}(x - x^2/2 + x^3/3 + \ldots)^3 + \ldots.$$

Since $N^{l_j} = 0$ (zero matrix) for all $k \geq m$, we must have

$$e^{L_j} = I_j + \frac{N_j}{\lambda_j}$$

or

$$B_j = e^{\text{Ln}(\lambda_j)I_j + L_j},$$

where we have loosely used the fact that $B_j = e^{\text{Ln}(B_j)}$. This completes the proof. $\quad\square$

Now we are ready to state and prove the Floquet theorem.

Theorem 4.3.3. *(Floquet theorem) Every fundamental matrix of (4.3.1) has the form*

$$\Phi(t) = P(t)e^{Ct},$$

where $P(t)$ is a T-periodic matrix, and C is a constant $n \times n$ matrix.

Proof. Let $\Phi(t)$ be a fundamental matrix of (4.3.1). Then

$$(\Phi^{-1}(t)\Phi(t+T))' = (\Phi^{-1}(t))'\Phi(t+T) + \Phi^{-1}(t)\Phi'(t+T)$$
$$= -\Phi^{-1}(t)A(t)\Phi(t+T) + \Phi^{-1}(t)A(t)\Phi(t+T)$$
$$= 0.$$

Thus

$$\Phi^{-1}(t)\Phi(t+T) = B \text{ (constant matrix)},$$

or

$$\Phi(t+T) = \Phi(t)B.$$

Since $\Phi^{-1}(t)$ and $\Phi(t+T)$ are nonsingular, B is also nonsingular. Thus by Lemma 4.1 $B = e^{CT}$ for some constant matrix C. Left to show that $P(t)$ is periodic with period T. We define $P(t) = \Phi(t)e^{-Ct}$. Then

$$P(t+T) = \Phi(t+T)e^{-C(t+T)}$$
$$= \Phi(t)e^{CT}e^{-C(t+T)}$$
$$= \Phi(t)e^{-Ct}$$
$$= P(t).$$

This completes the proof. $\quad\square$

Example 4.8. Let us find the Floquet multipliers and exponents of

$$x'(t) = \begin{pmatrix} -1 & 1 \\ 0 & 1 + \cos(t) - \frac{\sin(t)}{2 + \cos(t)} \end{pmatrix} \begin{pmatrix} x_1 \\ x_2 \end{pmatrix}.$$

We notice that the corresponding coefficient matrix is periodic with period 2π. We easily solve

$$x_2' = \left(1 + \cos(t) - \frac{\sin(t)}{2 + \cos(t)}\right) x_2$$

using separation of variables and obtain

$$x_2 = c_2 e^{t + \sin(t)} (2 + \cos(t)).$$

Substituting x_2 into $x_1' = -x_1 + x_2$ and then using the variation of parameters formula, we obtain

$$x_1 = c_2 e^{t + \sin(t)} + c_1 e^{-t}.$$

Setting $c_1 = c_2 = 1$, we obtain the fundamental matrix

$$\Phi(t) = \begin{pmatrix} e^{t + \sin(t)} & e^{-t} \\ e^{t + \sin(t)} (2 + \cos(t)) & 0 \end{pmatrix}.$$

From Definition 4.3.1, we have

$$B := \Phi^{-1}(0)\Phi(T) = \Phi^{-1}(0)\Phi(2\pi) = \begin{pmatrix} 0 & \frac{1}{3} \\ 1 & -\frac{1}{3} \end{pmatrix} \begin{pmatrix} e^{2\pi} & e^{-2\pi} \\ 3e^{2\pi} & 0 \end{pmatrix} = \begin{pmatrix} e^{2\pi} & 0 \\ 0 & e^{-2\pi} \end{pmatrix}.$$

Then the eigenvalues of B, $\rho_1 = e^{2\pi}$ and $\rho_2 = e^{-2\pi}$ are the Floquet multipliers with Floquet exponents $\mu_1 = 2\pi$ and $\mu_2 = -2\pi$.

Example 4.9. Find the Floquet multiplier and exponent of

$$x'(t) = (2 - \cos(t))x(t).$$

Using separation of variables, we find the solution

$$x(t) = K e^{2t - \sin(t)}$$

for some constant K. Set $K = 1$. Then the Floquet multiplier is $\rho = x^{-1}(0)x(2\pi) = 1e^{4\pi}$, and the Floquet exponent is $\mu = 4\pi$.

One of the main purposes of Floquet theory is to study the stability of the zero solution of (4.3.1). It is evident from Theorem 4.3.2 that the behavior of a solution depends on the Floquet exponent μ. In fact, from Theorem 4.3.2 we have that

$$x(t) = e^{\mu t} p(t),$$

where $p(t)$ is a periodic $n \times 1$ vector. Let x_j be the jth component of the vector solution. Then from Theorem 4.3.2 we have that

$$x_j(t) = \rho_j x_j(t)$$

and

$$x_j(t + NT) = \rho_j^N x_j(t) \to 0 \text{ as } N \to \infty.$$

Now we can easily see that each Floquet multiplier falls into one of the following categories:

(a) If $|\rho| < 1$, then $\mathrm{Re}(\mu) < 0$ and so $x(t) \to 0$ as $t \to \infty$;
(b) If $\rho = \pm 1$, then $\mathrm{Re}(\mu) = 0$, and so the solution is periodic with period T;
(c) If $|\rho| > 1$, then $\mathrm{Re}(\mu) > 0$, and so $x(t) \to \infty$ as $t \to \infty$.

We have the following general theorem concerning the stability of the zero solution of (4.3.1).

Theorem 4.3.4. *The zero solution of* (4.3.1) *is*

(i) *stable if and only if the Floquet multipliers ρ satisfy $|\rho| \le 1$ and there is a complete set of eigenvectors for any multiplier of modulus 1;*
(ii) *asymptotically stable if and only if $|\rho| < 1$ for every ρ.*

Proof. Let $\Phi(t)$ be the fundamental matrix of (4.3.1). Then by the Floquet theorem, we have that $\Phi(t) = P(t)e^{Ct}$. Then the solutions of (4.3.1) are given by $x(t) = P(t)e^{Ct}x_0$, where $x_0 \in \mathbb{R}^n$. We make two claims:

(1) the stability of the zero solution of (4.3.1) is the same as for

$$y'(t) = Cy(t), \tag{4.3.11}$$

which has solutions of the form $y(t) = e^{Ct}x_0$. To see this, let M be a positive constant. Since $P(t)$ is continuous and periodic, we have that $||P(t)|| \le M$ for all $t \in \mathbb{R}$. Thus $|x(t)| = |P(t)e^{Ct}x_0| \le M||e^{Ct}x_0|| = M|y(t)|$, and this completes the proof of the first claim.
(2) $y(t) = P^{-1}(t)x(t)$ is a solution of $y'(t) = Cy(t)$. To see this, let $y(t) = P^{-1}(t)x(t)$. Then

$$y(t) = P^{-1}(t)\Phi(t)x_0$$
$$= P^{-1}(t)P(t)e^{Ct}x_0$$
$$= e^{Ct}x_0,$$

which is a solution of $y'(t) = Cy(t)$. This completes the proof of the second claim.

Next, we address the proof of stability.

If $|\rho| < 1$ for all Floquet multipliers, then $\mathrm{Re}(\mu) < 0$ for all Floquet exponents, and the zero solution is AS for (4.3.11) and hence for (4.3.1).

Now we assume that $|\rho| \leq 1$ for all ρ and there are no generalized eigenvectors (see (3.2.26)) of e^{CT} corresponding to Floquet multipliers with $|\rho| = 1$. Then $\text{Re}(\mu) \leq 0$ for all eigenvalues of C, and there are no generalized eigenvectors of C corresponding to eigenvalues with $\text{Re}(\mu) = 0$. As a consequence, we have that $|e^{Ct}|$ is uniformly bounded. Thus the zero solution is stable for (4.3.11) and hence for (4.3.1).

If $\rho > 1$ for some ρ or if there is a generalized eigenvector of e^{CT} with $\rho = 1$, then either C has an eigenvalue with $\text{Re}(\mu) > 1$ or C has a generalized eigenvector with $\text{Re}(\mu) = 0$. In both cases, $|y(t)| \to \infty$ as $t \to \infty$. This implies that the zero solution of (4.3.11) is unstable. On the other hand, since $P(t)$ is continuous and periodic, we have that $P^{-1}(t)$ is also continuous and periodic. Thus

$$y(NT) = P^{-1}(NT)x(NT)x_0$$

for a positive integer N. Therefore

$$\infty = \lim_{N \to \infty} |y(NT)| \leq ||P^{-1}(NT)|| \lim_{N \to \infty} |x(NT)|,$$

and the zero solution is also unstable for (4.3.1). This completes the proof. $\qquad\square$

We provide the following example.

Example 4.10. Consider the periodic system with period π

$$x'(t) = \begin{pmatrix} -1 + \frac{1}{2}\sin^2(t) & -1 - \frac{1}{2}\cos(t)\sin(t) \\ 1 - \frac{1}{2}\cos(t)\sin(t) & -1 + \frac{1}{2}\cos^2(t) \end{pmatrix} \begin{pmatrix} x_1 \\ x_2 \end{pmatrix}.$$

By inspection we found the fundamental matrix to be

$$\Phi(t) = \begin{pmatrix} -e^{-t/2}\sin(t) & e^{-t}\cos(t) \\ e^{-t/2}\cos(t) & e^{-t}\sin(t) \end{pmatrix}.$$

We encourage the reader to verify that

$$\Phi'(t) = A(t)\Phi(t).$$

Then

$$B := \Phi^{-1}(0)\Phi(T) = \Phi^{-1}(0)\Phi(\pi)$$

$$= \begin{pmatrix} 0 & 1 \\ 1 & 0 \end{pmatrix} \begin{pmatrix} 0 & -e^{-\pi} \\ -e^{-\pi/2} & 0 \end{pmatrix} = \begin{pmatrix} -e^{-\pi/2} & 0 \\ 0 & -e^{-\pi} \end{pmatrix},$$

and thus the eigenvalues of B are given by $\rho_1 = -e^{-\pi/2}$ and $\rho_2 = -e^{-\pi}$ with $|\rho_i| < 1$, $i = 1, 2$. By Theorem 4.3.4 the origin $(0, 0)$ is asymptotically stable.

4.3.1 Mathieu's equation

We begin by considering the 2×2 real entries and constant matrix

$$D = \begin{pmatrix} a & b \\ c & d \end{pmatrix}.$$

Let $L(\lambda) = \det(D - \lambda I)$. Then $L(\lambda)$ can be easily computed to be

$$L(\lambda) = \lambda^2 - (a + d)\lambda + ad - bc = \lambda^2 - \text{tr}(D)\lambda + \det(D). \qquad (4.3.12)$$

On the other hand, if λ_1 and λ_2 are the eigenvalues of D, not necessarily distinct, then

$$L(\lambda) = (\lambda - \lambda_1)(\lambda - \lambda_2) = \lambda^2 - (\lambda_1 + \lambda_2)\lambda + \lambda_1 \lambda_2. \qquad (4.3.13)$$

A quick comparison of (4.3.12) with (4.3.13) gives

$$\text{tr}(D) = \lambda_1 + \lambda_2 \qquad (4.3.14)$$

and

$$\det(D) = \lambda_1 \lambda_2. \qquad (4.3.15)$$

Next, we consider Mathieu's equation

$$x''(t) + \big(\alpha + \beta \cos(t)\big)x(t) = 0 \qquad (4.3.16)$$

with parameters α and β. Eq. (4.3.16) was discovered by Émile Léonard Mathieu, who encountered it while studying vibrating elliptical drumheads. It has applications in many fields of the physical sciences, such as optics, quantum mechanics, and general relativity. The equation tends to occur in problems involving periodic motion or in the analysis of partial differential equations and boundary value problems possessing elliptic symmetry. Using the transformation $x_1 = x$, $x_2 = x'$, we write (4.3.16) in the system form

$$x'(t) = \begin{pmatrix} 0 & 1 \\ -\alpha - \beta \cos(t) & 0 \end{pmatrix} \begin{pmatrix} x_1 \\ x_2 \end{pmatrix}.$$

It is clear that the system is a Floquet system with a period of 2π. Let $\Phi(t)$ be a fundamental matrix. Then the Floquet multipliers ρ_1 and ρ_2 are the eigenvalues of the matrix

$$B = \Phi^{-1}(0)\Phi(T).$$

Since $\text{tr}(A) = 0$, where

$$A(t) = \begin{pmatrix} 0 & 1 \\ -\alpha - \beta \cos(t) & 0 \end{pmatrix},$$

by Theorem 4.3.1, we have that

$$\det(B) = \prod_{j=1}^{n} \rho_j = e^{\int_0^T \operatorname{tr}\left(A(s)\right)ds} = e^0 = 1.$$

Thus by (4.3.15),

$$\rho_1 \rho_2 = 1.$$

As consequence of (4.3.12)–(4.3.15), the eigenvalues or Floquet multipliers must satisfy the quadratic equation $\rho^2 - (\rho_1 + \rho_2)\rho + 1 = 0$. Letting $\phi := \phi(\alpha, \beta) = \rho_1 + \rho_2$, we arrive at the characteristic equation

$$\rho^2 - \phi\rho + 1 = 0. \tag{4.3.17}$$

Eq. (4.3.17) has the roots

$$\rho_{1,2} = \frac{1}{2}\left(\phi \pm \sqrt{\phi^2 - 4}\right).$$

Despite the fact that we cannot solve for ϕ, we can still deduce some information about the solutions by considering particular values for ϕ. Below we consider several cases that depend on assigned values of ϕ.

1. **Case 1: $\phi > 2$.**
   Since $\rho_{1,2} = \frac{1}{2}\left(\phi \pm \sqrt{\phi^2 - 4}\right)$, we have $\rho_1 > 1$, and since $\rho_1 \rho_2 = 1$, we have $\rho_1 > 1 > \rho_2 > 0$. As $\rho_2 = \frac{1}{\rho_1}$, we conclude that $\mu_2 = -\mu_1$. Thus according to Theorem 4.3.2, the solution must be of the form

   $$x(t) = c_1 e^{\mu_1 t} p_1(t) + c_2 e^{-\mu_1 t} p_2(t),$$

   where $p_1(t)$ and $p_2(t)$ are both periodic with period 2π. Moreover, the zero solution is unstable, and solutions are unbounded.

2. **Case 2: $\phi = 2$.**
   Then $\rho_1 = \rho_2 = 1$. In this case, Theorem 4.3.2 guarantees one solution of the form $e^{\mu t} p(t)$. On the other hand, if B has two corresponding linearly independent eigenvectors, then we can find two linearly independent 2π-periodic functions $p_1(t)$ and $p_2(t)$, so that the general solution has the form $x(t) = c_1 p_1(t) + c_2 p_2(t)$, since $\rho = 1$ implies $\mu = 0$. Now if B has only one eigenvector, then we end up with a solution of the form $p_1(t)$ and another of the form $t p_1(t) + p_2(t)$. This is due to the fact that the Jordan canonical form of the matrix B is $\begin{pmatrix} \lambda & 1 \\ 0 & \lambda \end{pmatrix}$, and the solution takes the form

$$\Phi(t) = P(t)e^{\begin{pmatrix} \lambda & 1 \\ 0 & \lambda \end{pmatrix}t} = P(t)\begin{pmatrix} e^{\lambda t} & te^{\lambda t} \\ 0 & e^{\lambda t} \end{pmatrix} = \begin{pmatrix} p_1(t) \\ p_2(t) \end{pmatrix}\begin{pmatrix} e^{\lambda t} & te^{\lambda t} \\ 0 & e^{\lambda t} \end{pmatrix}.$$

In conclusion, we have a solution of the form

$$x(t) = (c_1 + c_2 t)p_1(t) + c_2 p_2(t),$$

which is unbounded, and, moreover, the zero solution is unstable.

3. Case 3: $\phi = -2$.

Then $\rho_1 = \rho_2 = -1$. This is similar to Case 3. We know that $\rho = e^{\mu T} = -1$, which implies that $\mu = \frac{i\pi}{T}$, since

$$e^{\mu T} = e^{\frac{i\pi}{T}T} = e^{i\pi} = -1.$$

The solution is given by

$$x(t) = (c_1 + c_2 t)p_1(t)e^{\frac{i\pi}{T}t} + c_2 p_2(t)e^{\frac{i\pi}{T}t},$$

which is a combination of a periodic solution with period 4π and a nonperiodic solution. Again, solutions are unbounded, and the zero solution is unstable.

4. Case 4: $-2 < \phi < 2$.

Without loss of generality, we may define η by $\phi = 2\cos(\eta T)$, where $0 < \eta T < \pi$, so that

$$\rho = \frac{1}{2}\left(\phi \pm \sqrt{\phi^2 - 4}\right)$$
$$= \cos(\eta T) \pm i \, \sin(\eta T)$$
$$= e^{\pm i\eta T}.$$

Since $\zeta = 0$, Eqs. (4.3.9) and (4.3.10) reduce to

$$x_R = \text{Re}\left[e^{i\eta t}p(t)\right] = q(t)\cos(\eta t) - r(t)\sin(\eta t),$$
$$x_I = \text{Im}\left[e^{i\eta t}p(t)\right] = q(t)\sin(\eta t) + r(t)\cos(\eta t),$$

where $p(t) = q(t) + i \, r(t)$ is periodic with period 2π. With this in mind, the solution takes the form

$$x(t) = c_1\Big(q(t)\cos(\eta t) - r(t)\sin(\eta t)\Big) + c_2\Big(q(t)\sin(\eta t) + r(t)\cos(\eta t)\Big).$$

We easily see that the zero solution is stable and solutions are bounded, but it is not periodic in general as there are two frequencies 2π and η. In such a case, the solution is said to be pseudoperiodic. Next, we examine the period of η. Now $e^{i\eta t}$ has the period $\bar{T} := \frac{2\pi}{\eta}$. Since $0 < 2\cos(\eta T) < \pi$, or, $\phi \neq \pm 2$, we must have $\eta T \neq j\pi$, $j = 1, 2, \ldots,$ or $\frac{2\pi T}{\bar{T}} \neq j\pi$, which implies that $\frac{2T}{j} \neq \bar{T}$. This gives $\bar{T} \neq 2T, T, \frac{2}{3}T, \ldots.$

Note that in order that $\bar{T} = nT$, we must have

$$\eta = \frac{2\pi}{nT}$$

for $n \neq 1, 2$.

4.3.2 Applications to Mathieu's equation

4.3.2.1 Inverted pendulum

An application of Mathieu's equation is the inverted pendulum depicted in Fig. 4.3. We assume that the pendulum is frictionless and has a massless rod of length l. If F_{pivot} is the force acting on the pivot, τ_{net} is the net force to external torque, and I is the moment of inertia, then by Newton's second law, we have

$$F_{\text{pivot}} = ma = m\frac{d^2y}{dt^2}, \tag{4.3.18}$$

$$\tau_{\text{net}} = I\frac{d^2\theta}{dt^2} = ml^2\frac{d^2\theta}{dt^2}. \tag{4.3.19}$$

If F is the force acting on the particle and τ is the torque, then

$$\tau = rF\sin(\theta),$$

where r is the distance from the axis of rotation to the particle. So the gravitational torque is

$$\tau_{\text{grav}} = rF_{\text{grav}}\sin(\theta) = mgl\sin(\theta).$$

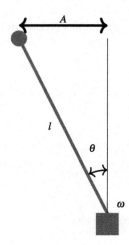

FIGURE 4.3

Inverted pendulum plotted in xy-coordinates.

Since the harmonic motion of the pendulum can be expressed as $y(t) = A\cos(\omega t)$, Eq. (4.3.18) becomes

$$F_{\text{pivot}} = ma = m\omega^2 A \cos(\omega t).$$

Now the torque exerted by the pivot, τ_{pivot} is given by

$$\tau_{\text{pivot}} = r F_{\text{pivot}} \sin(\theta) = -ml\omega^2 A \cos(\omega t) \sin(\theta).$$

Thus the total torque $\tau_{\text{net}} = \tau_{\text{grav}} + \tau_{\text{pivot}}$. As a result, expression (4.3.19) yields the expression

$$ml^2 \frac{d^2\theta}{dt^2} = mgl \sin(\theta) - ml\omega^2 A \cos(\omega t) \sin(\theta)$$

or

$$\frac{d^2\theta}{dt^2} + \left(\frac{g}{l} - \frac{\omega^2 A}{l} \cos(t)\right) \sin(\theta),$$

setting $\sin(\theta) \approx \theta$ for small oscillations, the above expression reduces to

$$\theta'' + \big(\alpha + \beta \cos(t)\big) = 0$$

with $\alpha = \frac{g}{l}$, $\beta = -\frac{\omega^2 A}{l}$, and period $T = 2\pi$.

4.4 Exercises

Exercise 4.1. Discuss the flow of solutions around the equilibrium solutions as we did in Fig. 4.1 for the following systems:

(a)

$$x'(t) = (1 - x^2)xe^x, \quad x(0) = x_0 \neq 0;$$

(b)

$$x'(t) = x^3 + x^2 + 6x, \quad x(0) = x_0 \neq 0.$$

Exercise 4.2. Discuss the stability as we did in Example 4.2 for the following IVPs:

(a)

$$(t^2 + 9)x'(t) + tx = 0, \quad x(0) = x_0 \neq 0;$$

(b)

$$(1 + t)x'(t) = -x, \quad x(t_0) = x_0, \quad t \geq t_0 > -1.$$

Exercise 4.3. Show that the zero solution of

$$x'(t) = -x^3, \ x(t_0) = x_0 \neq 0,$$

is UAS.

Exercise 4.4. The solution of

$$x'(t) = a(t)x(t), \ x(t_0) = x_0 \neq 0, \ t \geq t_0 \geq 0, \quad (4.4.1)$$

is given by

$$x(t) = x_0 e^{\int_{t_0}^{t} a(s)ds},$$

where $a(t)$ is continuous for $t \geq 0$. Show that the zero solution of (4.4.1) is

(a) stable if and only if there exists a positive constant M such that

$$e^{\int_{t_0}^{t} a(s)ds} < M, \ t \geq 0;$$

(b) asymptotically stable if and only if

$$\int_0^t a(s)ds \to -\infty \text{ as } t \to \infty;$$

(c) uniformly stable if and only if there exists a positive constant M such that $\int_{t_1}^{t_2} a(s)ds \leq M$ for all $0 < t_1 < t_2 < \infty$;

(d) uniformly asymptotically stable if and only if there exist positive constants M and α such that $\int_{t_0}^{t} a(s)ds \leq -\alpha(t - t_0) + M$.

Exercise 4.5. Find the fundamental matrix and study the stability of the zero solution of the given system for $t > 0$:

(a)

$$x'(t) = \begin{pmatrix} 0 & 1 \\ -\frac{1}{2t^2} & -\frac{5}{2t} \end{pmatrix} \begin{pmatrix} x_1 \\ x_2 \end{pmatrix},$$

(b)

$$x'(t) = \begin{pmatrix} 0 & 1 \\ -\frac{8}{t^2} & -\frac{10}{t} \end{pmatrix} \begin{pmatrix} x_1 \\ x_2 \end{pmatrix}.$$

Exercise 4.6. Find the Floquet multipliers and exponents of

$$x'(t) = \begin{pmatrix} \frac{\cos(t)+\sin(t)}{2+\sin(t)-\cos(t)} & 0 \\ 1 & 1 \end{pmatrix} \begin{pmatrix} x_1 \\ x_2 \end{pmatrix}.$$

Exercise 4.7. Find the Floquet multipliers and exponents of

$$x'(t) = \begin{pmatrix} 1 + \frac{\cos(t)}{2+\sin(t)} & 0 \\ 1 & -1 \end{pmatrix} \begin{pmatrix} x_1 \\ x_2 \end{pmatrix}.$$

Exercise 4.8. Find the Floquet multipliers and exponents of

(a)

$$x'(t) = \cos(3t)x(t);$$

(b)

$$x'(t) = \Big(\sin(3t) + \cos(3t) \Big) x(t);$$

(c)

$$x'(t) = \Big(-1 + \cos(2t) \Big) x(t).$$

Exercise 4.9. Prove that $B_j = e^{\mathrm{Ln}(B_j)}$ in the proof of Lemma 4.1.

Exercise 4.10. Carry out a similar discussion as in Section 4.3.1 for the second-order Hill equation

$$x''(t) + a(t)x(t) = 0,$$

where the function a is continuous and periodic with period T.

Exercise 4.11. Find the logarithm of the rotation matrix

$$A = \begin{pmatrix} \cos(t) & -\sin(t) \\ \sin(t) & \cos(t) \end{pmatrix}.$$

Exercise 4.12. Show that all solutions of (4.2.1) are bounded if and only if its zero solution is stable.

Qualitative analysis of linear systems

This chapter is exclusively devoted to the study of the stability of linear systems, near-linear systems, perturbed systems, and autonomous systems in the plane.

5.1 Preliminary theorems

In Section 4.2 of Chapter 4, we discussed the stability, uniform stability, asymptotic stability, and uniform asymptotic stability of the zero solution of the non-autonomous linear system

$$x' = A(t)x$$

by imposing conditions on its fundamental and principal matrices. In this section, we give explicit criteria that satisfy the conditions that were imposed only for the linear system with constant coefficients

$$x' = Ax, \tag{5.1.1}$$

where A is an $n \times n$ matrix with constant entries. We begin with the following lemma.

Lemma 5.1. *Let*

$$r(t) = t^m e^{-\alpha t}, \ t \ge 0,$$

where m is a non-negative integer, and α is a positive constant. Then there exists a positive constant D depending on m and α such that $r(t) \le D$.

Proof. If $m = 0$ then $r(t) \le 1 := D$. On the other hand, if $m > 0$, then by repeated application of L'Hôpital's rule we have $\lim\limits_{t \to \infty} t^m e^{-\alpha t} = 0$. Moreover, since $r(0) = 0$, $r(t)$ is continuous, and $r(t) \ge 0$, we have that $r(t) \le D$ for some positive constant D. This completes the proof. $\square$

Definition 5.1.1. Let A be a square matrix. We define the *spectrum* of A as

$$\sigma(A) = \{\lambda : \det(A - \lambda I) = 0\}.$$

Thus the inequality

$$Re(\sigma(A)) < \zeta$$

for some real number ζ means that all the eigenvalues of A have real parts less than ζ.

Advanced Differential Equations. https://doi.org/10.1016/B978-0-32-399280-0.00011-5

The next theorem is concerned with eigenvalues of multiplicity m.

Theorem 5.1.1. *Let λ be an eigenvalue of a matrix A with $Re\big(\sigma(A)\big) < \zeta$. Then for any integer $m \geq 0$, there exists a positive constant D such that*

$$t^m e^{\lambda t} \leq D e^{\zeta t} \text{ for all } t \geq 0.$$

Proof. Let $\lambda = \alpha + i\beta$ with $\alpha, \beta \in \mathbb{R}$. If $\beta = 0$, that is, λ is real, then since $Re\big(\sigma(A)\big) < \zeta$, we have that $\alpha - \zeta < 0$. So by Lemma 5.1

$$t^m e^{(\alpha - \zeta)t} \leq D \text{ for all } t \geq 0.$$

Multiplying this inequality by $e^{\zeta t}$, we obtain

$$t^m e^{\lambda t} = t^m e^{\alpha t} \leq D e^{\zeta t} \text{ for all } t \geq 0.$$

Now if $\beta \neq 0$, that is, λ is complex, then

$$|t^m e^{\lambda t}| = |t^m e^{(\alpha + i\beta)t}| = t^m e^{\alpha t} \leq D e^{\zeta t} \text{ for all } t \geq 0,$$

since $\alpha = Re\big(\sigma(A)\big) < \zeta$. This completes the proof. $\square$

Theorem 5.1.2. *Let λ be an eigenvalue of a matrix A with $Re\big(\sigma(A)\big) < \zeta$. Then for some positive constant D, any solution x of (5.1.1) satisfies*

$$|x(t)| \leq D e^{\zeta t} \text{ for all } t \geq 0.$$

Proof. Let $\lambda_1, \lambda_2, \ldots, \lambda_k$ be eigenvalues of A. We can form a set of fundamental solution, $x_1, x_2, \ldots, x_n$ and find an integer $N \geq 0$ such that each solution is of the form $x_j = t^m e^{\lambda_N t}$, $N = 1, 2, \ldots, k$, for integer m. Then by Theorem 5.1.1, we have that $|x_j(t)| \leq V_j e^{\zeta t}$ for positive constants V_j. If x is an arbitrary solution of (5.1.1), then it can be written in the form

$$x(t) = c_1 x_1(t) + c_2 x_2(t) + \ldots + c_n x_n(t).$$

Then for $t \geq 0$,

$$
\begin{aligned}
|x(t)| &\leq |c_1||x_1(t)| + |c_2||x_2(t)| + \ldots + |c_n||x_n(t)| \\
&\leq |c_1| V_1 e^{\zeta t} + |c_2| V_2 e^{\zeta t} + \ldots + |c_n| V_n e^{\zeta t} \\
&= \Big(|c_1| V_1 + |c_2| V_2 + \ldots + |c_n| V_n \Big) e^{\zeta t}.
\end{aligned}
$$

This completes the proof. $\square$

The next theorem plays an important role in the study of stability, and its proof depends on the Putzer algorithm, Theorem 3.2.14.

Theorem 5.1.3. *Let λ_i, $i = 1, 2, \ldots, k$, be the distinct eigenvalues of A, where λ_i has multiplicity n_i, and $n_1 + n_2 + \ldots + n_k = n$. If $Re\big(\sigma(A)\big) < \zeta$, then there is a positive constant D such that*

$$e^{At} \leq De^{\zeta t} \text{ for all } t \geq 0.$$

Proof. From Theorem 3.2.14 we know that

$$e^{At} = \sum_{j=0}^{n-1} \mu_{j+1}(t) P_j, \tag{5.1.2}$$

where $\mu_1(t), \mu_2(t), \ldots, \mu_n(t)$ satisfy the corresponding initial value problems

$$\mu_1' = \lambda_1 \mu_1, \ \mu_1(0) = 1; \ \mu_j' = \lambda_j \mu_j + \mu_{j-1}(t), \ \mu_j(0) = 0, \ j = 2, 3, \ldots, n,$$

and the matrices $P_0, P_1, \ldots, P_n$ are given by

$$P_0 = I, \ P_j = \prod_{k=1}^{j} (A - \lambda_j I), \ j = 1, 2, \ldots, n.$$

Then we may easily verify that

$$\mu_1(t) = e^{\lambda_1 t}, \ \mu_{j+1}(t) = e^{\lambda_{j+1} t} \int_0^t e^{-\lambda_{j+1} t} \mu_j(s) ds$$

for $j = 1, 2, \ldots, n - 1$. Without loss of generality, we may choose the order in $\{\lambda_j\}$ so that the real parts do not decrease, that is, if $\lambda_j = \rho_j + i\sigma_j$, then $\rho_j \leq \rho_{j+1}$. We claim that

$$|\mu_j(t)| \leq \frac{t^{j-1}}{(j-1)!} e^{\rho_j t} \text{ for } t > 0, \ j = 1, 2, \ldots, n. \tag{5.1.3}$$

Obviously, (5.1.3) holds for $j = 1$, since $\mu_1(t) = e^{\lambda_1 t}$. Thus let it hold for j. Now from the above integral defining $\mu_{j+1}(t)$ we have

$$\mu_{j+1}(t) = e^{\rho_{j+1} t} \int_0^t e^{-\rho_{j+1} s} \frac{s^{j-1}}{(j-1)!} e^{\rho_j s} ds.$$

Since $e^{(\rho_j - \rho_{j+1})s} \leq 1$, the above integral reduces to

$$\mu_{j+1}(t) = e^{\rho_{j+1} t} \frac{1}{(j-1)!} \int_0^t s^{j-1} ds = \frac{t^j}{j!} e^{\rho_{j+1} t}.$$

By our setup, we have ρ^n is the largest of the real parts, and hence

$$\mu_{j+1}(t) \leq \frac{t^j}{j!} e^{\rho_n t}.$$

For any exponent j and any $\rho^* > 0$, we have

$$t^j < h e^{\rho^* t}$$

for all $t \geq 0$ and some positive constant h. Thus

$$\mu_{j+1}(t) \leq h e^{(\rho^* + \rho_n)t}, \quad t \geq 0. \tag{5.1.4}$$

Now substitute (5.1.3) into (5.1.2) and then take the matrix norm to get

$$||e^{At}|| \leq \sum_{j=0}^{n-1} |\mu_{j+1}(t)| \, ||P_j|| \leq \Big[h \sum_{j=0}^{n-1} ||P_j|| \Big] e^{\zeta t} := D e^{\zeta t}. \tag{5.1.5}$$

This completes the proof. $\qquad\square$

As a direct consequence of Theorem 5.1.3, we have the following lemma.

Lemma 5.2. *Suppose all the eigenvalues of A have negative real parts. Then there are positive constants h, K, and ρ such that*

$$\mu_{j+1}(t) \leq h e^{-\rho t} \text{ and } ||e^{At}|| \leq K e^{-\rho t}, \, t \geq 0. \tag{5.1.6}$$

Theorem 5.1.4. **(i)** *Assume that every eigenvalue of the matrix A has a nonpositive real part and that the eigenvalues with zero real parts are simple. Then there is a positive constant D such that*

$$||e^{At}|| \leq D, \, t \geq 0.$$

Moreover, if x is a solution of (5.1.1), then it is bounded, and the zero solution is stable.

(ii) *If $Re(\sigma(A)) < 0$, then the zero solution is AS.*

(iii) *If there is an eigenvalue of A with positive real part, then the zero solution is unstable.*

Proof. If $\lambda = i\beta$ is a simple pure imaginary, then it contributes a bounded function of the form $e^{i\beta t}$ to a fundamental set of solutions. Thus we get a fundamental set of solutions with all elements bounded on $[0, \infty)$. Since e^{At} is formed from such a set, an argument similar to the proof of Theorem 5.1.3 shows that

$$||e^{At}|| \leq D$$

for some positive constant D. Now if x is a solution of (5.1.1) with $x(t_0) = x_0$, then

$$x(t) = e^{A(t-t_0)} x_0 \text{ for } t \geq t_0.$$

Let $\varepsilon > 0$ and set $\delta = \varepsilon/D$. Then for $|x_0| < \delta$, we have

$$|x(t)| = ||e^{A(t-t_0)}|| \, |x_0| \leq D\delta = \varepsilon.$$

This proves that the solution is bounded, and the zero solution is stable. The proof of (i) is complete.

As for (ii), the zero solution is stable by (i) since $Re(\sigma(A)) < 0$. Thus we may choose a real number $\zeta < 0$ such $Re(\sigma(A)) < \zeta$. By Theorem 5.1.3 there is a positive constant D such that $||e^{At}|| \leq De^{\zeta t}$ for all $t \geq 0$. Hence, for any initial condition $x(0) = x_0$, the solution x of (5.1.1) satisfies

$$|x(t, 0, x_0)| = |e^{At}x_0| \leq D|x_0|e^{\zeta t} \to 0 \text{ as } t \to \infty.$$

This completes the proof of (ii). □

Remark 5.1. Consider two matrices

$$A = \begin{pmatrix} 0 & 0 \\ 0 & 0 \end{pmatrix} \text{ and } B = \begin{pmatrix} 0 & 1 \\ 0 & 0 \end{pmatrix}.$$

It is clear that each matrix has an eigenvalue 0 with multiplicity 2 (not simple). The zero solution of the corresponding system $x' = Ax$ is stable, whereas the zero solution of $x' = Bx$ is unstable. Therefore careful examination of the stability is required when dealing with eigenvalues with zero real parts of multiplicity higher than one.

Theorem 5.1.5. *If all the eigenvalues of A are distinct with $Re(\sigma(A)) < -\zeta$, where $\zeta > 0$, then the zero solution of (5.1.1) is UAS.*

Proof. By Lemma 5.2, we have that

$$||e^{At}|| \leq Ke^{-\zeta t}, \ t \geq 0,$$

and hence

$$||e^{A(t-s)}|| \leq Ke^{-\zeta(t-s)}, \ t \geq s \geq 0.$$

Now if x is a solution of (5.1.1) with $x(s) = x_0$, then

$$x(t) = e^{A(t-s)}x_0 \text{ for } t \geq s,$$

and the results follow from Theorem 4.2.3, since

$$e^{A(t-s)} = \Phi(t)\Phi^{-1}(s).$$

This completes the proof. □

5.2 Near-constant systems

In this section, we examine when solutions of a linear non-autonomous system behave like solutions of an autonomous linear system near zero. We have already seen that the stability of zero solution of the system

$$x'(t) = A(t)x(t)$$

does not depend on the sign of the eigenvalues of the matrix $A(t)$. For example, the system

$$x'(t) = \begin{pmatrix} -1 + \frac{3}{2}\cos^2(t) & 1 - \frac{3}{2}\cos(t)\sin(t) \\ -1 - \frac{3}{2}\cos(t)\sin(t) & -1 + \frac{3}{2}\sin^2(t) \end{pmatrix} x(t) \qquad (5.2.1)$$

has the eigenvalues

$$\lambda = \frac{-1 \pm i\sqrt{7}}{4}$$

and the principal matrix

$$\Psi(t) = \begin{pmatrix} e^{t/2}\cos(t) & e^{-t}\sin(t) \\ -e^{t/2}\sin(t) & e^{-t}\cos(t) \end{pmatrix}.$$

This tells us that solutions are unbounded and the zero solution is unstable. Thus the question that we wish to examine is that if, in some sense, the matrix $A(t)$ is close to a constant matrix A, then how closely do solutions of the non-autonomous system

$$x'(t) = A(t)x(t) \qquad (5.2.2)$$

behave like solutions of the autonomous system

$$x'(t) = Ax(t)? \qquad (5.2.3)$$

We assume that both matrices are $n \times n$ and $A(t)$ has continuous entries on some interval that includes the initial time t_0.

Theorem 5.2.1. *Assume that every eigenvalue of the matrix A has a nonpositive real part and that the eigenvalues with zero real parts are simple. If*

$$\int_0^\infty ||A(t) - A||dt \leq E \qquad (5.2.4)$$

for some positive constant E, then solutions of (5.2.2) are bounded, and its zero solution is uniformly stable.

Proof. By Theorem 5.1.4 there is a positive constant D such that

$$e^{At} \leq D.$$

Rewrite (5.2.2) as

$$x'(t) = Ax(t) + \left(A(t) - A\right)x(t) \qquad (5.2.5)$$

and set $g(t) = \left(A(t) - A\right)x(t)$. If x is a solution of (5.2.2) with $x(t_0) = x_0$, then it satisfies (5.2.5), and by the variation of parameters formula (3.3.6) we have that

$$x(t) = e^{A(t-t_0)}x_0 + \int_{t_0}^t e^{A(t-s)}\left(A(s) - A\right)x(s)ds. \qquad (5.2.6)$$

Taking the appropriate norms of both sides of (5.2.6) and estimating give, for $t \geq t_0 \geq 0$,

$$|x(t)| = ||e^{A(t-t_0)}|||x_0| + \int_{t_0}^{t} ||e^{A(t-s)}|| \, ||A(s) - A|| \, |x(s)| ds$$

$$\leq D|x_0| + \int_{t_0}^{t} D||A(s) - A|| \, |x(s)| ds$$

$$\leq D|x_0| e^{\int_{t_0}^{t} D||A(s)-A|| \, ds} \quad \text{(by Gronwall's inequality)} \qquad (5.2.7)$$

$$\leq D|x_0| e^{DE}. \qquad (5.2.8)$$

It follows that solutions are bounded and the zero solution is uniformly stable by choosing $\delta = \frac{\varepsilon}{De^{DE}}$ for any $\varepsilon > 0$ with $|x_0| < \delta$.   □

Theorem 5.2.2. *Assume that all the eigenvalues of the matrix A of system (5.2.3) are distinct with negative real parts and there is a positive number (to be identified later in the proof) θ such that*

$$||A(t) - A|| < \theta \qquad (5.2.9)$$

for all $t \geq t_0 \geq 0$. Then the zero solution of (5.2.2) is uniformly asymptotically stable.

Proof. Since $Re(\sigma(A)) < 0$, we may choose a real number $\zeta > 0$ such $Re(\sigma(A)) < -\zeta$. By Lemma 5.2, there is a positive constant K such that $||e^{At}|| \leq Ke^{-\zeta t}$ for all $t \geq 0$. If x is a solution of (5.2.2) with $x(t_0) = x_0$, then it satisfies (5.2.5), and by the variation of parameters formula (3.3.6), we have that

$$x(t) = e^{A(t-t_0)}x_0 + \int_{t_0}^{t} e^{A(t-s)}(A(s) - A)x(s)ds.$$

Let

$$0 < \theta < \frac{\zeta}{K}. \qquad (5.2.10)$$

Taking the appropriate norms of both sides gives, for $t \geq 0$,

$$|x(t)| = ||e^{A(t-t_0)}|||x_0| + \int_{t_0}^{t} ||e^{A(t-s)}|| \, ||A(s) - A|| \, |x(s)| ds$$

$$\leq Ke^{-\zeta(t-t_0)}|x_0| + \int_{t_0}^{t} Ke^{-\zeta(t-s)} \, \theta \, |x(s)| ds$$

or

$$|x(t)|e^{\zeta(t-t_0)} \leq K|x_0| + \int_{t_0}^{t} Ke^{\zeta(s-t_0)} \, \theta \, |x(s)| ds.$$

Let $z(t) = e^{\zeta(t-t_0)}|x(t)|$. Then the above inequality reduces to

$$z(t) \leq K|x_0| + k\theta \int_{t_0}^t z(s)ds.$$

An application of Gronwall's inequality leads to

$$z(t) \leq k|x_0|e^{k\theta(t-t_0)} \text{ for all } t \geq t_0 \geq 0,$$

and, consequently,

$$x(t)| \leq k|x_0|e^{(k\theta-\zeta)(t-t_0)} \text{ for all } t \geq t_0 \geq 0,$$

from which we arrive at the uniform stability and uniform asymptotic stability of the zero solution by imitating the proof of Theorem 4.2.3. This completes the proof. $\square$

Remark 5.2. There is no standard way for forming (5.2.5). It depends on what we want from the eigenvalues of A and conditions (5.2.4) and (5.2.9).

Example 5.1. We show that the zero solution of the non-autonomous system

$$x'(t) = \begin{pmatrix} \frac{e^{-t}}{3} & 1 - \frac{te^{-t}}{6} \\ -2 + \frac{1}{16(1+t)^2} & -3 + \frac{e^{-t}}{16} \end{pmatrix} x \qquad (5.2.11)$$

is uniformly asymptotically stable. We write the system in the form

$$x'(t) = \begin{pmatrix} 0 & 1 \\ -2 & -3 \end{pmatrix} x + \left[\begin{pmatrix} \frac{e^{-t}}{3} & 1 - \frac{te^{-t}}{6} \\ -2 + \frac{1}{16(1+t)^2} & -3 + \frac{e^{-t}}{16} \end{pmatrix} - \begin{pmatrix} 0 & 1 \\ -2 & -3 \end{pmatrix} \right] x$$

$$= Ax + (A(t) - A)x.$$

Now the eigenvalues of A are -2 and -1. We easily verify that

$$e^{A(t-t_0)} = \begin{pmatrix} e^{-(t-t_0)} & e^{-3(t-t_0)} \\ -e^{-(t-t_0)} & -3e^{-(t-t_0)} \end{pmatrix}, \ t \geq t_0 \geq 0,$$

and

$$||e^{A(t-t_0)}||_\infty \leq 4e^{-(t-t_0)}, \ t \geq t_0 \geq 0.$$

Moreover,

$$||A(t) - A||_\infty = \max\{\frac{1}{3} + \frac{1}{6e}, \frac{1}{16} + \frac{1}{16}\} = \frac{1}{8} \text{ for all } t \geq t_0 \geq 0,$$

and condition (5.2.10) is satisfied with $K = 4$, $\zeta = 1$, and $\theta = 2$. Thus by Theorem 5.2.2 the zero solution of (5.2.11) is uniformly asymptotically stable.

5.3 **Perturbed linear systems**

Suppose we know about the stability of the zero solution of the linear system

$$x'(t) = A(t)x(t), \ t \geq 0. \tag{5.3.1}$$

The question is, what can we say about the stability of the zero solution of the *perturbed* linear system

$$x'(t) = A(t)x(t) + B(t)x(t), \ t \geq 0, \tag{5.3.2}$$

where A and B are $n \times n$ continuous matrices on the interval $[0, \infty)$? In other words, under what conditions on $B(t)$ is the perturbed system (5.3.2) stable? It turns out that if, in some sense, the norm of the matrix B is small, then a type of stability of (5.3.2) can be deduced. Note that in (5.3.2), $B(t)$ is known as the *perturbation term*.

Theorem 5.3.1. *Assume that the zero solution of* (5.3.1) *is uniformly sable and that*

$$\int_0^\infty ||B(t)|| dt \leq E \tag{5.3.3}$$

for some positive constant E. Then the zero solution of (5.3.2) *is uniformly stable.*

Proof. Let $\Phi(t)$ be the fundamental matrix of (5.3.1). Due to the US of the zero solution, for some positive constant K, we have

$$||\Phi(t)\Phi^{-1}(s)|| \leq K.$$

By the variation of parameters formula (3.3.4), we have that

$$x(t) = \Phi(t)\Phi^{-1}(t_0)x_0 + \int_{t_0}^t \Phi(t)\Phi^{-1}(s)B(s)ds$$

or

$$|x(t)| = ||\Phi(t)\Phi^{-1}(t_0)|| |x_0| + \int_{t_0}^t ||\Phi(t)\Phi^{-1}(s)|| \, ||B(s)|| \, |x(s)| ds$$

$$\leq K|x_0| + \int_{t_0}^t K||B(s)|| \, |x(s)| ds$$

$$\leq K|x_0|e^{\int_{t_0}^t K||B(s)|| ds} \quad \text{(by Gronwall's inequality)}$$

$$\leq K|x_0|e^{KE}, \tag{5.3.4}$$

which implies the uniform stability. This completes the proof. $\qquad \square$

The next example shows that the integrability condition (5.3.3) for $||B(t)||$ is necessary.

Example 5.2. Letting

$$x_1' = x_2,$$

the second-order differential equation

$$x'' + x = 0$$

can be written in the matrix form

$$x'(t) = \begin{pmatrix} 0 & 1 \\ -1 & 0 \end{pmatrix} \begin{pmatrix} x_1 \\ x_2 \end{pmatrix}$$

with

$$x = \begin{pmatrix} x_1 \\ x_2 \end{pmatrix}.$$

Thus its zero solution is uniformly stable. On the other hand, Theorem 5.3.1 says nothing about the perturbed system

$$x'' - \frac{2}{t} x' + x = 0, \; t > 0,$$

which is equivalent to

$$x'(t) = \begin{pmatrix} 0 & 1 \\ -1 & 0 \end{pmatrix} x + \begin{pmatrix} 0 & 0 \\ 0 & \frac{2}{t} \end{pmatrix} x.$$

Here $B(t) = \begin{pmatrix} 0 & 0 \\ 0 & \frac{2}{t} \end{pmatrix}$ is the perturbed matrix in the sense that $|B(t)| = \frac{2}{t} \to 0$ as $t \to \infty$. Moreover,

$$\int_{t^*}^{\infty} ||B(t)|| dt = \int_{t^*}^{\infty} \frac{2}{t} dt = \infty$$

for any $t^* > 0$, and hence $B(t)$ does not satisfy condition (5.3.3).

Note that the perturbed system has two linearly independent solutions

$$\begin{pmatrix} x_1 \\ x_2 \end{pmatrix} = \begin{pmatrix} \sin(t) - t\cos(t) \\ t\sin(t) \end{pmatrix}, \quad \begin{pmatrix} x_1 \\ x_2 \end{pmatrix} = \begin{pmatrix} \cos(t) + t\sin(t) \\ t\cos(t) \end{pmatrix},$$

and consequently its solutions are unbounded, and as a result, its zero solution is unstable.

Theorem 5.3.2. *Assume that the zero solution of* (5.3.1) *is uniformly asymptotically stable, that is, if* $\Phi(t)$ *is the fundamental matrix of* (5.3.1), *then by* (4.2.3) *there are positive constants K and ξ such that*

$$||\Phi(t)\Phi^{-1}(s)|| \leq Ke^{\xi(t-s)}, \; t \geq s \geq 0.$$

Suppose

$$\int_0^\infty ||B(t)||dt \le \gamma(t - t_0) + \beta, \quad t \ge t_0 \ge 0, \tag{5.3.5}$$

for positive constants γ and β. Then the zero solution of (5.3.2) is uniformly asymptotically stable, provided that

$$\gamma < \frac{\xi}{K}.$$

Proof. Imitate the proof of Theorem 5.3.1 and get

$$|x(t)|e^{\xi(t-t_0)} \le K|x_0| + \int_{t_0}^t Ke^{\xi(s-t_0)} ||B(s)|| \, |x(s)|ds.$$

Using Gronwall's inequality along with condition (5.3.5), we arrive at

$$|x(t)| \le Ke^{K\beta}|x_0|e^{K(\gamma - \frac{\xi}{K})(t-t_0)}. \tag{5.3.6}$$

The rest of the proof follows along the lines of Theorem 5.2.2 using inequality (5.3.6). This completes the proof. $\qquad\qquad\square$

5.4 Autonomous systems in the plane

In most applications, the corresponding modeling systems can be written as two-dimensional autonomous systems in the form

$$x' = P(x, y), \ y' = Q(x, y), \tag{5.4.1}$$

where P and Q are continuous on some subset of $\mathbb{R}^2$. Recall that the system is called *autonomous* if P and Q do not depend explicitly on t. System (5.4.1) can be written in the form $z' = f(z)$, where

$$z = \begin{pmatrix} x \\ y \end{pmatrix} \text{ and } f(z) = \begin{pmatrix} P(z) \\ Q(z) \end{pmatrix}.$$

Thus the second-order differential equation

$$x'' = g(x, x')$$

can be written in the form (5.4.1) as the system

$$x' = y, \ y' = g(x, y).$$

We begin with the following lemma. We denote the solution of (5.4.1) by the pair $(x(t), y(t))$, $t \in (a, b)$.

Lemma 5.3. *Let $(x(t), y(t))$ be a solution of (5.4.1). Then for any constant k, the pair $(x(t + k), y(t + k))$ is also a solution of (5.4.1).*

Proof. Let $x_1 = x(t + k)$ and $x_2 = y(t + k)$. Then by the chain rule we have

$$x_1' = x'(t + k) = P(x(t + k), y(t + k)) = P(x_1, x_2),$$
$$x_2' = y'(t + k) = Q(x(t + k), y(t + k)) = Q(x_1, x_2),$$

which imply that the pair $(x_1(t), x_2(t))$ is a solution of (5.4.1) for $t \in (a - k, b - k)$. This completes the proof. $\square$

The results of Lemma 5.3 do not hold for non-autonomous systems. To see this, we consider the system

$$x' = ty, \ y' = y.$$

It readily follows that

$$x(t) = (t - 1)e^t, \ y(t) = e^t$$

is a solution of the system. However, for any nonzero constant k, we have $x(t + k) = (t + k - 1)e^{t+k}$ and $x'(t + k) = (t + k)e^{t+k} \neq ty(t)$.

Definition 5.4.1. The solution $(x(t), y(t))$ of (5.4.1) as t varies describes parametrically a curve in the plane, which we call a *trajectory* (or orbit).

Theorem 5.4.1. *Different trajectories of (5.4.1) cannot intersect.*

Proof. Let C_1 and C_2 be two trajectories represented by $(x_1(t), y_1(t))$ and $(x_2(t), y_2(t))$, respectively, with a common point (x_0, y_0). Then there are two different times (otherwise, uniqueness is violated), say t_1 and t_2, such that

$$(x_0, y_0) = (x_1(t_1), y_1(t_1)) = (x_2(t_2), y_2(t_2)).$$

By Lemma 5.3

$$x(t) = x_1(t + t_1 - t_2), \ y(t) = y_1(t + t_1 - t_2)$$

is a solution. Since $(x(t_2), y(t_2)) = (x_0, y_0)$, we have that $x(t)$ and $y(t)$ must agree respectively with $x_2(t)$ and $y_2(t)$ due to the uniqueness. Thus C_1 and C_2 must coincide. This completes the proof. $\square$

Remark 5.3. Be aware that a trajectory is a curve represented parametrically by one or more solutions. Thus $x(t), y(t)$ and $x(t + k), y(t + k)$ for $k \neq 0$ represent distinct solutions but the same trajectory.

Example 5.3. The damped harmonic motion represented by the second-order differential equation

$$u'' + 2u' + 2u = 0$$

can be easily transformed into the planar autonomous system

$$x' = y,$$
$$y' = -2x - 2y.$$

The eigenvalues of the system are found to be $\lambda = -1 \pm i$. Hence the solution is

$$x = e^{-t} (c_1 \cos(t) + c_2 \sin(t)),$$
$$y = e^{-t} (-c_1 \sin(t) + c_2 \cos(t)).$$

To describe the trajectories, we find that

$$x^2 + y^2 = (c_1^2 + c_2^2)e^{-2t},$$

which describes a family of spirals.

We begin a detailed study of the linear plane autonomous system

$$X' = AX, \tag{5.4.2}$$

where

$$A = \begin{pmatrix} a & b \\ c & d \end{pmatrix}$$

is a nonsingular matrix, and

$$X = \begin{pmatrix} x \\ y \end{pmatrix}.$$

The characteristic equation for this system is

$$\lambda^2 - (a+d)\lambda + ad - bc = 0.$$

By introducing

$$p = a + d,$$
$$q = ad - bc,$$

the characteristic equation becomes

$$\lambda^2 - p\lambda + q = 0.$$

Let λ_1 and λ_2 be the roots of this equation,

$$\lambda_{1,2} = \frac{1}{2}\left[p \pm \sqrt{p^2 - 4q} \right]. \tag{5.4.3}$$

It is obvious that the signs of $\lambda_{1,2}$ and hence the stability of the zero solution depend on the discriminant

$$\Delta = p^2 - 4q = (\text{tr}(A))^2 - 4\det(A).$$

Based on Theorem 5.1.4, we can now state the following stability theorem of the zero solution of (5.4.2).

Theorem 5.4.2. *The zero solution of* (5.4.2) *is*

a) asymptotically stable if any of the followings occurs:
i)

$$\Delta > 0, \ q > 0, \ and \ p < 0,$$

ii)

$$\Delta = 0, \ p < 0,$$

iii)

$$\Delta < 0, \ p < 0;$$

b) stable if

$$\Delta < 0, \ p = 0;$$

c) unstable if any of the followings occurs:
i)

$$\Delta > 0, \ q > 0, \ and \ p > 0,$$

ii)

$$\Delta = 0, \ p > 0,$$

iii)

$$\Delta < 0, \ p > 0.$$

Remark 5.4. In the case where $p = 0$ and $q = 0$, further analysis is required to determine the stability of the zero solution.

Note that according to Theorem 5.4.2, the zero solution $(0,0)$ of the system in Example 5.3 is asymptotically stable.

Next, we consider the behavior of the trajectories of system (5.4.2) for various cases of the eigenvalues. In what follows, we use the matrix transformation

$$Y = BX, \ \det(B) \neq 0,$$

so that the essential behavior near the critical point $(0,0)$ remains unchanged.

1. *Real and distinct roots.* If we let

$$B = \begin{pmatrix} c & \lambda_1 - a \\ c & \lambda_2 - a \end{pmatrix},$$

then system (5.4.2) transforms into the system

$$x' = \lambda_1 x,$$
$$y' = \lambda_2 y,$$

where, for simplicity, x and y are again used as the new coordinates. For some arbitrary constants c_1 and c_2, the solutions of the system are

$$x(t) = c_1 e^{\lambda_1 t}, \quad y(t) = c_2 e^{\lambda_2 t}. \tag{5.4.4}$$

Eliminating the time t in (5.4.4) gives

$$y = c x^{\frac{\lambda_2}{\lambda_1}}, \tag{5.4.5}$$

where c is an arbitrary constant. When λ_1 and λ_2 have the same sign, then Eq. (5.4.5) represent parabolic curves that are tangent to the origin as shown in Fig. 5.1(a). The origin is called a *proper node*. We have the following three subcases.

1.(a) If both λ_1, λ_2 are negative, then the origin is asymptotically stable. In this case, we have a *stable node*.

1.(b) If λ_1, λ_2 are positive, then the origin is unstable. In this case, we have an *unstable node*.

1.(c) If λ_1, λ_2 have opposite signs, then (5.4.5) represents hyperbolic curves as shown in Fig. 5.1(b). The origin is unstable and is called a *saddle point*.

2. *Real and equal roots.* In this case, $\Delta = (a - d)^2 + 4bc = 0$, which implies that

$$\lambda_1 = \lambda_2 = \frac{(a + d)}{2} := \lambda.$$

2. (a) If b or $c = 0$ and $a = d$, then system (5.4.2) reduces to

$$x' = \lambda x,$$
$$y' = \lambda y.$$

As before, solving the system and eliminating t lead to the trajectories that are lines through the origin given by

$$y = \frac{c_2}{c_1} x.$$

Thus, if $\lambda < 0$, then the origin is asymptotically stable and is a *proper node*, as shown in Fig. 5.1(c). If $\lambda > 0$, then the origin is unstable.

(b) This is a more complicated case. If we let

$$B = \begin{pmatrix} \frac{a-d}{2b} & 1 \\ \frac{1}{b} & 0 \end{pmatrix}, \ b \neq 0,$$

then system (5.4.2) transforms into the system

$$x' = \lambda x,$$
$$y' = x + \lambda y.$$

For some arbitrary constants c_1 and c_2, the solutions of the system are

$$x(t) = c_1 e^{\lambda t}, \ \ y(t) = (c_1 t + c_2) e^{\lambda t}. \tag{5.4.6}$$

The trajectories are shown in Fig. 5.1(d), and the origin is an *improper node*; it is asymptotically stable when $\lambda < 0$ and unstable when $\lambda > 0$.

3. *Complex roots.* In this case, we let

$$\lambda_1 = \alpha + i\beta \ \text{ and } \ \lambda_2 = \alpha - i\beta$$

and choose the matrix

$$B = \begin{pmatrix} c & \alpha - a \\ 0 & \beta \end{pmatrix}$$

so that system (5.4.2) transforms into

$$x' = \alpha x - \beta y,$$
$$y' = \beta x + \alpha y.$$

3. (a) For $\alpha = 0$, this system becomes

$$x' = -\beta y,$$
$$y' = \beta x,$$

with solutions

$$x(t) = c_1 \cos(\beta t) + c_2 \sin(\beta t), \ \ y(t) = c_1 \sin(\beta t) + c_2 \cos(\beta t)$$

for arbitrary constants c_1 and c_2. It is easy to verify that the trajectories are circles given by

$$x^2 + y^2 = c_1^2 + c_2^2$$

and are shown in Fig. 5.1(e). In this case the origin is stable but not asymptotically stable.

3. (b) For $\alpha \neq 0$, the above system has the solutions

$$x(t) = e^{\alpha t}(c_1 \cos(\beta t) + c_2 \sin(\beta t)), \quad y(t) = e^{\alpha t}(c_1 \sin(\beta t) + c_2 \cos(\beta t))$$

for arbitrary constants c_1 and c_2. The trajectories are spirals given by

$$x^2 + y^2 = (c_1^2 + c_2^2)e^{2\alpha t}$$

and are shown in Fig. 5.1(f). In this case the origin is a *focal point*, which is asymptotically stable when $\alpha < 0$ and unstable when $\alpha > 0$.

The solutions of a planar system of linear differential equations can be classified according to the trace and the determinant of the coefficient matrix. This diagram schematically shows the different types of solutions.

Poincaré Diagram: Classification of Phase Portraits in the $(\det A, \mathrm{Tr} A)$-plane

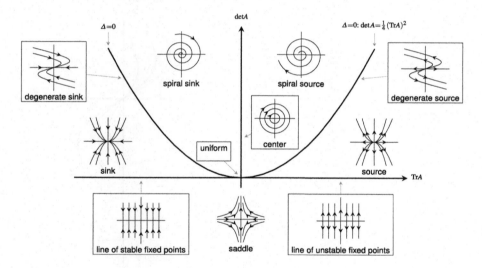

5.5 Hamiltonian and gradient systems

Consider the pendulum with friction discussed in Example 3.5:

$$x_1' = x_2,$$

$$x_2' = -\frac{g}{L}\sin(x_1) - \frac{b}{m}x_2.$$

Define the *energy function* $H : [t_0, \alpha) \times \mathbb{R}^2 \to [0, \infty)$ by

$$H(x_1, x_2) = \frac{1}{2}x_2^2(t) + \frac{g}{L}\Big(1 - \cos(x_1(t))\Big).$$

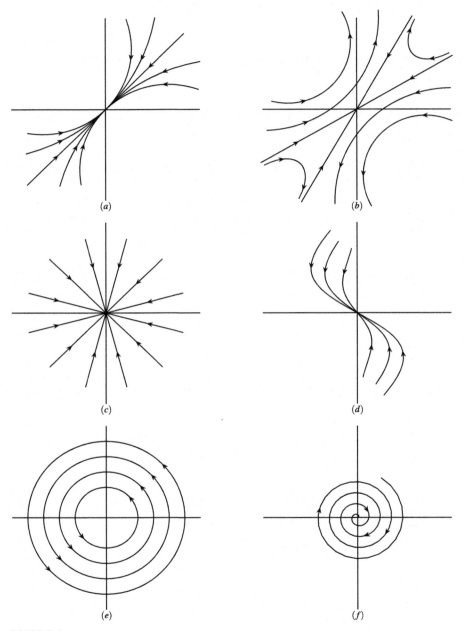

FIGURE 5.1

A phase portrait gallery.

Then along the solutions of (3.2.7) we have that

$$\frac{d}{dt}H(x_1, x_2) = x_2\, x_2' + \frac{g}{L}\sin(x_1)\, x_1'$$

$$= x_2\left(-\frac{g}{L}\sin(x_1) - \frac{b}{m}x_2\right) + \frac{g}{L}\sin(x_1)x_2$$

$$= -\frac{b}{m}x_2^2.$$

If the friction $b > 0$, then $\frac{d}{dt}H(t) < 0$, and the *energy is dissipated*. On the other hand, if $b = 0$, then the energy is *conserved*, and theoretically, this solution lies on a *level curve*

$$H(x_1(t), x_2(t)) = \text{constant} := c$$

of the function $H(x_1, x_2)$ with variables x_1 and x_2. Thus, in the case $b > 0$, this solution will not stay on a level curve. Hence centers are possible for (3.2.5).

When the energy is conserved, we obtain an example of a *Hamiltonian system*. In our situation the system

$$\begin{cases} x_1' = x_2, \\ x_2' = -\frac{g}{L}\sin(x_1) \end{cases} \tag{5.5.1}$$

can be written in the form

$$x_1' = x_2 = \frac{\partial H}{\partial x_2},$$

$$x_2' = -\frac{g}{L}\sin(x_1) = -\frac{\partial H}{\partial x_1}.$$

Since the Hamiltonian function conserves the energy, it is constant along solutions. The level curves are given by

$$\frac{1}{2}x_2^2(t) - \frac{g}{L}\cos(x_1(t)) = c.$$

System (5.5.1) has the equilibrium solutions $(\pm n\pi, 0)$, $n = 0, 1, \ldots$. It will be shown in Chapter 6, when there is no damping or friction, that is, $b = 0$, then $(\pm n\pi, 0)$ are centers for $n = 2, 4, 6, \ldots$ and unstable saddles for $n = 1, 3, 5, \ldots$. Phase portraits and level curves are shown in Fig. 5.2 for different values of c.

Thus we have the following definition.

Definition 5.5.1. A planar system of differential equations is called a *Hamiltonian system* if there exists a differentiable function $H(x_1, x_2) : \mathbb{R}^2 \to \mathbb{R}$ such that

$$\begin{cases} x_1'(t) = \frac{\partial H}{\partial x_2}, \\ x_2'(t) = -\frac{\partial H}{\partial x_1}. \end{cases} \tag{5.5.2}$$

Such a function $H(x_1, x_2)$ is called a *Hamiltonian function* of system (5.5.2).

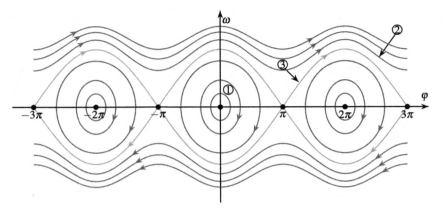

FIGURE 5.2

Phase portraits and level curves for (5.5.1) for different values of c.

Consider the second-order differential equation of the form

$$x'' - f(x)x' - g(x) = 0.$$

In system form, we have

$$x_1' = x_2,$$
$$x_2' = g(x_1) + f(x_1)x_2.$$

Let

$$H(x_1, x_2) = \frac{1}{2}x_2^2(t) + \left(-\int_0^{x_1} g(s)ds\right),$$

where $\frac{1}{2}(x')^2$ corresponds to the *kinetic energy*, and $-\int_0^{x(t)} g(s)ds$ represents the *potential energy*. Then it is easy to verify that

$$x_1' = x_2 = \frac{\partial H}{\partial x_2},$$
$$x_2' = g(x_1) + f(x_1)x_2 = -\frac{\partial H}{\partial x_1},$$

only if $f(x) = 0$. We conclude that the *Newtonian system*

$$x'' = g(x)$$

with g continuously differentiable on $\mathbb{R}$ is a Hamiltonian system.

For the Hamiltonian system (5.5.2), the energy is always conserved. To see this, let $(x_1(t), x_2(t))$ be any solution of (5.5.2). Then

$$\frac{d}{dt}H(x_1(t), x_2(t)) = \frac{\partial H}{\partial x_1}x_1' + \frac{\partial H}{\partial x_2}x_2'$$
$$= \frac{\partial H}{\partial x_1}\frac{\partial H}{\partial x_2} + \frac{\partial H}{\partial x_2}\left(-\frac{\partial H}{\partial x_1}\right) = 0. \qquad (5.5.3)$$

That is why a Hamiltonian function is sometimes called an *energy function*. In applications, the Hamiltonian function is defined by

$$H(x, y) = K(x, y) + V(x, y),$$

where K is the kinetic energy, and V is the potential energy.

Next, we establish conditions for the planar system

$$x' = f(x, y), \quad y' = g(x, y) \qquad (5.5.4)$$

to be a Hamiltonian system. If (5.5.4) is a Hamiltonian system, then there is a function H satisfying

$$\frac{\partial H}{\partial y} = f(x, y) \text{ and } -\frac{\partial H}{\partial x} = g(x, y).$$

Assuming that the function H has continuous second-order partial derivatives, we arrive at

$$\frac{\partial f}{\partial x} = \frac{\partial^2 H}{\partial x \partial y} = \frac{\partial^2 H}{\partial y \partial x} = -\frac{\partial g}{\partial y}. \qquad (5.5.5)$$

Thus we have shown that condition (5.5.5) is a sufficient condition for system (5.5.4) to be a Hamiltonian system.

The system

$$x' = x + y^2,$$
$$y' = -y - x^2$$

is Hamiltonian since $\frac{\partial f}{\partial x} = 1 = -\frac{\partial g}{\partial y} = -(-1)$.

The next theorem provides necessary and sufficient conditions for the existence of Hamiltonian function.

Theorem 5.5.1. *The planar system* (5.5.4) *is a Hamiltonian system if and only if*

$$\frac{\partial f}{\partial x} = -\frac{\partial g}{\partial y}. \qquad (5.5.6)$$

Proof. We already know that if system (5.5.4) is Hamiltonian, then (5.5.6) is true. It remains to show that if (5.5.4) holds, then there is $H(x, y)$ satisfying

$$x'(t) = \frac{\partial H}{\partial y} = f(x, y)$$

and

$$y'(t) = -\frac{\partial H}{\partial x} = g(x, y).$$

Let us begin with $\frac{\partial H}{\partial x} = -g(x, y)$. Then a direct integration with respect to x keeping y fixed gives

$$H(x, y) = -\int g(x, y)dx + C_1(y),$$

where the constant of integration C_1 may depend on y since H is a function of two variables x and y. Once we determine C_1, we can obtain H. Differentiating the last expression in H with respect to y and setting it equal to $f(x, y)$ leads to

$$\frac{\partial H}{\partial y} = \frac{\partial}{\partial y}\left(-\int g(x, y)dx + C_1(y)\right) = f(x, y) \text{ by definition}$$

or

$$C_1'(y) = \frac{\partial}{\partial y}\int g(x, y)dx + f(x, y).$$

Then integrating the last expression with respect to y gives the desired function $C_1(y)$, that is,

$$C_1(y) = \int \left(\frac{\partial}{\partial y}\int g(x, y)dx\right)dy + \int f(x, y)dy.$$

The proof will be accomplished if we can show $C_1(y)$ does not depend on x. Equivalently, we can show that the right side of the above expression is independent of x:

$$\frac{\partial}{\partial x}\left[\int f(x, y)dy + \int\left(\frac{\partial}{\partial y}\int g(x, y)dx\right)dy\right]$$
$$= \int \frac{\partial f}{\partial x}dy + \int \frac{\partial}{\partial x}\left(\frac{\partial}{\partial y}\int g(x, y)dx\right)dy$$
$$= \int \frac{\partial f}{\partial x}dy + \int \frac{\partial}{\partial y}\left(\frac{\partial}{\partial x}\int g(x, y)dx\right)dy$$
$$= \int \frac{\partial f}{\partial x}dy + \int \frac{\partial g}{\partial y}dy$$
$$= \int \frac{\partial f}{\partial x}dy - \int \frac{\partial f}{\partial x}dy = 0.$$

This completes the proof.   □

Example 5.4. We already know that the system

$$x' = x + y^2, \quad y' = -y - x^2$$

is Hamiltonian since $\frac{\partial f}{\partial x} = 1 = -\frac{\partial g}{\partial y} = -(-1)$. Let us now try to find the Hamiltonian function H. It follows from Theorem 5.5.1 that

$$H(x, y) = -\int g(x, y)dx + C_1(y) = -\int(-y - x^2)dx + c_1(y) = xy + \frac{x^3}{3} + C_1(y),$$

where

$$C_1'(y) = \frac{\partial}{\partial y}\int g(x, y)dx + f(x, y) = \frac{\partial}{\partial y}\left(\int(-y - x^2)dx\right) + x + y^2.$$

It follows that $C_1(y) = \frac{y^3}{3}$. Finally,

$$H(x, y) = xy + \frac{x^3}{3} + \frac{y^3}{3}.$$

Next, we try to explore the stability of an equilibrium solution of a Hamiltonian system. We have seen that the autonomous system (5.5.4) is called Hamiltonian or conservative if there exists a Hamiltonian function $H : \Omega \to \mathbb{R}$ that is not constant on any open set in Ω, but it follows from (5.5.3) that it is constant on solutions (orbits).

Theorem 5.5.2. *Suppose (5.5.4) is a Hamiltonian system. Then none of its equilibrium solutions can be asymptotically stable.*

Proof. Suppose q is an asymptotically stable equilibrium solution or a point of the Hamiltonian system given by (5.5.4). Then there is a neighborhood Λ of q such that for $p \in \Lambda$, $\phi(t, p) \to q$ as $t \to \infty$, where $\phi(t, p)$ is an orbit or trajectory through p. However, if the system is assumed conservative with Hamiltonian function H, then H is constant on orbits, so $H(p) = H(\phi(t, p)) = H(q)$ for each $p \in \Lambda$. This implies that H is constant on an open set, which is a contradiction. This completes the proof. $\square$

When a Hamiltonian system has an equilibrium point at which the integral has a minimum, then an inference of stability can be made.

Definition 5.5.2. A function $E : \Omega \to \mathbb{R}$ is said to have a *strong minimum* at q if there is a neighborhood Λ of q such that $E(x) > E(q)$ for every $x \in \Lambda$ except for $x = q$.

The name *strong* is implied from the strict inequality.

Theorem 5.5.3. *Suppose (5.5.4) is a Hamiltonian system with equilibrium solution q and Hamiltonian function H. If H has a strong minimum at q, then q is stable.*

Proof. Shift the equilibrium solution q to the origin $(0, 0)$. Let Λ be a neighborhood of $(0, 0)$. Let $\varepsilon > 0$ be small enough such that the set $A = \{d : ||d|| = \varepsilon\} \subset \Lambda$. Set

$$V(d) = H(d) - H(0).$$

Denote the minimum of V on the set A by V_ε. Now we chose $\delta > 0$ with $\delta < \varepsilon$, so that we have $V(d) < \varepsilon$ if $||d|| < \delta$. Then we have $||d(t)|| < \varepsilon$ for all $t > 0$. This implies that the origin is stable. This completes the proof. □

Next, we state parallel results concerning strong maximums and saddle points.

Definition 5.5.3. A function $E : \Omega \to \mathbb{R}$ is said to have a *strong maximum* at q if there is a neighborhood Λ of q such that $E(q) > E(x)$ for every $x \in \Lambda$ except for $x = q$.

The name *strong* is implied from the strict inequality.

Theorem 5.5.4. *Suppose* (5.5.4) *is a Hamiltonian system with equilibrium solution q and Hamiltonian function H. If H has a strong maximum at q, then q is an unstable saddle.*

Remark 5.5. Note that if (x_0, y_0) is a critical point of (5.5.4), then it will also be a critical point of the Hamiltonian function H. We can use the second derivative test again to determine whether or not $H(x, y)$ has a saddle at (x_0, y_0) or a maximum or a minimum there. The Jacobian at the critical point is

$$J(x_0, y_0) = \begin{pmatrix} H_{xy}(x_0, y_0) & H_{yy}(x_0, y_0) \\ -H_{xx}(x_0, y_0) & -H_{xy}(x_0, y_0) \end{pmatrix}.$$

Furthermore,

$$\det J(x_0, y_0) = -\left(H_{xy}(x_0, y_0)\right)^2 + H_{xx}(x_0, y_0)H_{yy}(x_0, y_0).$$

Recall from Section 4.3.1 that if λ_1 and λ_2 are eigenvalues of $J(x_0, y_0)$, not necessarily distinct, then $\det J(x_0, y_0) = \lambda_1 \lambda_2$. Thus if we assume that none of the eigenvalues of the $J(x_0, y_0)$ is zero, then we have $\det J(x_0, y_0) \neq 0$.

If $\det J(x_0, y_0) > 0$, which is the same as saying that the function H has a strict minimum or maximum at (x_0, y_0), then (x_0, y_0) is a stable center for (5.5.4) if both eigenvalues are negative. This is the case since by Theorem 5.5.2 no critical point of a Hamiltonian system can be asymptotically stable. If the both eigenvalues are positive, then (x_0, y_0) is unstable for (5.5.4).

If $\det(x_0, y_0) < 0$, the equilibrium point (x_0, y_0) is unstable saddle for (5.5.4).

Consider the Hamiltonian system given in Example 5.4. We have already found the Hamiltonian function

$$H(x, y) = xy + \frac{x^3}{3} + \frac{y^3}{3}.$$

The two equilibrium solutions are $(0, 0)$ and $(-1, -1)$. Define

$$D(x, y) = H_{xx}H_{yy} - \left(H_{xy}\right)^2.$$

Then

$$D(x, y) = 4xy - 1.$$

Now $D(0, 0) = -1$, and $(0, 0)$ is an unstable saddle. On the other hand, $D(-1 - 1) = 3 > 0$, and since $H_{xx}(-1, -1) = -2 < 0$, the Hamiltonian function has a strong maximum at the equilibrium solution $(-1, -1)$. Thus by Theorem 5.5.4 $(-1, -1)$ is unstable.

Next, we discuss dissipative systems that we call *gradient systems*.

Definition 5.5.4. A planar system of differential equations is called *gradient* if there exists a differentiable function $V(x, y) : \mathbb{R}^2 \to \mathbb{R}$ such that

$$\begin{cases} x'(t) = -\frac{\partial V}{\partial x}, \\ x'_2(t) = -\frac{\partial V}{\partial y}. \end{cases} \tag{5.5.7}$$

The function $V(x, y)$ is called the *gradient function* of system (5.5.7).

The name *gradient* comes from the fact that system (5.5.7) can be written in the form

$$X' = -\nabla V,$$

where

$$X = \begin{pmatrix} x \\ y \end{pmatrix}, \quad \nabla V = \begin{pmatrix} \frac{\partial V}{\partial x} \\ \frac{\partial V}{\partial y} \end{pmatrix}.$$

For the gradient system (5.5.7), the energy is dissipated. To see this, assume that V is an energy function for system (5.5.7). Then if $(x(t), y(t))$ is any solution of (5.5.7), then

$$\frac{d}{dt} V(x(t), y(t)) = \frac{\partial V}{\partial x} x' + \frac{\partial V}{\partial y} y'$$
$$= -\left[\left(\frac{\partial V}{\partial x} \right)^2 + \left(\frac{\partial V}{\partial y} \right)^2 \right] < 0, \tag{5.5.8}$$

provided that $\left(\frac{\partial V}{\partial x} \right)^2 + \left(\frac{\partial V}{\partial y} \right)^2 \neq 0$.

Next, we establish conditions for the planar system

$$x' = f(x, y), \quad y' = g(x, y) \tag{5.5.9}$$

to be a gradient system. If (5.5.9) is a gradient system, then there is a function V satisfying

$$-\frac{\partial V}{\partial x} = f(x, y) \quad \text{and} \quad -\frac{\partial V}{\partial y} = g(x, y).$$

Assuming that the function V has continuous second-order partial derivatives, we arrive at

$$\frac{\partial f}{\partial y} = -\frac{\partial^2 V}{\partial x \partial y} = -\frac{\partial^2 V}{\partial y \partial x} = \frac{\partial g}{\partial x}. \tag{5.5.10}$$

Thus we have shown that condition (5.5.10) is a sufficient condition for system (5.5.9) to be a gradient system. The system of differential equations

$$x' = -2xy, \quad y' = -(x^2 - 1)$$

is gradient since $\frac{\partial f}{\partial y} = -2x = \frac{\partial g}{\partial x}$. Following parallel steps as in Theorem 5.5.1, we easily find the gradient function

$$V(x, y) = x^2 y - y.$$

Note that

$$\frac{dV}{dt} = 2xyx' + (x^2 - 1)y' = -[4x^2 y^2 + (x^2 - 1)^2] = -[V_x^2 + V_y^2] < 0$$

for $V_x^2 + V_y^2 \neq 0$. As in the case of a Hamiltonian system, if (x_0, y_0) is a critical point of (5.5.9), then it will also be a critical point of the gradient function V. Now we state parallel theorems to Theorems 5.5.3 and 5.5.4.

Theorem 5.5.5. *Suppose (5.5.9) is a gradient system with equilibrium solution q and gradient function H. If H has a strong minimum at q, then q is stable.*

Theorem 5.5.6. *Suppose (5.5.9) is a gradient system with equilibrium solution q and gradient function H. If H has a strong maximum at q, then q is an unstable saddle.*

Example 5.5. Consider the system

$$x' = -8x^3 + 2x, \quad y' = -2y + 2,$$

which is gradient with gradient function

$$V(x, y) = 2x^4 + y^2 - x^2 - 2y.$$

The system has three equilibrium solutions given by $(-1/2, 1)$, $(0, 1)$, $(1/2, 1)$. Using the second derivative test, we find that V has a strong minimum at $(-1/2, 1)$ and $(1/2, 1)$. Thus by Theorem 5.5.5 the points are stable for the system. The gradient function V has neither a maximum nor a minimum at $(0, 1)$ and hence nothing can be concluded about the stability of $(0, 1)$. We will see in Chapter 6 that $(0, 1)$ is unstable.

5.6 Exercises

Exercise 5.1. Assume that every eigenvalue of an $n \times n$ constant matrix A has a non-positive real part and that the eigenvalues with zero real part are simple. Suppose that $g : [0, \infty) \to \mathbb{R}^n$ is continuous with

$$\int_0^t |g(s)| ds < \infty.$$

Then show that every solution of

$$x'(t) = Ax(t) + g(t)$$

is bounded on $[0, \infty)$.

Exercise 5.2. Prove (iii) of Theorem 5.1.4.

Exercise 5.3. Examine the stability of the zero solution of the given system.

(a)

$$x'(t) = \begin{pmatrix} 0 & 2 & 1 \\ -2 & 0 & -1 \\ 0 & 0 & -1 \end{pmatrix} x;$$

(b)

$$x'(t) = \begin{pmatrix} 1 & 2 & 1 \\ 6 & -1 & 0 \\ -1 & -2 & -1 \end{pmatrix} x;$$

(c)

$$x'(t) = \begin{pmatrix} 6 & -1 \\ 5 & 2 \end{pmatrix} x.$$

Exercise 5.4. Show that the zero solution of the non-autonomous system

$$x'(t) = \begin{pmatrix} -1 + te^{-t} & 2 + e^{-t} \\ 2 + \frac{1}{(1+t)^2} & -4 + (1+t)^{\frac{-1}{2}} \end{pmatrix} x$$

is stable.

Exercise 5.5. Assume all the eigenvalues of the matrix A of system (5.2.3) have negative real parts and there is a positive number θ such that $||A(t) - A|| < \theta$ for t is sufficiently large. Show that solutions of (5.2.2) are bounded and its zero solution is asymptotically stable.

Exercise 5.6. Let A be a constant matrix with $Re(\rho(A)) < 0$. Choose constants $K > 0$ and $\alpha < 0$ such that

$$||e^{At}|| \le K e^{\alpha t}, \ t \ge 0.$$

Let C be a continuous matrix-valued function such that for positive constant β, we have

$$||C(t)|| \le \frac{\beta}{K}, \ t \ge 0.$$

Show that the zero solution of

$$x' = [A + C(t)]x(t)$$

is

1. stable if $\beta = -\alpha$,
2. unstable if $\beta > -\alpha$, and
3. asymptotically stable if $0 < \beta < -\alpha$.

Exercise 5.7. a) Let A be a constant matrix with $Re(\rho(A)) < 0$. Let C be a continuous matrix-valued function on $[0, \infty)$. Show that if

$$\int_0^\infty ||C(t)|| dt < \infty,$$

then the zero solution of

$$x' = [A + C(t)]x(t)$$

is asymptotically stable.

b) Use the results of part a) to show that the zero solution of

$$u''(t) + \left(\frac{1}{t^2 + 1} + 1 \right) u'(t) + \left(e^{-t} + 1 \right) u(t) = 0$$

is asymptotically stable for $t \ge 0$.

Exercise 5.8. Give an example as an application of Theorem 5.3.2.

Exercise 5.9. Show that all trajectories of the system

$$\begin{pmatrix} x \\ y \end{pmatrix}' = \begin{pmatrix} 0 & 1 \\ -4 & 0 \end{pmatrix} \begin{pmatrix} x \\ y \end{pmatrix}$$

are periodic and determine their periods.

Exercise 5.10. Describe the type of stability of the origin of each system and sketch the trajectories.
(a) $x' = x$, $y' = 2x + 2y$;
(b) $x' = -x$, $y' = x - y$;
(c) $x' = -x + 2y$, $y' = x - y$;
(d) $x' = -x + y$, $y' = 2x$;
(e) $x' = 2x - 8y$, $y' = x - 2y$;
(f) $x' = -3x + 2y$, $y' = -2x$.

Exercise 5.11. Consider the damped harmonic motion represented by the second-order differential equation

$$mx'' + ax' + kx = 0,$$

where m, a, and k are positive constants. By changing the equation into a system discuss the nature and stability of the origin in the following cases: $a = 0$; $a^2 - 4mk = 0$; $a^2 - 4mk < 0$; $a^2 - 4mk > 0$. Interpret the results physically.

Exercise 5.12. Decide whether each system is Hamiltonian or not; in the case it is, find the Hamiltonian function. Next, find the equilibrium solutions (if any) and classify their stabilities when possible.
(a) $x'' + x + x^3$;
(b) $\theta'' + \sin(\theta) = 0$;
(c) $x'' + x^2 - 1$;
(d) $x'' + x^2 + 1$.

Exercise 5.13. Decide whether each system is gradient or not; in the case it is, find the gradient function. Next, find the equilibrium solutions (if any) and classify their stabilities when possible.
(a) $x' = -4x(x - 1)(x - 2)$, $y' = -2y$;
(b) $x' = 1 - 2x$, $y' = -7 - 3y$;
(c) $x' = 3 - 2y^2x$, $y' = -2yx^2 - 4$;
(d) $x' = 4x^3 + 4xy$, $y' = 2x^2 + 2y - 1$.

Nonlinear systems

In general, nonlinear differential equations are important in applications, and their solutions are not known explicitly. Sometimes these equations may be linearized by an expansion process in which nonlinear terms are discarded. When nonlinear terms make vital contributions to the solution, this cannot be done, but sometimes it is enough to retain a few "small" ones. In this chapter, we look for different ways to obtain vital information regarding the behavior of solutions.

6.1 Bifurcations in scalar systems

In this section, we limit our discussion to the scalar differential equation

$$x' = f(x), \tag{6.1.1}$$

where $f : \mathbb{R} \to \mathbb{R}$ is continuously differentiable. We assume that x^* is an equilibrium solution of (6.1.1). We always assume that the equilibria are isolated, that is, if x^* is an equilibrium, then there is an open interval containing x^* but no other equilibria.

Fig. 6.1 shows a generic plot of $f(x)$ for $x^* = a, b, c$. The equilibrium solution a is an *attractor* (stable) since arrows to its left and right point toward it. The equilibrium solution b is a *repeller* (unstable) since the arrow to its left and the arrow to its right point away from it. Finally, the equilibrium solution c is *semi-stable* since the arrow to its left points toward it and the one to its right points away from it. In the case of the equilibrium solution a, we may say that it is *locally asymptotically stable* with respect to small perturbation or deviation from the equilibrium solution. To better illustrate the notion, consider the scenario of a population of foxes in a protected refuge that is in a locally asymptotically stable state a. Suppose a weather event caused a small number of the foxes to die. Local asymptotic stability means that the system will eventually return to the original state a. However, if the number of dead foxes is big, then the perturbation is not small, and in this event, there is no guarantee that the population of foxes will return to the original state a. In the case of the population return to state a for all perturbations, no matter how large, the state a is called *globally asymptotically stable*. Such topics will be considered in detail in Chapter 7.

Now we examine the dynamics of (6.1.1) in the neighborhood of its equilibrium solution x^*. If (6.1.1) is in an equilibrium state x^* and $\eta(t)$ represents a small pertur-

Advanced Differential Equations. https://doi.org/10.1016/B978-0-32-399280-0.00012-7

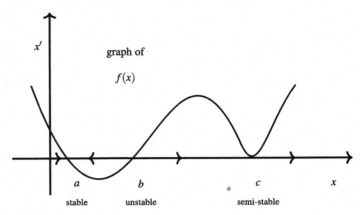

FIGURE 6.1

Generic phase line diagram.

bation from that state, then $x(t) = x^* + \eta(t)$. It follows that $x'(t) = \eta'(t)$. Thus using Taylor series expansion about x^*, we have

$$x'(t) = \eta'(t) = f(x^* + \eta)$$
$$= f(x^*) + \frac{f'(x^*)}{1!}(x^* + \eta - x^*) + \frac{f''(x^*)}{2!}(x^* + \eta - x^*)^2 + \cdots$$
$$= f'(x^*)\eta + \frac{f''(x^*)}{2!}\eta^2 + \cdots$$
$$= f'(x^*)\eta + O(\eta^2) + \cdots .$$

If $f'(x^*) \neq 0$, then the term $|f'(x^*)\eta| \gg |\frac{f''(x^*)}{2!}\eta^2|$. Thus we may neglect the term $O(\eta^2)$ and higher-order terms for that same reason. Then we arrive at the linearization of the system about the equilibrium state x^* given by

$$\eta'(t) = s\eta, \tag{6.1.2}$$

where $s = f'(x^*)$ is the slope of $f(x)$ at x^*. Eq. (6.1.2) has the solution

$$\eta(t) = ce^{st} \tag{6.1.3}$$

for some constant c. It is clear from (6.1.3) that if $s > 0$, then solutions grow exponentially, and the equilibrium solution x^* is unstable. On the other hand, if $s < 0$, then the solutions decay exponentially to zero, and the equilibrium solution x^* is asymptotically stable and hence is stable. If $s = 0$, then nothing can be said about the stability. Note that different stabilities may be obtained for different systems when

$s = 0$. To see this, we consider the scalar differential equations

$$x' = x^2, \ x' = -x^2, \ x' = x^3, \ \text{and} \ x' = -x^3.$$

All they share the equilibrium solution $x^* = 0$ and $s = f'(0) = 0$. Consider the first equation with $f(x) = x^2$. Then $f'(x) = 2x$, which is negative for $x < 0$ and positive for $x > 0$. Thus by the previous discussion the arrow to its left points toward $x^* = 0$, and the one to it right points away from it. It follows that $x^* = 0$ is semi-stable. By similar arguments we find that $x^* = 0$ is semi-stable for the second equation, unstable for the third equation, and stable for the fourth equation.

Example 6.1. If $x(t)$ represents a population at time t, $r > 0$ is the growth rate, and $K > 0$ is the carrying capacity, then the logistics model of the population growth is given by

$$x' = rx(1 - \frac{x}{K}). \tag{6.1.4}$$

Setting the right hand of (6.1.4) equal to zero, we arrive at two equilibrium solutions

$$x^* = 0, \ K.$$

Let $f(x) = rx(1 - \frac{x}{K})$. Then $f'(x) = r - \frac{2r}{K}x$. It follows from (6.1.2) that

$$s = f'(0) = r \ \text{and} \ s = f'(K) = -r.$$

We conclude that $x^* = 0$ is unstable and $x^* = K$ is stable.

Although the dynamics of scalar differential equations is simple, since both solutions converge to a constant or become unbounded as $t \to \infty$, they can have some interesting behaviors if solutions depend on some embedded parameters. To be specific, when the parameters are varied and pass some critical values, the systems may experience some abrupt changes or undergo some qualitative changes. These qualitative changes are called *bifurcations*, and the parameter values at which the bifurcations occur are called *bifurcation points*. We are interested in scalar differential equations of the form

$$x' = f(x, \mu), \tag{6.1.5}$$

where f is continuous in both arguments, and $\mu \in \mathbb{R}$ is a parameter.

Thus the system is said to undergo a bifurcation as μ crosses some critical value μ_0 if, in some way, the nature of the solutions changes qualitatively between the regions $\mu < \mu_0$ and $\mu > \mu_0$.

Three most important one-dimensional equilibrium bifurcations are described locally by the following differential equations:

$$x' = \mu - x^2, \quad \text{saddle-node};$$
$$x' = \mu x - x^2, \quad \text{transcritical; and}$$
$$x' = \mu x - x^3, \quad \text{pitchfork}.$$

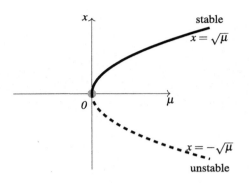

FIGURE 6.2

Bifurcation diagram determined by $x' = \mu - x^2$.

We will study each of these in more detail. We begin by considering *saddle node bifurcation*, which is the basic mechanism for creation and destruction of fixed points. As parameter varies, two fixed points move toward each other, collide, and mutually annihilate, that is, the saddle-node bifurcation results in fixed points being created or destroyed.

Example 6.2. Consider the scalar differential equation with parameter μ

$$x' = \mu - x^2. \tag{6.1.6}$$

Clearly, there are no equilibrium solutions for $\mu < 0$, and $x^* = \pm\sqrt{\mu}$ for $\mu > 0$. It is clear that $s = f'(\sqrt{\mu}) = -2\sqrt{\mu} < 0$ and $s = f'(-\sqrt{\mu}) = 2\sqrt{\mu} > 0$, which imply that the solid bold branch $x = \sqrt{\mu}$ is stable and the dashed branch $x = -\sqrt{\mu}$ is unstable. The bifurcation diagram is shown in Fig. 6.2, and since there is an exchange of stability at $\mu = 0$, we have a bifurcation at $\mu = 0$. This type of bifurcation is known by the names *saddle node bifurcation*, *fold bifurcation*, *turning point*, and *blue-sky bifurcation*. Note that at the bifurcation point $\mu = 0$, $s = f'(0) = 0$, and hence the linearization term disappears when equilibrium solutions coalesce.

Remark 6.1. In (6.1.6) of Example 6.2, if we change the term $-x^2$ to x^2, then

$$x' = \mu + x^2.$$

There are no equilibrium solutions for $\mu > 0$, and the curve $y = \mu + x^2$ does not intersect the x-axis. As $\mu \searrow 0$, the graph $y = \mu + x^2$ intersects the origin tangentially, and we have one equilibrium solution. As μ becomes negative, two equilibrium solutions appear at $x = \pm\sqrt{-\mu}$. Then, using the stability indicator s, we see that $x = -\sqrt{-\mu}$ is stable and $x = \sqrt{-\mu}$ is unstable. We conclude there is a *saddle node bifurcation* at $(0, 0)$ as indicated in the bifurcation diagram shown in Fig. 6.3.

Next, we discuss a class of bifurcations that occur as stability changes when going from one equilibrium solution to another. This type of bifurcation is called *transcritical bifurcation*.

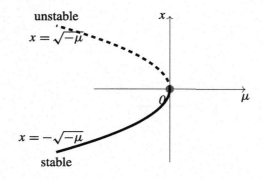

FIGURE 6.3

Bifurcation diagram determined by $x' = \mu + x^2$.

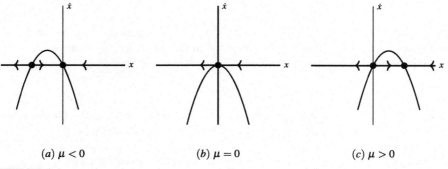

$(a)\ \mu < 0$ $(b)\ \mu = 0$ $(c)\ \mu > 0$

FIGURE 6.4

Phase portraits for different values of μ.

Example 6.3. To illustrate the idea, we consider the population growth model with carrying capacity

$$x' = \mu x - x^2.$$

The system has the equilibrium solutions $x^* = 0, \mu$. Using $s = f'(x^*)$, it follows that $x^* = 0$ is stable if $\mu < 0$ and unstable if $\mu > 0$. Similarly, $x^* = \mu$ is stable for $\mu > 0$ and unstable for $\mu < 0$. Fig. 6.4 shows the phase portraits for different values of μ. Also, Fig. 6.5 shows the bifurcation diagram. The stable branches are shown by bold solid lines, and the unstable branches are shown by dashed lines.

Next, we discuss a type of bifurcation called *pitchfork bifurcation*. The name comes from the shape of the bifurcation diagram that resembles a pitchfork. This type of bifurcation is common in problems that have a *symmetry*, for example, the buckling beam. In the case of pitchfork bifurcation the equilibrium solutions tend to appear and disappear symmetrically, creating a type of bifurcations that are different

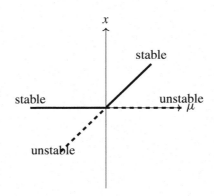

FIGURE 6.5

Bifurcation diagram that shows $(\mu, x) = (0, 0)$ is a transcritical bifurcation point.

from those that we have already discussed. Pitchfork bifurcations can come in one of two types, in the *supercritical bifurcation* and in the *subcritical bifurcation*.

Supercritical bifurcation is typified by equations of the form

$$x' = \mu x - x^3.$$

For $\mu < 0$, the origin is the only equilibrium solution, and it is stable. On the other hand, for $\mu > 0$, the origin is unstable, and the other two equilibrium solutions $\pm\sqrt{\mu}$ are stable.

Similarly, subcritical bifurcation is typified by equations of the form

$$x' = \mu x + x^3.$$

For $\mu < 0$, the origin is stable, and the other two equilibrium solutions $\pm\sqrt{\mu}$ are unstable. On the other hand, for $\mu > 0$, the origin is the only equilibrium solution, and it is unstable.

Example 6.4. We consider the scalar differential equation

$$x' = \mu x - x^3.$$

It is clear that $x^* = 0$ is an equilibrium solution regardless of the sign of μ and $x^* = \pm\sqrt{\mu}$ only if $\mu \geq 0$. Note that for any fixed value of μ, the function $y = \mu x - x^3$ is an odd function in x. Hence critical points will appear and disappear symmetrically with respect to the origin, or the system has the left–right symmetry. Using $s = f'(x^*) = \mu - 3(x^*)^2$, it follows $x^* = 0$ is stable for $\mu \leq 0$ and unstable for $\mu \geq 0$. In addition, $s = f(\pm\sqrt{\mu}) = -2\mu < 0$ implies that the equilibrium solutions $\pm\sqrt{\mu}$ are stable. The bifurcation diagram is displayed in Fig. 6.6.

Fig. 6.6 indicates that two stable equilibrium solutions or branches are created at the bifurcation point $(0, 0)$ and still exist after the bifurcation. A similar remark can be drawn regarding the subcritical bifurcation.

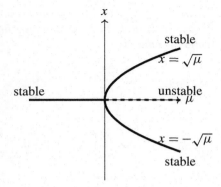

FIGURE 6.6

Bifurcation diagram determined for $x' = \mu x - x^3$; supercritical bifurcation at $(0, 0)$.

The next theorem is the scalar version of the *implicit function theorem*.

Theorem 6.1.1. *For scalar system* (6.1.5), *suppose that* $f : \mathbb{R} \times \mathbb{R} \to \mathbb{R}$ *with continuous first partial derivatives. Suppose x_0 is an equilibrium solution at μ_0, meaning that $f(x_0, \mu_0) = 0$. Assume that*

$$f(x_0, \mu_0) = 0, \quad \frac{\partial f}{\partial x}(x_0, \mu_0) \neq 0.$$

Then there exist $\delta, \varepsilon > 0$ and a C^1- function

$$\tilde{x} : (\mu_0 - \varepsilon, \mu_0 + \varepsilon) \to \mathbb{R}$$

such that $x = \tilde{x}$ is the unique solution of

$$f(x, \mu) = 0$$

with $|x - x_0| < \delta$ and $|\mu - \mu_0| < \varepsilon$.

For bifurcation theory, we have the following important result.

Corollary 6.1. *Suppose that $f : \mathbb{R} \times \mathbb{R} \to \mathbb{R}$ has continuous first partial derivatives. Suppose x_0 is an equilibrium solution at μ_0 for* (6.1.5). *Then a necessary condition for a solution (x_0, μ_0) of* (6.1.5) *to be a bifurcation point of equilibria is that*

$$\frac{\partial f}{\partial x}(x_0, \mu_0) = 0. \tag{6.1.7}$$

The next theorem provides necessary conditions that guarantee the different types of bifurcation points we have discussed so far.

Theorem 6.1.2. *Suppose that $f : \mathbb{R} \times \mathbb{R} \to \mathbb{R}$ is continuous in both arguments and has continuous first and second partial derivatives. Suppose that at (x_0, μ_0), f satisfies the necessary bifurcation conditions*

$$f(x_0, \mu_0) = 0, \quad \frac{\partial f}{\partial x}(x_0, \mu_0) = 0.$$

(1) If

$$\frac{\partial f}{\partial \mu}(x_0, \mu_0) \neq 0, \quad \frac{\partial^2 f}{\partial x^2}(x_0, \mu_0) \neq 0,$$

then a saddle node bifurcation occurs at (x_0, μ_0).
(2) If

$$\frac{\partial f}{\partial \mu}(x_0, \mu_0) = 0, \quad \frac{\partial^2 f}{\partial x \partial \mu}(x_0, \mu_0) \neq 0, \quad \frac{\partial^2 f}{\partial x^2}(x_0, \mu_0) \neq 0,$$

then a transcritical bifurcation occurs at (x_0, μ_0).
(3) If

$$\frac{\partial f}{\partial \mu}(x_0, \mu_0) = 0, \quad \frac{\partial^2 f}{\partial x^2}(x_0, \mu_0) = 0,$$

$$\frac{\partial^2 f}{\partial x \partial \mu}(x_0, \mu_0) \neq 0, \quad \frac{\partial^3 f}{\partial x^3}(x_0, \mu_0) \neq 0,$$

then a pitchfork bifurcation occurs at (x_0, μ_0). In addition,
(i) if $\frac{\partial^3 f}{\partial x^3}(x_0, \mu_0) < 0$, then there is a supercritical bifurcation at (x_0, μ_0), and
(ii) if $\frac{\partial^3 f}{\partial x^3}(x_0, \mu_0) > 0$, then there is a subcritical bifurcation at (x_0, μ_0).

Example 6.5. We consider the scalar system

$$x' = \mu x - e^x.$$

Then its bifurcation points must satisfy

$$\mu x - e^x = 0.$$

Plotting the graphs $y = \mu x$ and $y = e^x$, we see that the line $y = x$ is tangent to the curve at $(x, y) = (1, e)$. In other words, if we set $f(x, \mu) = \mu x - e^x$, then $f(1, e) = 0$. This says that $x_0 = 1$ is an equilibrium solution for $\mu = e$. Moreover, $\frac{\partial f}{\partial x}(1, e) = 0$. In addition, $\frac{\partial f}{\partial \mu}(x_0, \mu_0) = 1 \neq 0$, and $\frac{\partial^2 f}{\partial x^2}(x_0, \mu_0) = -e \neq 0$, and so by item 1 of Theorem 6.1.2 there is a saddle node bifurcation at $(1, e)$.

Remark 6.2. Suppose bifurcation for (6.1.5) occurs at (x_0, μ_0). We may translate to the origin $(0, 0)$ by the change of variables $x \to x - x_0$, $\mu \to \mu - \mu_0$. Moreover, if

$x = \tilde{x}(\mu)$ is a solution branch, then $x \to x - \tilde{x}(\mu)$ maps the branch to $x = 0$. Suppose we have a saddle node bifurcation. For $(0, 0)$ to be a bifurcation point, we must have

$$f(x_0, \mu_0) = 0, \quad \frac{\partial f}{\partial x}(x_0, \mu_0) = 0.$$

Further, suppose that

$$\frac{\partial f}{\partial \mu}(0, 0) = a \neq 0, \quad \frac{\partial^2 f}{\partial x^2}(0, 0) = b \neq 0.$$

Then if in the Taylor expansion of $f(x, \mu)$, we neglect quadratic terms in μ and cubic terms in x, then near $(x, \mu) = (0, 0)$, we obtain

$$f(x, \mu) = \mu \frac{\partial f}{\partial \mu}(0, 0) + x \frac{\partial f}{\partial x}(0, 0) + \frac{x^2}{2} \frac{\partial^2 f}{\partial x^2}(0, 0) + \cdots$$
$$= a\mu + bx^2 + \cdots. \tag{6.1.8}$$

The form $a\mu + bx^2$ in (6.1.8) is called a *normal form* or *representative* for scalar saddle node bifurcations. Thus the original system may be approximated near the origin by

$$x' = a\mu + bx^2.$$

The signs of a and b determine the type of bifurcations and the shape of the bifurcation diagrams as shown in Example 6.2 and Remark 6.1.

We end the section with the following application.

Example 6.6. (*Bucking of a rod*) The potential energy of the system of two rigid rods of length L connected by a torsional spring with spring constant k is given by the function

$$R(x) = \frac{1}{2}kx^2 + 2\lambda L\big(\cos(x) - 1\big),$$

where λ is the strength of compressive force, and x is the angle of the rod to the horizontal axis. The term $\frac{1}{2}kx^2$ is the energy required to compress the spring by the angle $2x$, and $2\lambda L\big(\cos(x) - 1\big)$ is the work done on the system by the external force. Equilibrium solutions satisfy $R'(x) = 0$ or

$$x - \mu \sin(x) = 0, \quad \mu = \frac{2\lambda L}{k},$$

where μ is a dimensionless force parameter. The above equation has the unbuckled equilibrium solution $x = 0$. Set

$$f(x, \mu) = R'(x) = \mu \sin(x) - x.$$

Then the necessary conditions for an equilibrium bifurcation are

$$\frac{\partial f}{\partial x} = \mu - 1 = 0,$$

which occurs at $\mu = 1$. By finding the Taylor series of f at $(0, 1)$ we arrive at

$$f(x, \mu) = (\mu - 1)x - \frac{1}{6}x^3 + \cdots.$$

We can see from Example 6.4 or verifying the conditions of item 3 of Theorem 6.1.2 that there is a pitchfork bifurcation at $(x, \mu) = (0, 1)$. The bifurcation equilibria near this point are given for $0 < \mu - 1 \ll 1$ by

$$x = \sqrt{6(\mu - 1)} + \cdots.$$

This can also be seen by plotting the graphs $y = x$ and $y = \mu \sin(x)$.

This simple one-dimensional problem is an illustration for the buckling of an elastic beam, one of the first bifurcation problems which was originally studied by Euler in 1757.

For the rest of this chapter, if $x = \begin{pmatrix} x_1 \\ x_2 \\ \vdots \\ x_n \end{pmatrix}$, then $|x| = \sum\limits_{i=1}^{n} |x_i|$, and for an $n \times n$

matrix, $A = a_{ij}$, $|A| = \sum\limits_{i,j=1}^{n} |a_{ij}|$.

6.2 Stability of systems by linearization

Consider the general system of ordinary differential equations

$$\begin{aligned} x_1' &= f_1(x_1, \ldots, x_n), \\ x_2' &= f_2(x_1, \ldots, x_n), \\ &\vdots \\ x_n' &= f_n(x_1, \ldots, x_n). \end{aligned}$$

Using the vector notations

$$x = \begin{pmatrix} x_1 \\ x_2 \\ \vdots \\ x_n \end{pmatrix}$$

and

$$f(x) = \begin{pmatrix} f_1(x) \\ f_2(x) \\ \vdots \\ f_n(x) \end{pmatrix},$$

the above system can be written in the vector form

$$x' = f(x), \tag{6.2.1}$$

where $f : \mathbb{R}^n \to \mathbb{R}^n$ with continuously differentiable components. We assume that the vector $x^* = (x_1^*, x_2^*, \ldots, x_n^*)$ is a fixed point of (6.2.1).

Define the linear part of f at x^* by $J(x^*)$, where

$$J(x^*) := \begin{pmatrix} \frac{\partial f_1}{\partial x_1}(x^*) & \frac{\partial f_1}{\partial x_2}(x^*) & \cdots & \frac{\partial f_1}{\partial x_n}(x^*) \\ \frac{\partial f_2}{\partial x_1}(x^*) & \frac{\partial f_2}{\partial x_2}(x^*) & \cdots & \frac{\partial f_2}{\partial x_n}(x^*) \\ \vdots & \vdots & \ddots & \vdots \\ \frac{\partial f_n}{\partial x_1}(x^*) & \frac{\partial f_n}{\partial x_2}(x^*) & \cdots & \frac{\partial f_n}{\partial x_n}(x^*) \end{pmatrix}, \tag{6.2.2}$$

which is called the *Jacobian* of f at x^*. Since $f \in C^1(\mathbb{R}^n, \mathbb{R}^n)$, Taylor's theorem for functions of several variables says that

$$f(x) = J(x^*)(x - x^*) + g(x) \quad \text{(we have used } f(x^*) = 0), \tag{6.2.3}$$

where g is a function that is small in the neighborhood of x^* in the sense that

$$\lim_{x \to x^*} \frac{|g(x)|}{|x - x^*|} = 0. \tag{6.2.4}$$

Note that since $f(x^*) = 0$, (6.2.3) implies that $g(x^*) = 0$. Our hope is to prove that if $Re(\rho(J(x^*))) < 0$ or the origin for

$$y' = J(x^*)y$$

is asymptotically stable, then the equilibrium point x^* of (6.2.1) is asymptotically stable. Note that by introducing the transformation $y = x - x^*$ we may assume without loss of generality that the fixed point is at the origin. Thus we have the following general theorem.

Theorem 6.2.1. *Assume that $Re(\rho(A)) < 0$ where A is an $n \times n$ constant matrix. Let U be an open set of $\mathbb{R} \times \mathbb{R}^n$ that contains $[0, \infty) \times B_r(0)$, where $B_r(0) = \{x \in \mathbb{R}^n : |x - 0| \le r\}$ for positive constant r. Assume that $g : U \to \mathbb{R}^n$ is continuous in both arguments and Lipschitz in the second argument such that*

$$\lim_{x \to 0} \frac{|g(t, x)|}{|x|} = 0 \ \text{uniformly for} \ t \in [0, \infty) \tag{6.2.5}$$

with $g(t, 0) = 0$. *Then the origin for the perturbed nonlinear system*

$$x' = Ax + g(t, x) \tag{6.2.6}$$

is asymptotically stable.

Proof. Since $Re(\rho(A)) < 0$, we can find constants $K > 0$ and $\xi < 0$ such that

$$\|e^{At}\| \leq Ke^{\xi t}, \quad t > 0.$$

Let $x(t)$ be a solution of (6.2.6) on some interval (a, b) containing 0 with $x(0) = x_0$. Then the variation of parameters formula (3.3.6) yields

$$x(t) = e^{A(t-t_0)}x_0 + \int_{t_0}^{t} e^{A(t-s)}g(s, x(s))ds. \tag{6.2.7}$$

Taking the appropriate norms of both sides of (6.2.7) and estimating give, for $t \geq 0$,

$$|x(t)| = K|x_0|e^{\xi t} + \int_0^t Ke^{\xi(t-s)}|g(s, x(s))|ds, \quad t \in [0, b).$$

Multiplying both sides by $e^{-\xi t}$, we arrive at the expression

$$|x(t)|e^{-\xi t} = K|x_0| + \int_0^t Ke^{-\xi s}|g(s, x(s))|ds, \quad t \in [0, b).$$

Chose $\eta > 0$ such that $\eta K < -\xi$. Thus by (6.2.5) we have that

$$|g(t, x)| \leq \eta|x| \text{ for } (t, x) \in [0, \infty) \times \bar{B}_r(0).$$

Set $\delta^* = \delta/(2K + 1) < \delta$. Suppose that $|x_0| < \delta^*$, and let (a, b) be the maximal interval of existence with $x(0) = x_0$.

Let $[0, c] \subseteq (a, b)$ be an interval such that $|x(t)| \leq \delta$ for all $t \in [0, c]$. This is possible due to the continuity of x. By the above inequality of g we have that

$$|g(t, x)| \leq \eta|x| \text{ for } t \in [0, c].$$

Then we have

$$|x(t)|e^{-\xi t} = K|x_0| + \int_0^t Ke^{-\xi s}\eta|x(s)|ds, \quad t \in [0, c].$$

Use Gronwall's inequality and then evaluate the integral to get

$$|x(t)|e^{-\xi t} \leq K^2|x_0|\eta\frac{1}{K\eta}[e^{K\eta t} - 1]$$
$$= K|x_0|e^{K\eta t}.$$

Since $|x_0| < \delta^*$, $K|x_0| < \delta$, and so $|x(t)|e^{-\xi t} \le \delta e^{K\eta t}$. It follows that

$$|x(t)| \le \delta e^{(\xi + K\eta)t}, \quad t \in [0, c], \tag{6.2.8}$$

where $\xi + K\eta < 0$.

Of course, we only showed that inequality (6.2.8) holds for $|x(t)| \le \delta$ on $[0, c]$. It is left as an exercise to show that the inequality holds on $[0, \infty)$. This completes the proof. $\qquad\square$

Remark 6.3. (1) If the origin is stable for the linear system $x' = Ax$ but not asymptotically stable, then it is not necessary that the origin is stable for the nonlinear system (6.2.6).
(2) If the origin is unstable for the linear system $x' = Ax$, then it is unstable for the nonlinear system (6.2.6).

Remark 6.4. Let $D \subset \mathbb{R}^n$ containing the origin. Consider the autonomous system

$$x' = f(x), \tag{6.2.9}$$

where $f : D \to \mathbb{R}^n$ with continuously differentiable components and $f(0) = 0$. Let $A = J(0)$ be the Jacobian matrix of f at $x = 0$. Rewrite (6.2.9) as

$$x' = Ax + g(x), \tag{6.2.10}$$

where $g(x) = f(x) - Ax$. The differentiability of f implies $g(x) = o(|x|)$ as $x \to \infty$. Therefore $g(x)$ satisfies condition (6.2.5).

The next theorem provides results concerning the UAS of the zero solution. Its proof is left as an exercise.

Theorem 6.2.2. *Assume that $Re(\rho(A)) < 0$, where A is an $n \times n$ constant matrix (or the zero solution of the linear part is AS). Assume that a continuous function $g(t, x)$ satisfies Lipschitz or local Lipschitz condition in the second argument for $x \in D$. Suppose that*

$$\lim_{x \to 0} \frac{|g(t, x)|}{|x|} = 0 \ \textit{uniformly for } t \in [0, \infty) \tag{6.2.11}$$

with $g(t, 0) = 0$. Then the origin for the perturbed nonlinear system

$$x' = Ax + g(t, x)$$

is uniformly asymptotically stable.

We have the following example.

Example 6.7. The zero solution of the nonlinear non-autonomous system

$$x' = -x + \frac{t}{t+1}x^2$$

is UAS. It is obvious that the zero solution of the linear part $x' = -x$ is AS. We must show that condition (6.2.11) holds. Let $g(t, x) = \frac{t}{t+1}x^2$. Then g is continuous, and for any open subset $D = (\alpha, \beta) \subset \mathbb{R}$ containing the origin, we have

$$|g(t, x) - g(t, y)| \leq 2\beta|x - y| \text{ for all } (t, x), (t, y) \in [0, \infty) \times D.$$

Moreover,

$$\lim_{x \to 0} \frac{|g(t, x)|}{|x|} \leq \lim_{x \to 0} \frac{x^2}{|x|} = 0 \text{ uniformly for } t \in [0, \infty).$$

Thus by Theorem 6.2.2 the zero solution is UAS.

The next example shows the needs for a different approach when the eigenvalues of the matrix A fail to have negative real parts.

Example 6.8. Consider the planar autonomous system

$$x' = -y + x\sqrt{x^2 + y^2}, \quad y' = x + y\sqrt{x^2 + y^2}.$$

It is clear that $(0, 0)$ is the only equilibrium point. We rewrite the system in the form

$$\begin{pmatrix} x \\ y \end{pmatrix}' = \begin{pmatrix} 0 & -1 \\ 1 & 0 \end{pmatrix} \begin{pmatrix} x \\ y \end{pmatrix} + \begin{pmatrix} x\sqrt{x^2 + y^2} \\ y\sqrt{x^2 + y^2} \end{pmatrix}.$$

The eigenvalues of

$$A = \begin{pmatrix} 0 & -1 \\ 1 & 0 \end{pmatrix}$$

are $\pm i$. Thus Theorem 6.2.1 does not apply. Instead, we use the polar coordinates. Namely,

$$x = r\cos(\theta), \quad y = r\sin(\theta), \quad \text{where } r^2 = x^2 + y^2,$$

and $r = r(t)$ and $\theta = \theta(t) = \tan^{-1}(\frac{y}{x})$. Thus

$$\frac{d}{dt}(x^2 + y^2) = \frac{d}{dt}r^2$$

or

$$xx' + yy' = rr'.$$

A substitution of x' and y' yields

$$rr' = x(-y + x\sqrt{x^2 + y^2}) + y(x + y\sqrt{x^2 + y^2})$$
$$= (x^2 + y^2)\sqrt{x^2 + y^2}$$
$$= r^2 r.$$

Thus we have the first-order differential equation

$$r' = r^2,$$ (6.2.12)

which implies that the function r is increasing. Thus trajectories are spiraling away from the origin, and hence the origin is unstable. In addition,

$$\theta' = \frac{xy' - x'y}{x^2 + y^2} = 1.$$

Choosing $\theta(0) = 0$, we have $\theta(t) = t$, which indicates trajectories spiral away from the origin counterclockwise. Note that using separation of variables, (6.2.12) can be easily solved to find $r(t) = \frac{1}{k-t}$ for some suitable constant k. Notice that $r(t) \to \infty$ as $t \to k^-$.

Example 6.9. In Example 3.5, we considered a pendulum with friction and arrived at the planar autonomous system

$$x_1' = x_2,$$
$$x_2' = -\frac{g}{L}\sin(x_1) - \frac{b}{m}x_2.$$ (6.2.13)

It is clear that system (6.2.13) has infinitely many equilibrium points $(n\pi, 0)$, $n = 0, 1, \ldots$, which correspond to vertical positions with zero velocity. We will only analyze two equilibrium points $(0, 0)$ and $(\pi, 0)$. For the equilibrium point $(0, 0)$, we write system (6.2.13) in the form

$$\begin{pmatrix} x_1 \\ x_2 \end{pmatrix}' = \begin{pmatrix} 0 & 1 \\ -\frac{g}{L} & -\frac{b}{m} \end{pmatrix}\begin{pmatrix} x_1 \\ x_2 \end{pmatrix} + \frac{g}{L}\begin{pmatrix} 0 \\ x_1 - \sin(x_1) \end{pmatrix}.$$

The linear part has the eigenvalues

$$\lambda_1 = -\frac{b}{2m} + \frac{1}{2}\left(\frac{b^2}{m^2} - 4\frac{g}{L}\right)^{1/2}$$

and

$$\lambda_1 = -\frac{b}{2m} - \frac{1}{2}\left(\frac{b^2}{m^2} - 4\frac{g}{L}\right)^{1/2}.$$

If $\frac{b^2}{m^2} - 4\frac{g}{L} < 0$ (when damping is weak), then the eigenvalues are complex with negative real parts, and the origin is a stable spiral. If $\frac{b^2}{m^2} - 4\frac{g}{L} \geq 0$ (when the damping is strong), that is,

$$\frac{b^2}{m^2} > 4\frac{g}{L},$$

we have that $\lambda_2 \leq \lambda_1 < 0$, and hence the equilibrium solution $(0, 0)$ for system (6.2.13) is asymptotically stable by Theorem 6.2.1.

The physical and geometrical interpretation for $(x_1, x_2') = (0, 0)$ being a stable spiral corresponds to the pendulum hanging vertically downward ($x_1 = 0$) with zero velocity ($x_2' = 0$), and thus, when the damping is weak, a perturbation will cause the pendulum to spiral and approach the vertically downward position.

For the equilibrium point $(\pi, 0)$, we use the transformation

$$\begin{pmatrix} x_1 \\ x_2 \end{pmatrix} = \begin{pmatrix} y_1 \\ y_2 \end{pmatrix} + \begin{pmatrix} \pi \\ 0 \end{pmatrix}$$

in (6.2.13), obtaining

$$\begin{pmatrix} y_1 \\ y_2 \end{pmatrix}' = \begin{pmatrix} y_2 \\ \frac{g}{L}\sin(y_1) - \frac{b}{m}y_2 \end{pmatrix}.$$

To examine this system near $(0, 0)$, we write the above new nonlinear system in y in the form

$$\begin{pmatrix} y_1 \\ y_2 \end{pmatrix}' = \begin{pmatrix} 0 & 1 \\ \frac{g}{L} & -\frac{b}{m} \end{pmatrix} \begin{pmatrix} y_1 \\ y_2 \end{pmatrix} - \frac{g}{L} \begin{pmatrix} 0 \\ y_1 - \sin(y_1) \end{pmatrix}.$$

The linear parts have the eigenvalues

$$\lambda_1 = -\frac{b}{2m} + \left(\frac{b^2}{m^2} + 4\frac{g}{L}\right)^{1/2},$$

$$\lambda_1 = -\frac{b}{2m} - \left(\frac{b^2}{m^2} + 4\frac{g}{L}\right)^{1/2}.$$

It is clear that $\lambda_2 < 0 < \lambda_1$ and the equilibrium point $(0, 0)$ and hence $(\pi, 0)$ are unstable.

Note that we could calculate the Jacobian given by (6.2.2) at $(\pi, 0)$ and obtain

$$J((\pi, 0)) = \begin{pmatrix} 0 & 1 \\ \frac{g}{L} & -\frac{b}{m} \end{pmatrix},$$

instead of using transformation.

If we compute $J((\pm n\pi, 0))$, then we will see that $(\pm n\pi, 0)$ are stable for $n = 2, 4, \ldots$ and unstable for $n = 1, 3, \ldots$.

Example 6.10. (*Lotka–Volterra predator–prey model*) In Example 4.1 we considered the Lotka–Volterra system

$$\frac{dx}{dt} = ax - bxy,$$

$$\frac{dy}{dt} = -cy + dxy \tag{6.2.14}$$

and found that the equilibrium solutions are

$$x_1^* = (0,0) \text{ and } x_2^* = (c/d, a/b).$$

The first equilibrium solution x_1^* effectively represents the extinction of both species. If both populations are at 0, then they will continue to be so indefinitely. The second equilibrium solution x_2^* represents a fixed point at which both populations sustain their current, non-zero numbers, and do so indefinitely. The levels of population at which this equilibrium is achieved depend on the chosen values of the parameters a, b, c, and d. Now we will further discuss the stability of both equilibrium solutions x_1^* and x_2^* and hope that x_1^* is unstable and x_2^* is stable. The Jacobian matrix is

$$J(x,y) = \begin{pmatrix} a - by & -bx \\ dy & -c + dx \end{pmatrix}.$$

For $x_1^* = (0,0)$, we have

$$J(0,0) = \begin{pmatrix} a & 0 \\ 0 & -c \end{pmatrix}$$

with eigenvalues $a > 0$ and $-c < 0$. Thus the equilibrium point $(0,0)$ is a saddle and hence unstable.

For $x_2^* = (c/d, a/b)$,

$$J(c/d, a/b) = \begin{pmatrix} 0 & -\frac{bc}{d} \\ \frac{da}{b} & 0 \end{pmatrix},$$

which has purely imaginary eigenvalues. Hence linearization is inconclusive. Note that the use of polar coordinates will not work here in the sense that it is hard, if not impossible, to obtain any information from it. We look at the vector fields surrounding the equilibrium point by dividing the region of interest into four quadrants as depicted in Fig. 6.1. We do so by drawing vertical and horizontal lines through the equilibrium point $(c/d, a/b)$. Those lines are called the x and y *nullclines*. We begin by writing system (6.2.14) in the form

$$x' = x(a - by), \quad y' = y(-c + dx). \tag{6.2.15}$$

The x nullclines for system (6.2.15) where $x' = 0$ are $x = 0$ and $y = \frac{a}{b}$. Trajectories cross these two lines vertically. Similarly, trajectories cross the y nullclines where $y' = 0$ horizontally. We see that the equilibria are the intersections of the x and y nullclines. On the ray to the left of the equilibrium and below the line $y = a/b$, we have $x < \frac{c}{d}$ and $y = \frac{a}{b}$. We know that the vector field is vertical, and so we only need to check the sign of y'. From (6.2.15) we have

$$y' = y(-c + dx) = \frac{a}{b}(-c + dx) < \frac{a}{b}\left(-c + d\frac{c}{d}\right) = 0.$$

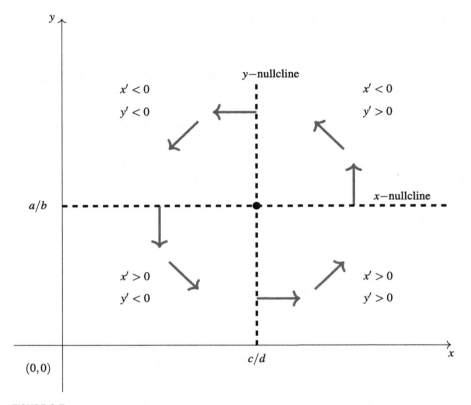

FIGURE 6.7

Nullclines and vector field.

Thus in that region, we have that $x' > 0$ and $y' < 0$. Similarly, if we consider the region to the left of the vertical line $x = \frac{c}{d}$ and above the horizontal line $y = c/d$, then we have that $x = \frac{c}{d}$ and $y > \frac{a}{b}$. Again, the vector field is horizontal, and we only need to check the sign of x'. From (6.2.15) we have

$$x' = x(a - by) < \frac{c}{d}(a - b\frac{a}{b}) = 0.$$

Thus in that particular region, we have $x' < 0$, $y' < 0$. This task can be easily performed for the other two regions. The complete vector field around the equilibrium is displayed in Fig. 6.7. To obtain the trajectories, we write the system in the form

$$\frac{dy}{dx} = \frac{y(-c + d\,x)}{x(a - by)}.$$

After rearranging the terms, we get

$$\frac{(a - by)}{y}dy = \frac{(-c + d\,x)}{x}dx.$$

Simplifying and then integrating both sides yield

$$a\mathrm{Ln}(y) - by = -c\mathrm{Ln}(x) + d\,x + K \text{ for some constant } K.$$

Exponentiating both sides gives

$$y^a e^{-by} = x^{-c}e^{d\,x}e^K.$$

We consider the y nullcline where the value of x is fixed at c/d. We choose a specific trajectory by assigning a fixed value for K. Then the right side of the above relation is positive β, so that

$$y^a = \beta e^{by}. \tag{6.2.16}$$

The left and right sides of (6.2.16) are the power and exponential functions, respectively. Therefore if we plot them on the same graph, then there can be at most two crossings, or Eq. (6.2.16) can have at most two solutions for y. Thus along the y nullcline, there can be at most two crossings. That means that an orbit or trajectory cannot spiral into or out from an equilibrium point. Otherwise, many values of y would be required, which is not the case here. This determines that the equilibrium point $(c/d, a/b)$ is encircled with periodic trajectories, and hence it is stable. Note that the periodic trajectories encircling the equilibrium point $(c/d, a/b)$ indicate that a small perturbation from the equilibrium solution puts the populations on a periodic orbit that stays near the equilibrium and the system never returns to equilibrium. A phase diagram is shown in Fig. 6.8.

6.3 An SIR epidemic model

Here is the perfect place to apply what we have studied so far to an *SIR* model, where

$S = S(t)$ is the number of susceptible individuals (not yet having the disease but capable of getting it)
$I = I(t)$ is the number of infected individuals, and
$R = R(t)$ is the number of recovered or removed individuals.

We assume that the total population is constant and everyone is susceptible. Then

$$S + I + R = N \text{ (constant)},$$

that is,

$$S + I + R = 1 \text{ (by normalization)}.$$

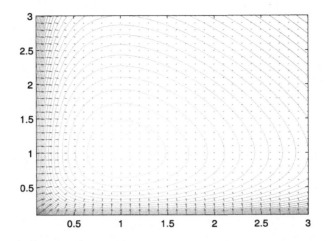

FIGURE 6.8

Prey–predator dynamics as described by the level curves. The arrows describe the velocity and direction of solutions. We chose $a = b = c = d = 1$.

Let

μ = birth rate = death rate (because the number of people is constant),

γ = removal rate, and

λ = infection rate.

We assume that all newborns are susceptible and arrive at the epidemic model

$$\begin{cases} S'(t) = (S + I + R)\mu - \mu S - (SI)\lambda, \\ I'(t) = -\mu I - \gamma I + \lambda SI, \\ R'(t) = -\mu R + \gamma I. \end{cases}$$

We have $(S + I + R)\mu = \mu$, since the population is constant and normalized at 1. The term of μS represents the death of susceptible, and the negative implies loss or removal from being susceptible. On the other hand, the term $(SI)\lambda$ denotes those who got infected, and the minus represents that they are infected and not susceptible anymore. Similarly, the term μI represents those who die from the infection, and the minus indicates a loss for those who are infected already. In addition, the term γI indicates those who are removed from the population due to beating the epidemic or got vaccinated. Finally, the term μR denotes those who are removed due to death. Keep in mind that *minus* means loss from a certain class for S, I, and R, and *plus* implies gain for the same classes. Thus by only analyzing $S \geq 0$ and $I \geq 0$ we arrive at the 2 × 2-planar system

$$\begin{cases} S' = \mu - \mu S - \lambda SI, \\ I' = -(\mu + \gamma)I + \lambda SI. \end{cases} \tag{6.3.1}$$

Note that system (6.3.1) is nonlinear and $(0, 0)$ is not an equilibrium solution. Also, the system is valid in the region $0 \leq S + I \leq 1$ since we are neglecting R. It is important to note that the system does not allow those who recover from the epidemic to become susceptible again. Suppose $S(t_0) = S_0$ and $I(t_0) = I_0$. Since the right side of (6.3.1) is continuous with continuous first partial derivatives, we know that the solutions $S(t, t_0, S_0)$ and $I(t, t_0, I_0)$ exist and are unique. Moreover, $S(t, t_0, S_0)$ and $I(t, t_0, S_0)$ are continuous and differentiable with respect to all variables. Next, we examine how big the interval on which the solutions exist. Due to our constraints, we are interested in the triangular region of existence

$$T = \{(S, I) : S \geq 0, I \geq 0, \ S + I < 1\}.$$

Along $S = 0$, we have $S' = \mu > 0$, $0 \leq I < 1$, and $I' = -(\mu + \gamma)I < 0$ if $I > 0$. Moreover,

$$\frac{dI}{dS} = \frac{I'}{S'} < 0.$$

Thus any trajectory that starts on the I-axis moves downward inside T in the positive direction of S.

Now, along $I = 0$, we have $S' = \mu(\mu - S) > 0$, $0 \leq I < 1$, and $I' = 0$. Thus any trajectory that starts on the S-axis moves toward the point $(1, 0)$, at which the line $S + I = 1$ intersects the S-axis. Note that $(1, 0)$ is an equilibrium solution.

Along the line $S + I = 1$, we have

$$(S + I)' = S' + I' = \mu - \mu(S + I) - \gamma I,$$

and since $S + I = 1$, this expression reduces to

$$(S + I)' = -\gamma I < 0.$$

This tells us that any trajectory starting at the line $S + I = 1$ moves to the left of the line toward the inside of the triangle T. We conclude that the triangular region T is positively invariant, that is, if $(S_0, I_0) \in T$, then $S(t, t_0, S_0)$, $I(t, t_0, I_0) \in T$ for all $t \geq t_0$. Thus the maximum interval of existence for solutions in T is $[t_0, \infty)$. From now on we refer to the triangular region T as the feasible region. Next, we search for the equilibrium solutions (S^*, I^*) in the feasible region. To do so, we must solve for S and I by setting the right sides of (6.3.1) to zero, that is, $S' = 0$, and $I' = 0$:

$$\mu - \mu S - \lambda SI = 0, \quad \text{and} \quad -(\mu + \gamma)I + \lambda SI = 0.$$

From the second equation we obtain

$$(-\mu - \gamma + \lambda S)I = 0.$$

Thus if $I \neq 0$, then we arrive at

$$S^* = \frac{\mu + \gamma}{\lambda}.$$

Clearly, $S^* \geq 0$, and in order for S^* to be in the feasible region, we must have

$$\mu + \gamma < \lambda.$$

Substituting $S^* = \frac{\mu+\gamma}{\lambda}$ into the equation $\mu - \mu S^* - \lambda S^* I^* = 0$, we arrive at

$$I^* = \frac{\mu}{\mu + \gamma} - \frac{\mu}{\lambda}.$$

It is clear $I^* > 0$, since

$$I^* = \frac{\mu}{\mu + \gamma} - \frac{\mu}{\lambda} > \frac{\mu}{\lambda} - \frac{\mu}{\lambda} = 0.$$

Next, we show that (S^*, I^*), belongs to the feasible region:

$$0 < S^* + I^* = \frac{\mu + \gamma}{\lambda} + \frac{\mu}{\mu + \gamma} - \frac{\mu}{\lambda}$$
$$= \frac{\gamma}{\lambda} + \frac{\mu}{\mu + \gamma}$$
$$< \frac{\gamma}{\mu + \gamma} + \frac{\mu}{\mu + \gamma}$$
$$= 1.$$

Thus the two equilibrium solutions $(1, 0)$ and (S^*, I^*) belong to the feasible region T. Next, we use Theorem 6.2.1 to study the stability of the two equilibrium solutions. We do not want solutions to approach $(1, 0)$ and stay there. Instead, we rather have that solutions approach (S^*, I^*), that is, we hope for $(1, 0)$ to be unstable.

We begin with $(1, 0)$. The Jacobian matrix is

$$J(S, I) = \begin{pmatrix} -\mu - \lambda I & -\lambda S \\ \lambda I & -(\mu + \gamma) + \lambda S \end{pmatrix}.$$

Thus

$$J(1, 0) = \begin{pmatrix} -\mu & -\lambda \\ 0 & -(\mu + \gamma) + \lambda \end{pmatrix}$$

with two eigenvalues

$$d_1 = -\mu \quad \text{and} \quad d_2 = -(\mu + \gamma) + \lambda.$$

Clearly, $d_1 < 0$, and $d_2 > 0$ as long as $\mu + \gamma < \lambda$. Thus by Theorem 6.2.1 the equilibrium solution $(1, 0)$ is unstable. Recall that if

$$A = \begin{pmatrix} a & b \\ c & d \end{pmatrix}$$

is a nonsingular matrix, then the characteristic equation is

$$\lambda^2 - (a+d)\lambda + ad - bc = 0.$$

By introducing

$$p = a + d,$$
$$q = ad - bc,$$

the characteristic equation becomes

$$\lambda^2 - p\lambda + q = 0.$$

The roots of this equation are

$$\lambda_{1,2} = \frac{1}{2}\left[p \pm \sqrt{p^2 - 4q}\right]. \tag{6.3.2}$$

Now the Jacobian at (S^*, I^*) is

$$J(S^*, I^*) = \begin{pmatrix} -\mu - \lambda I^* & -\lambda S^* \\ \lambda I^* & 0 \end{pmatrix}.$$

Using (6.3.2), we arrive at the two eigenvalues

$$d_{1,2} = \frac{-(\mu + \lambda I^*) \pm \sqrt{(\mu + \lambda I^*)^2 - 4\lambda^2 S^* I^*}}{2}$$

with negative real parts. Hence by Theorem 6.2.1 the equilibrium solution (S^*, I^*) is asymptotically stable.

6.4 Limit cycle

A central issue in the theory of nonlinear systems in differential equations is to determine the existence of closed orbits or limit cycles. Those closed orbits or limit cycles represent periodic solutions of the system. We consider the two-dimensional autonomous system

$$x' = P(x, y), \ y' = Q(x, y), \tag{6.4.1}$$

where P and Q are continuous on some subset of $\mathbb{R}^2$. Systems given by (6.4.1) may admit solutions that commonly tend to a limiting finite periodic solution. This limiting closed curve in the phase plane is called a *limit cycle*.

Definition 6.4.1. A limit cycle of system (6.4.1) is a closed curve, and no other solution that is a closed curve exists in its neighborhood.

Note that every neighboring trajectory spirals and approaches the limit cycle from the inside or from the outside as $t \to \infty$ or as $t \to -\infty$.

Definition 6.4.2. If every neighboring trajectory approaches a limit cycle as $t \to \infty$ or as $t \to -\infty$, then the limit cycle is said to be *stable*.

It is hard, if not impossible, to determine if a nonlinear system admits a limit cycle. However, the next result, known as Bendixson–Dulac criteria should provide a criterion for the *nonexistence* of limit cycles or closed trajectories of the system given by (6.4.1).

Theorem 6.4.1. *(Bendixson–Dulac theorem) Suppose the first partial derivatives of $P(x, y)$ and $Q(x, y)$ are continuous on some connected subset U in $\mathbb{R}^2$. If*

$$\frac{\partial P}{\partial x} + \frac{\partial Q}{\partial y}$$

is not identically zero and has the same sign in any open set of U, then system (6.4.1) has no closed trajectory in U.

Proof. Suppose the contrary, that is, there is a closed curve C in U. Then by Green's theorem we have

$$\int_C P(x, y)dy - Q(x, y)dx = \int\int_R (\frac{\partial P}{\partial x} + \frac{\partial Q}{\partial y})dxdy, \qquad (6.4.2)$$

where R is the region enclosed by C. Let $x = x(t)$, $y = y(t)$ be a parameterization of C. Then

$$\int_C P(x, y)dy - Q(x, y)dx = \int_0^T (P\frac{dy}{dt} - Q\frac{dx}{dt})dt,$$

where T is the period of C. Using (6.4.1), we arrive at

$$\int_C P(x, y)dy - Q(x, y)dx = \int_0^T (PQ - QP)dt = 0.$$

Thus from (6.4.2) we have

$$\int\int_R (\frac{\partial P}{\partial x} + \frac{\partial Q}{\partial y})dxdy = 0,$$

which can only hold if $\frac{\partial P}{\partial x} + \frac{\partial Q}{\partial y}$ is zero or changes sign, which is a contradiction. Thus C cannot be close, and the system does not admit a limit cycle. This completes the proof. $\qquad \square$

Example 6.11. Let g and h be differentiable functions. Consider the two-dimensional system

$$x' = g(y), \quad y' = (1 + x^2)y + h(x).$$

Then

$$\frac{\partial P}{\partial x} + \frac{\partial Q}{\partial y} = 1 + x^2 > 0, \quad x \in \mathbb{R},$$

and by Theorem 6.4.1 the system does not admit a periodic solution.

The next theorem generalizes Theorem 6.4.1. For its proof, we refer the reader to Exercise 6.20.

Theorem 6.4.2. *(Generalization of Bendixson–Dulac theorem) Suppose the first partial derivatives of $P(x, y)$, $Q(x, y)$, and $h(x, y)$ are continuous on some connected subset U in $\mathbb{R}^2$. If*

$$\frac{\partial(hP)}{\partial x} + \frac{\partial(hQ)}{\partial y}$$

is not identically zero and of one sign in any open set of U in $\mathbb{R}^2$, then system (6.4.1) cannot have a closed trajectory in U.

We have the following example as an application.

Example 6.12. For constants a, b, c, k, consider the two-dimensional system

$$x' = (y - a)x, \quad y' = bx + cy + ky^2.$$

Then

$$\frac{\partial P}{\partial x} + \frac{\partial Q}{\partial y} = y - a + c + 2ky,$$

and hence Theorem 6.4.1 is not applicable. Let us try Theorem 6.4.2 with $h(x, y) = x^{-(2k+1)}$. Then

$$\frac{\partial(hP)}{\partial x} + \frac{\partial(hQ)}{\partial y} = (c + 2ka)x^{-(2k+1)} \neq 0$$

if $c + 2ka \neq 0$, and by Theorem 6.4.2 the system does not admit a periodic solution in the half-plane $x > 0$ or $x < 0$.

In fact, we can do better. If $x = 0$, then starting from any point on the y-axis, the equation has a solution that lies on the y-axis. Thus we conclude that the equation has no closed cycle or trajectory and hence a periodic solution in $\mathbb{R}^2$, since otherwise the uniqueness will be violated.

So far we have discussed results concerning the nonexistence of periodic solutions, or periodic trajectories, or limit cycles. Next, we look at the existence of limit cycles.

Theorem 6.4.3. *A closed orbit or limit cycle of (6.4.1) surrounds at least one equilibrium point of the system. In other words, if no equilibrium point of the system exists in a region, then there can be no periodic orbits in that region.*

Let $(0, 0)$ be the only equilibrium solution of (6.4.1). Moreover, suppose any solution of (6.4.1) that starts near the origin will spiral away from the origin. Since there are no equilibrium solutions, our first thought might be that all solutions of (6.4.1) have trajectories that spiral out to infinity. However, the next theorem, due to Poincaré and Bendixson, presents us a different scenario. It shows, under certain conditions, the existence of a limit cycle. Its proof requires high-level arguments, and we omit it.

Theorem 6.4.4. *(Poincaré–Bendixson) Suppose R is a closed bounded region of the plane enclosed by two simple curves μ_1 and μ_2. If R is free of equilibrium solutions of (6.4.1) and there is a trajectory C confined to R, then either C is a limit cycle, or it spirals toward a limit cycle.*

The main point of the theorem is that if we find one solution that exists for all t large enough (that is, we let t go to infinity) and stays within a bounded region, then we have found either a periodic orbit or a solution that spirals toward a limit cycle; that is, in the long term, the behavior will be very close to a periodic function. We should take the theorem more as a qualitative statement rather than something to help us in computations. In practice, it is hard to find solutions and therefore hard to show rigorously that they exist at all times. Another caveat to consider the theorem is that it only works in two dimensions. In three dimensions and higher, there is simply too much room.

As an application, we provide the following example.

Example 6.13.

$$x' = y + x[1 - x^2 - y^2], \quad y' = -x + y[1 - x^2 - y^2]. \tag{6.4.3}$$

It is clear that $(0, 0)$ is the only equilibrium solution of (6.4.3). Let us pass to polar coordinates. Namely,

$$x = r\cos(\theta), \quad y = r\sin(\theta), \quad \text{where } r^2 = x^2 + y^2,$$

and $r = r(t)$ and $\theta = \theta(t) = \tan^{-1}(\frac{y}{x})$. Thus

$$\frac{d}{dt}(x^2 + y^2) = \frac{d}{dt}r^2$$

or

$$xx' + yy' = rr'.$$

A substitution of x' and y' yields

$$r' = r(1 - r^2),$$

$$\theta' = \frac{xy' - x'y}{x^2 + y^2} = 1.$$

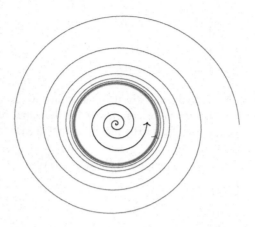

FIGURE 6.9

Limit cycle.

For constants c and k, the solutions are given by

$$r(t) = \frac{1}{\sqrt{1 + ce^{-2t}}}, \; c > -1, \; \theta(t) = -t + k, \; t \geq 0.$$

For $c = 0$, we have that $r = 0$ and $(x(t), y(t)) = (\cos(t), -\sin(t))$ is a periodic solution with trajectory given by the unit circle in red (light gray in print version) in Fig. 6.9. For $c > 0$, we have $0 < r(t) < 1$, and as t increases, $r \nearrow 1$, and the corresponding solution approaches the unit circle from inside of the circle in a spiral way as $t \to \infty$, as indicated in Fig. 6.9. For $-1 < c < 0$, we have $r(t) > 1$, and as t increases, $r \searrow 1$, and the corresponding solution approaches the unit circle from outside of the circle in a spiral way as $t \to \infty$, as indicated in Fig. 6.9. Therefore the unit circle $x^2 + y^2 = 1$, given parametrically by $(x(t), y(t)) = (\cos(t), -\sin(t))$ is a *limit cycle* for (6.4.3), and it is asymptotically stable. Note that $(0, 0)$ is the only equilibrium solution inside the unit circle. As for the construction of the region R, we let $\mu_1 < \mu < \mu_2$, where $0 < \mu_1 < 1$ and $1 < \mu_2 < \infty$, and define the annulus region

$$R = \{(x, y) : \mu_1 < x^2 + y^2 = 1 < \mu_2\}.$$

Clearly, R contains no equilibrium solution, and by Theorem 6.4.4, there is a periodic cycle in R.

6.5 Lotka–Volterra competition model

If two populations $x_1(t)$ and $x_2(t)$ were growing logistically and not affecting each other, their growth can be described by two uncoupled logistic equations

$$\begin{cases} x_1'(t) = r_1 x_1(t)\left(1 - \frac{x_1(t)}{K_1}\right), \\ x_2'(t) = r_2 x_2(t)\left(1 - \frac{x_2(t)}{K_2}\right), \\ x_1(0) > 0, \ x_2(0) > 0, \end{cases}$$

where $r_i, i = 1, 2$, are the growth rates (birth–death), and $K_i, i = 1, 2$, are positive constants. If the two populations were competing for the same resources, then the growth rate of each population is affected by the other. In that case the model becomes

$$\begin{cases} x_1'(t) = r_1 x_1(t)\left(1 - \frac{x_1(t)}{K_1} - \lambda_1 x_2(t)\right), \\ x_2'(t) = r_2 x_2(t)\left(1 - \frac{x_2(t)}{K_2} - \lambda_2 x_1(t)\right), \\ x_1(0) > 0, \ x_2(0) > 0, \end{cases}$$

where $\lambda_i, i = 1, 2$, are positive constants. To simplify the calculations, we convert the system by using non-dimensional variables by measuring x_1 in units of K_1, x_2 in units of K_2, and time in units of $\frac{1}{r_1}$. We let

$$x_1(t) = K_1 x(t), \ x_2(t) = K_2 y(t), \ \text{and} \ t = \frac{s}{r_1}.$$

By the chain rules, it follows that

$$K_1 \frac{dx}{ds} = \frac{dx_1}{dt}\frac{dt}{ds} = \frac{dx_1}{dt}\frac{1}{r_1}$$

or

$$x_1'(t) = r_1 K_1 \frac{dx}{ds}.$$

Substituting this back into the first equation of the model, we arrive at

$$r_1 K_1 \frac{dx}{ds} = r_1 K_1 x(s)\left(1 - x(s) - \lambda_1 K_2 y(s)\right),$$

from which we get

$$\frac{dx}{ds} = x(s)\left(1 - x(s) - \lambda_1 K_2 y(s)\right).$$

Similarly, we obtain

$$\frac{dy}{ds} = r y(s)\left(1 - y(s) - \lambda_2 K_1 x(s)\right),$$

where $r = \frac{r_2}{r_1}$.

Interchanging s with t and letting $\mu_1 = \lambda_1 K_2$ and $\mu_2 = \lambda_2 K_1$, we arrive at the dimensionless Lotka–Volterra competition model

$$\begin{cases} x'(t) = x(t)\big(1 - x(t) - \mu_1 y(t)\big), \\ y'(t) = ry(t)\big(1 - y(t) - \mu_2 x(t)\big), \\ x(0) = x_0 > 0, \; y(0) = y_0 > 0. \end{cases} \qquad (6.5.1)$$

Next, we show that solutions of (6.5.1) are positive and remain bounded for all $t > 0$.

Proposition 6.1. *If $x(t)$, $y(t)$ is a solution pair of (6.5.1) with positive initial conditions x_0 and y_0, respectively, then they are positive and bounded for all $t > 0$.*

Proof. Let $x(t)$, $y(t)$ be a solution of (6.5.1) with positive initial conditions x_0 and y_0. Dividing the first equation of (6.5.1) by $x(t)$ and the second equation by $y(t)$, respectively, and integrating, we get the integral relations

$$x(t) = x_0 e^{\int_0^t \big(1 - x(s) - \mu_1 y(s)\big) ds}$$

and

$$y(t) = y_0 e^{r \int_0^t \big(1 - y(s) - \mu_2 s(s)\big) ds}.$$

We see that if the initial conditions are positive, then the components of the solution are positive. It is left to show that the solutions remain bounded for all $t > 0$. Since x and y are positive, we have that

$$x(t)\big(1 - x(t) - \mu_1 y(t)\big) \le x(t)\big(1 - x(t)\big)$$

and

$$ry(t)\big(1 - y(t) - \mu_2 x(t)\big) \le ry(t)\big(1 - y(t)\big).$$

It follows that

$$x'(t) \le x(t)\big(1 - x(t)\big) \; \text{and} \; y'(t) \le ry(t)\big(1 - y(t)\big).$$

From now on, we write x for $x(t)$ and y for $y(t)$. Consider the differential equation

$$z' = z(1 - z), \; z(0) = x(0) = x_0.$$

Separating the variables and then integrating, we get $\int \frac{dz}{z(1-z)} = \int dt$. Using partial fractions on the left side integral, we obtain $\text{Ln}|\frac{z}{1-z}| = t + C$. Taking the exponential on both sides, solving for z, and then applying the initial condition, we arrive at the unique solution

$$z(t) = \frac{x_0 e^t}{1 + x_0(e^t - 1)} \to 1 \; \text{as} \; t \to \infty.$$

Since z is positive and has a limit, we conclude that z is bounded for all $t > 0$. Moreover,

$$z(t) \leq 1 \text{ as } t \to \infty.$$

From this discussion we arrive at

$$x(t) \leq z(t) \leq 1.$$

Similar results can be obtained for $y(t)$. This concludes the proof. $\square$

Setting the right sides of (6.5.1) equal to zero, we obtain the equilibrium solutions

$$E_0 = (0, 0), \ E_1 = (0, 1), \ E_2 = (1, 0),$$

and

E_3, which comprises of the solutions (if they exist) of $x + \mu_1 y = 1$, $\mu_2 x + y = 1$.

Next, we use Theorem 6.2.1 to analyze the stability of the equilibrium solutions. The Jacobian matrix is

$$J(x, y) = \begin{pmatrix} 1 - 2x - \mu_1 y & -\mu_1 x \\ -r\mu_2 y & r(1 - 2x - \mu_2 x) \end{pmatrix}.$$

For $E_0 = (0, 0)$, we have

$$J(0, 0) = \begin{pmatrix} 1 & 0 \\ 0 & r \end{pmatrix}$$

with eigenvalues $\lambda = 1, r$, and hence by Theorem 6.2.1 and Remark 6.3 E_0 is unstable or repeller. At E_1,

$$J(0, 1) = \begin{pmatrix} 1 - \mu_1 & 0 \\ -r\mu_2 & -r \end{pmatrix}$$

with eigenvalues $\lambda = 1 - \mu_1$ and $-r$, and hence by Theorem 6.2.1 and Remark 6.3 E_1 is either a saddle or asymptotically stable. Similarly, at E_2,

$$J(1, 0) = \begin{pmatrix} -1 & -\mu_1 \\ 0 & r(1 - \mu_2) \end{pmatrix}$$

with eigenvalues $\lambda = -1$ and $r(1 - \mu_2)$, and hence by Theorem 6.2.1 and Remark 6.3 E_2 is either a saddle or asymptotically stable. For, E_3 we solve the system

$$x + \mu_1 y = 1, \ \mu_2 x + y = 1.$$

After some calculations, we obtain

$$E_3 = (\frac{1 - \mu_1}{1 - \mu_1 \mu_2}, \frac{1 - \mu_2}{1 - \mu_1 \mu_2}).$$

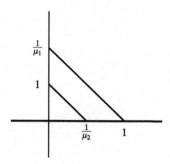 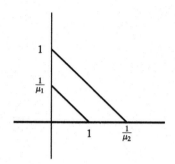

Case: (*a*) $\mu_1 < 1$, $\mu_2 > 1$ *Case*: (*b*) $\mu_1 > 1$, $\mu_2 < 1$

FIGURE 6.10

No equilibrium points in the first quadrant.

The stability of the equilibrium solution of E_3 is important to biologists since it shows the coexistence of both species. We will analyze the stability by considering different cases that are based on the eigenvalues.

Case (a): $\mu_1 < 1$, $\mu_2 > 1$.
In this case the lines $x + \mu_1 y = 1$ and $\mu_2 x + y = 1$ do not intersect in the first quadrant, and it follows that E_3 is not an equilibrium solution as depicted in Fig. 6.10. In addition, in Case (a), E_2 is asymptotically stable, and E_1 is a saddle point. Since the interior of the first quadrant does not contain any equilibrium solution, it follows from Theorem 6.4.3 that there is no periodic solution in the first quadrant. Recall that by Theorem 6.4.3 a closed orbit or limit cycle surrounds at least one equilibrium point. Thus by Theorem 6.4.3 the omega limit set of any trajectory is the equilibrium solution E_2, and hence E_2 is globally asymptotically stable; see Fig. 6.11. The biological meaning is that population x wins the competition. In other words, population y becomes instinct.

Case (b): $\mu_1 > 1$, $\mu_2 < 1$.
In this case the roles of E_1 and E_2 are reversed, but still only one competitor survives.

Next, we consider the cases where E_3 is in the first quadrant, that is, E_3 is an equilibrium solution.

It turned out that E_3 is an equilibrium solution in two cases.

Case (c): $\mu_1 < 1$, $\mu_2 < 1$ and **Case (d):** $\mu_1 > 1$, $\mu_2 > 1$.
Next, we make use of Theorem 6.4.2 (generalization of the Bendixson–Dulac theorem) to study the possibilities of the existence of a periodic solution that surrounds E_3. Let $h(x, y) = \frac{1}{xy}$. Then h is a continuously differentiable function in $G = \{(x, y) : x > 0, y > 0\}$ with

$$\frac{\partial}{\partial x}\left[\frac{1}{xy}x(1 - x - \mu_1 y)\right] = -\frac{1}{y}$$

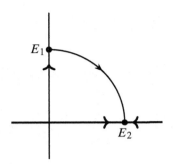

FIGURE 6.11

E_2 is globally asymptotically stable.

and

$$\frac{\partial}{\partial y}\Big[\frac{1}{xy}ry\big(1-y-\mu_2 x\big)\Big]=-\frac{r}{x}.$$

Thus

$$\frac{\partial}{\partial x}\Big[\frac{1}{xy}x\big(1-x-\mu_1 y\big)\Big]+\frac{\partial}{\partial y}\Big[\frac{1}{xy}ry\big(1-y-\mu_2 x\big)\Big]=-\Big(\frac{1}{y}+\frac{r}{x}\Big),$$

which is negative in G. Therefore there is no periodic solution in the first quadrant in both case (c) and case (d).

Let (x^*, y^*) be a solution of $x + \mu_1 y = 1$, $\mu_2 x + y = 1$, that is, $(x^*, y^*) = (\frac{1-\mu_1}{1-\mu_1\mu_2}, \frac{1-\mu_2}{1-\mu_1\mu_2})$. Then

$$
\begin{aligned}
J(x^*, y^*) &= \begin{pmatrix} 1-2x^*-\mu_1 y^* & -\mu_1 x^* \\ -r\mu_2 y^* & r(1-2y^*-\mu_2 x^*) \end{pmatrix} \\[4pt]
&= \begin{pmatrix} 1-2x^*-\mu_1\frac{1-\mu_2}{1-\mu_1\mu_2} & -\mu_1 x^* \\ -r\mu_2 y^* & r(1-2y^*-\mu_2\frac{1-\mu_1}{1-\mu_1\mu_2}) \end{pmatrix} \\[4pt]
&= \begin{pmatrix} -x^* & -\mu_1 x^* \\ -r\mu_2 y^* & -ry^* \end{pmatrix},
\end{aligned}
$$

and $\mathrm{tra}(J(x^*, y^*)) = -x^* - ry^* < 0$ for $x^*, y^* > 0$ implies that the sum of the eigenvalues is negative. Moreover,

$$\det(J(x^*, y^*)) = rx^* y^* - r\mu_1\mu_2 x^* y^* = rx^* y^*(1-\mu_1\mu_2).$$

As

$$\det(J(x^*, y^*) - \lambda I) = \lambda^2 - \mathrm{tra}(J(x^*, y^*))\lambda + \det(J(x^*, y^*)),$$

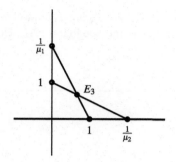

Case(c) : $\mu_1 < 1,\ \mu_2 < 1$

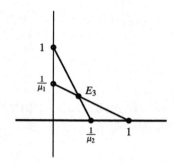

Case(d) : $\mu_1 > 1,\ \mu_2 > 1$

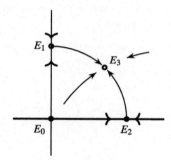

(e) : $\mu_1 < 1,\ \mu_2 < 1,\ 1 - \mu_1\mu_2 > 0$

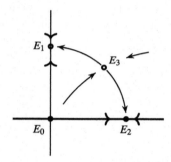

(f) : $\mu_1 > 1,\ \mu_2 > 1,\ 1 - \mu_1\mu_2 < 0$

FIGURE 6.12

Coexistence.

we have that

$$\lambda^2 + (x^* + ry^*)\lambda + rx^*y^*(1 - \mu_1\mu_2) = 0.$$

It follows that the eigenvalues are given by

$$\lambda = \frac{-x^* - ry^* \pm \sqrt{(x^* + ry^*)^2 - 4rx^*y^*(1 - \mu_1\mu_2)}}{2}.$$

So E_3 is asymptotically stable if $1 - \mu_1\mu_2 > 0$ and a saddle if $1 - \mu_1\mu_2 < 0$. Fig. 6.12(d) and (f).

If $\mu_1\mu_2 = 1$, then the two lines are parallel or coincident. If coincident, then there is a line of equilibrium solutions, and if parallel, then cases (a) and (b) apply. In case (c), $1 - \mu_1 > 0$ and $1 - \mu_2 > 0$, so E_1 and E_2 are saddle unstable points. Since $\mu_1\mu_2 < 1$, E_3 is an attractor, that is, all solutions tend to E_3. In that case the two populations coexist. See Fig. 6.12(d).

In case (d), $\mu_1 - 1 > 0$ and $\mu_2 - 1 > 0$, and hence E_1 and E_2 are asymptotically stable. Also, in case (d) $\mu_1\mu_2 > 1$, and so E_3 is a saddle point. See Fig. 6.12(f). Here the initial conditions determine the winner.

6.6 Bifurcation in planar systems

We consider the two-dimensional autonomous system

$$x' = P(x, y, \mu), \ \ y' = Q(x, y, \mu) \tag{6.6.1}$$

depending upon a real parameter μ. We assume that P and Q are continuous in all arguments on some subset of $\mathbb{R}^2$. As μ changes values or signs, the behavior of solutions changes and may result in bifurcation points. For example, as we move across a particular equilibrium solution, the stability may change from stable to unstable, and vice versa. We begin by considering the following example.

Example 6.14.　Consider the two-dimensional linear system

$$x' = x + \mu y, \ y' = x - y$$

for parameter μ. The matrix

$$\begin{pmatrix} 1 & \mu \\ 1 & -1 \end{pmatrix}$$

has the eigenvalues

$$\lambda = \pm\sqrt{1 + \mu}.$$

If $\mu > -1$, then the eigenvalues are real with opposite signs, and therefore the origin $(0, 0)$ is a saddle point (unstable). For the particular value $\mu = -1$, the system becomes

$$x' = x - y, \ y' = x - y$$

and results in the whole line $y = x$ of fixed points. If $\mu < -1$, then the eigenvalues are purely imaginary, and the origin $(0, 0)$ is a center (stable). As $\mu > -1$ decreases to $\mu < -1$, the nature of the solution changes at $\mu = -1$; as a result, the equilibrium solution $(0, 0)$ changes from an unstable saddle to a stable center as μ passes the critical value -1. Thus we have a *bifurcation* at $\mu = -1$.

Example 6.15.　Consider the planar autonomous system

$$x' = -y - x(x^2 + y^2 - \mu), \ \ y' = x - y(x^2 + y^2 - \mu).$$

It is clear that $(0, 0)$ is the only equilibrium point. We rewrite the system in the form

$$\begin{pmatrix} x \\ y \end{pmatrix}' = \begin{pmatrix} \mu & -1 \\ 1 & \mu \end{pmatrix}\begin{pmatrix} x \\ y \end{pmatrix} + \begin{pmatrix} -x(x^2 + y^2) \\ -y(x^2 + y^2) \end{pmatrix}.$$

The eigenvalues of

$$A = \begin{pmatrix} \mu & -1 \\ 1 & \mu \end{pmatrix}$$

are $\mu \pm i$.

(1) If $\mu < 0$, then the trajectories spiral toward the origin, and the origin is asymptotically stable by Theorem 6.2.1.

(2) If $\mu > 0$, then the origin is unstable for the nonlinear system.

(3) If $\mu = 0$, then Theorem 6.2.1 says nothing about the stability or instability about the origin of the nonlinear system.

Next we examine in more detail the trajectories of the system by using polar coordinates as we did in Example 6.8. By doing so we arrive at

$$r' = r(\mu - r^2) \text{ and } \theta' = 1. \tag{6.6.2}$$

Separating the variables and using partial fractions we arrive at

$$\int \frac{dr}{r(r - \sqrt{\mu})(r + \sqrt{\mu})} = t + k \text{ for some constant } k.$$

This leads to the solution

$$r(t) = \frac{\sqrt{\mu}}{\sqrt{1 + ce^{-2\mu t}}}. \tag{6.6.3}$$

Similarly, if we choose $\theta(0) = 0$, then we have $\theta(t) = t$.

Assume that $\mu > 0$ and consider the following three cases.

(3a) When $c = 0$, we have from (6.6.3) $r(t) = \sqrt{\mu}$, which represents a periodic solution with radius $\sqrt{\mu}$.

(3b) If $c < 0$, then (6.6.3) represents an orbit or trajectory that spirals counterclockwise toward the circle $r(t) = \sqrt{\mu}$ from the outside.

(3c) If $c > 0$, then (6.6.3) represents an orbit or trajectory that spirals counterclockwise toward the circle $r(t) = \sqrt{\mu}$ from the inside.

Let $0 < \mu_1 < \mu < \mu_2$ and define the annulus region

$$R = \{(x, y) : \mu_1 < x^2 + y^2 < \mu_2\}.$$

Clearly, R contains no equilibrium solution, and then cases (3a)–(3c) prove that the circle given by $r^2(t) = \mu$ or $x^2 + y^2 = \mu$ is a *stable limit cycle* and the origin is unstable.

Next, we examine the stability of the origin when $\mu = 0$.

In this case (6.6.2) gives

$$r' = -r^3,$$

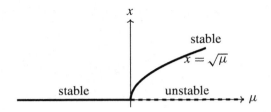

FIGURE 6.13

Bifurcation diagram for Example 6.15.

which has the solution

$$r(t) = \frac{1}{\sqrt{2t + K}}$$

for a suitable constant K. Clearly, $r(t) \to 0$ as $t \to \infty$, and hence the zero solution (or the origin) is asymptotically stable (spirally stable).

Finally, for $\mu < 0$, from (6.6.2) we have

$$r' = r(r^2 + l^2) \text{ and } \theta' = 1,$$

where $l^2 = -\mu > 0$. Separating the variables and using partial fractions, we arrive at

$$\int \frac{dr}{r(r^2 + l^2)} = -dt \text{ for some constant } k.$$

This leads to the solution

$$r(t) = \frac{cl^2 e^{-2l^2 t}}{1 - cl^2 e^{-2l^2 t}} \text{ for some constant } c.$$

Clearly, $r(t) \to 0$, as $t \to \infty$, and hence the zero solution (or the origin) is asymptotically stable (spirally stable). In conclusion, for $\mu \leq 0$, the origin $(0, 0)$ is stable, and for $\mu > 0$, $(0,0)$ is unstable, and $r = \sqrt{\mu}$ is stable. Thus as μ passes through the critical value $\mu = 0$ from negative to positive, the origin changes from stable spiral to unstable spiral, and there appears a new limit cycle that is a periodic solution. This type of bifurcation is common and known as *Hopf bifurcation*.

The bifurcation diagram is shown in Fig. 6.13.

The next theorem provides a necessary condition for the existence of Hopf bifurcation. The proof is complicated, and hence we omit it.

Theorem 6.6.1. *Consider the two-dimensional autonomous system*

$$x' = P(x, y, \mu), \; y' = Q(x, y, \mu) \tag{6.6.4}$$

depending upon a real parameter μ, where P and Q are differentiable in the neighborhood of $(0, 0, 0)$. Assume that for small μ,

$$P(0, 0, \mu) = Q(0, 0, \mu) = 0. \tag{6.6.5}$$

Let the Jacobian

$$J(0, 0, 0) = \begin{pmatrix} P_x(0, 0, 0) & P_y(0, 0, 0) \\ Q_x(0, 0, 0) & Q_y(0, 0, 0) \end{pmatrix} = \begin{pmatrix} 0 & -w \\ w & 0 \end{pmatrix} \text{ for some } w \neq 0. \tag{6.6.6}$$

Let

$$a = \frac{1}{16}\left(P_{xxx} + Q_{xxy} + P_{xyy} + Q_{yyy}\right)$$

$$+ \frac{1}{16w}\left(P_{xy}(P_{xx} + P_{yy}) - Q_{xy}(Q_{xx} + Q_{yy}) - P_{xx}Q_{xx} + P_{yy}Q_{yy}\right) \tag{6.6.7}$$

evaluated at $(0, 0, 0)$ with $a \neq 0$. Suppose that

$$a\left(P_{\mu x} + Q_{\mu y}\right) > 0. \tag{6.6.8}$$

Then for sufficiently small $\mu < 0$, there is a unique periodic solution with amplitude $A(\mu)$ such that

$$\lim_{\mu \to 0^-} \frac{A(\mu)}{\sqrt{|\mu|}} = c > 0.$$

If

$$a\left(P_{\mu x} + Q_{\mu y}\right) < 0, \tag{6.6.9}$$

then the periodic solution exists for $\mu > 0$. The stability of the periodic solution is determined by the sign of

$$P_{\mu x} + Q_{\mu y} \tag{6.6.10}$$

and is the opposite of the stability of the zero solution in the range of μ where the periodic solution exists.

Example 6.16. As an application of Theorem 6.6.1, we consider the system given in Example 6.14. It readily follows that the Jacobian

$$J(0, 0, 0) = \begin{pmatrix} \mu & -1 \\ 1 & \mu \end{pmatrix} = \begin{pmatrix} 0 & -1 \\ 1 & 0 \end{pmatrix}.$$

Thus condition (6.6.6) is satisfied with $w = 1$. After simple calculations, it follows that (6.6.7) evaluated at $(0, 0, 0)$ implies that $a = -1$. Moreover,

$$a\left(P_{\mu x} + Q_{\mu y}\right) = -1(1 + 1) < 0.$$

Thus by Theorem 6.6.1, the system possesses a periodic solution for $\mu > 0$ that is stable (spirally stable) since the origin is unstable in the region where μ is positive.

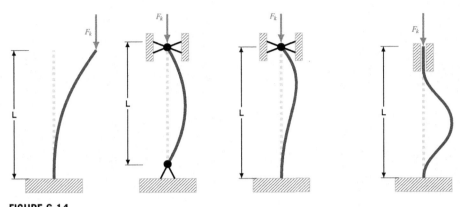

FIGURE 6.14

Buckling beam.

We end this section with an application to Euler's buckling beam.

Example 6.17. (Euler's buckling beam) One of the most famous applications to bifurcations is the so-called Euler's buckling beam. Consider different scenarios of beams in Fig. 6.14, where excessive pressure is applied to the beam. The beam can support a small force or pressure, but as the force is increased, the beam will go through a qualitative change in its vertical standing position. At one point, as we keep increasing the force, the beam position will get compromised and results in a so-called *buckling,* that is, there will be a critical value at which if the pressure is increased beyond it, then the beam will buckle. Fig. 6.14 provides different situations for such buckling beams along with supported ends. As for modeling, suppose we have a beam of length one unit and the beam is placed along the horizontal axis with one end supported by a fixed wall and a force $\mathbb{F} \geq 0$ applied along its axis. Let $\delta > 0$ be the viscous damping, and let $K > 0$ be the stiffness. The small deflection of the beam is denoted by $v(x, t)$, where x denotes the horizontal axes. After transforming the model from a partial differential equation to a second-order differential equation, it was shown in [27] that the amplitude u in variable time t satisfies the second-order differential equation

$$u''(t) + \delta u'(t) - \pi^2(\mathbb{F} - \pi^2)u(t) + \frac{1}{2}K\pi^4 u^3(t) = 0. \qquad (6.6.11)$$

To better study the differential equation, we put it in the system form

$$\begin{cases} x' = y, \\ y' = \pi^2(\mathbb{F} - \pi^2)x - \delta y - \frac{1}{2}K\pi^4 x^3, \end{cases} \qquad (6.6.12)$$

where the force $\mathbb{F}$ is the only parameter. The system has three equilibrium solutions

$$(0,0), \quad \left(\sqrt{\frac{2(\mathbb{F}-\pi^2)}{K\pi^2}},0\right), \quad \text{and} \quad \left(-\sqrt{\frac{2(\mathbb{F}-\pi^2)}{K\pi^2}},0\right).$$

Note that the equilibrium solutions

$$\left(\sqrt{\frac{2(\mathbb{F}-\pi^2)}{K\pi^2}},0\right) \text{ and } \left(-\sqrt{\frac{2(\mathbb{F}-\pi^2)}{K\pi^2}},0\right)$$

are only valid for $\mathbb{F} > \pi^2$.

Now

$$J(0,0) = \begin{pmatrix} 0 & 1 \\ \pi^2(\mathbb{F}-\pi^2) & -\delta \end{pmatrix}$$

with eigenvalues

$$\lambda = \frac{-\delta \pm \sqrt{\delta^2 + 4\pi^2(\mathbb{F}-\pi^2)}}{2},$$

and, as a consequence, $(0,0)$ is stable if $\mathbb{F} < \pi^2$ and unstable if $\mathbb{F} > \pi^2$. The physical implication is that the beam will buckle if $\mathbb{F} > \pi^2$.

Now we consider the equilibrium solution $\left(\sqrt{\frac{2(\mathbb{F}-\pi^2)}{K\pi^2}},0\right)$ for $\mathbb{F} > \pi^2$. We translate it to the origin with transformation

$$x_1 = x - \sqrt{\frac{2(\mathbb{F}-\pi^2)}{K\pi^2}}, \quad y_1 = 0$$

and arrive at the linearized matrix

$$J(0,0) = \begin{pmatrix} 0 & 1 \\ -2\pi^2(\mathbb{F}-\pi^2) & -\delta \end{pmatrix}$$

with eigenvalues

$$\lambda = \frac{-\delta \pm \sqrt{\delta^2 - 8\pi^2(\mathbb{F}-\pi^2)}}{2}.$$

It follows that $(0,0)$ is stable if $\mathbb{F} > \pi^2$ and that the equilibrium solution $\left(\sqrt{\frac{2(\mathbb{F}-\pi^2)}{K\pi^2}},0\right)$ is stable for $\mathbb{F} > \pi^2$. The physical implication is that the beam will buckle if $\mathbb{F} > \pi^2$.

Similarly, we can easily show that the equilibrium solution $\left(-\sqrt{\frac{2(\mathbb{F}-\pi^2)}{K\pi^2}},0\right)$ is also stable for $\mathbb{F} > \pi^2$. Thus we have a pitchfork bifurcation at $\mathbb{F} = \pi^2$. The bifurcation diagram in Fig. 6.15 shows that bending is symmetric and can take place in the negative direction as well.

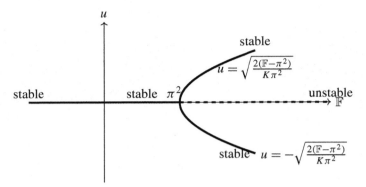

FIGURE 6.15

Bifurcation diagram for $u = \pm\sqrt{\frac{2(\mathbb{F}-\pi^2)}{K\pi^2}}$.

6.7 Manifolds and Hartman–Grobman theorem

In the study of dynamical systems and differential equations, the stable manifold theorem is an important result about the structure of the set of orbits approaching a given hyperbolic fixed point. Stable manifolds are directly related to the important Hartman–Grobman theorem, which states that the flow generated by a smooth vector field in a neighborhood of a hyperbolic equilibrium point is topologically conjugate with the flow generated by its linearization. Hartman's counterexample shows that, in general, the conjugacy cannot be taken to be C^1. However, the stable manifold theorem tells us that there are important structures for the two flows that can be matched up by smooth changes of variables.

Let $f : E \subseteq \mathbb{R}^n \to \mathbb{R}^n$ be C^1 and consider the autonomous nonlinear differential equation

$$x' = f(x). \tag{6.7.1}$$

Definition 6.7.1. An equilibrium point x_0 of (6.7.1) is called *hyperbolic* if none of the eigenvalues of the matrix $Df(x_0)$ has a zero real part, where throughout this section, we adopt the notation

$$Df(x_0) = J(x_0).$$

For example, the system

$$x' = y, \quad y' = -x - x^3 - 3y$$

has $(0,0)$ as its only equilibrium solution. The linearization at $(0,0)$ is

$$Df(0,0) = \begin{pmatrix} 0 & 1 \\ -1 & -3 \end{pmatrix},$$

which has the eigenvalues

$$\lambda = \frac{-2 \pm \sqrt{5}}{2}.$$

Thus $(0, 0)$ is a hyperbolic equilibrium solution since none of the eigenvalues has the zero real part.

Our aim is to show that the local behavior of the nonlinear system (6.7.1) near a hyperbolic equilibrium point x_0 is qualitatively determined by the behavior of the linear system

$$x' = Ax, \tag{6.7.2}$$

where the matrix $A = Df(x_0)$. Now we define the flow or trajectory of (6.7.1) that suits our notations for this section. Denote the maximal interval of existence (α, β) of the solution $\phi(t, x_0)$ of the IVP

$$x' = f(x), \quad x(0) = x^* \tag{6.7.3}$$

by $I(x^*)$ since α or β usually depend on x^*. If $x^* \in E$ and $t \in I(x^*)$, then the set of mappings ϕ_t defined by

$$\phi_t(x^*) = \phi(t, x^*)$$

is called the flow of (6.7.1). We define the set $\Omega \subseteq \mathbb{R} \times E$ as

$$\Omega = \{(t, x^*) \in \mathbb{R} \times E : t \in I(x^*)\}.$$

We leave it an exercise to show that the set Ω is an open subset of $\mathbb{R} \times E$ and $\phi \in C^1(\Omega)$.

Theorem 6.7.1. *Let $E \subseteq \mathbb{R}^n$ be open with $f \in C^1(E)$, and let $\phi_t : E \to E$ be the flow of (6.7.1) defined for all $t \in \mathbb{R}$.*
(i) The set $S \subseteq E$ is said to be invariant with respect to the flow ϕ_t if $\phi_t(S) \subseteq S$ for all $t \in \mathbb{R}$.
(ii) The set E is said to be positively (or negatively) invariant with respect to the flow ϕ_t if $\phi_t(S) \subseteq S$ for all $t > 0$ (or $t < 0$).

We furnish the following example.

Example 6.18. Consider the nonlinear system of differential equations

$$x' = \begin{pmatrix} x_1 + x_2^2 \\ -x_2 \end{pmatrix}$$

with

$$x(0) = \begin{pmatrix} a \\ b \end{pmatrix} = x^*.$$

Now $x_2' = -x_2$ has the solution $x_2 = be^{-t}$. Substituting this solution of x_2 into the first differential equation and using the variation of parameters formula, we arrive at $x_1 = ae^t + \frac{1}{3}b^2(e^t - e^{-2t})$. Thus

$$\phi_t(x^*) = \begin{pmatrix} ae^t + \frac{1}{3}b^2(e^t - e^{-2t}) \\ be^{-t} \end{pmatrix}.$$

Next, we show that the set

$$S = \{x \in \mathbb{R}^2 : x_1 = -x_2^2/3\}$$

is invariant under the flow ϕ_t. Let $x^* \in S$. Then $a = -b^2/3$, and it follows that

$$\phi_t(x^*) = \begin{pmatrix} -\frac{1}{3}b^2 e^{-2t} \\ be^{-t} \end{pmatrix} \in S.$$

Therefore $\phi_t(S) \subseteq S$ for all $t \in \mathbb{R}$. We will see later that the set S will be called the *stable manifold* for system (6.7.1).

We are ready to formally define stable and unstable manifolds.

Consider the linear system given by (6.7.2) and suppose the eigenvalues of A and their corresponding generalized eigenvectors are $\lambda_j = \alpha_j + i\beta_j$ and $k_j + i \, d_j$ for $j = 1, 2, \ldots, n$. Let those vectors be the columns of the matrix P. Then $P^{-1}AP = J$, where J is the Jordan form of A. Then the sets

$$E^s := Span\{k_j, d_j : \alpha_j < 0\},$$

$$E^u := Span\{k_j, d_j : \alpha_j > 0\},$$

$$E^c := Span\{k_j, d_j : \alpha_j = 0\}$$

are the *stable, unstable*, and *center* subspaces for system (6.7.2).

Definition 6.7.2. Suppose that x_0 is a hyperbolic equilibrium point of (6.7.1).
(i) The *stable manifold* of x_0 is the set

$$W^s(x_0) = \{x^* \in E : \lim_{t \uparrow \infty} \phi_t(x^*) = x_0\}.$$

(ii) The *unstable manifold* of x_0 is the set

$$W^u(x_0) = \{x^* \in E : \lim_{t \downarrow -\infty} \phi_t(x^*) = x_0\}.$$

We have the following example.

Example 6.19. Consider the system given in Example 6.18. Then the eigenvalues of $A = Df(0, 0)$ are $\lambda = \pm 1$. If $\lambda = 1$, then its corresponding eigenvector is given by $v_1 = (1, 0)^T$. Similarly, $v_2 = (0, 1)^T$ is the corresponding eigenvector for $\lambda = -1$. Thus

$$E^u = \{(x, y) \in \mathbb{R}^2 : y = 0\}$$

and

$$E^s = \{(x, y) \in \mathbb{R}^2 : x = 0\}.$$

Now $\lim_{t \uparrow \infty} \phi_t(x^*) = x_0$ if and only if $a + b^2/3 = 0$. Thus the stable manifold is

$$W^s(x_0) = \{x^* \in \mathbb{R}^2 : a = -b^2/3\}.$$

Similarly, $\lim_{t \downarrow -\infty} \phi_t(x^*) = (0, 0)$ if and only if $b = 0$. Thus the unstable manifold is

$$W^u(0, 0) = \{x^* \in \mathbb{R}^2 : b = 0\}.$$

6.7.1 The stable manifold theorem

The stable manifold theorem that we are about to state and give a sketch of its proof states that (6.7.1) has stable and unstable manifolds W^s and W^u that are tangent at x_0 to the stable and unstable subspaces E^s and E^u of the linearized system (6.7.2), where x_0 is a hyperbolic equilibrium solution of (6.7.1). Furthermore, W^s and W^u are of the same dimensions as E^s and E^u. Besides, W^s and W^u are positively and negatively invariant under ϕ_t, respectively, and satisfy

$$\lim_{t \uparrow \infty} \phi_t(x^*) = x_0 \quad \text{for all } x^* \in W^s,$$

$$\lim_{t \downarrow -\infty} \phi_t(x^*) = x_0 \quad \text{for all } x^* \in W^u.$$

We furnish the following example that should serve as warm up.

Example 6.20. Let $g \in C^1(E)$, where $E \subseteq \mathbb{R}^2$ is an open set containing $(0, 0)$. Consider the nonlinear system

$$\begin{cases} x' = -x, \\ y' = y + g(x), \end{cases}$$

where we assume that $g(0) := g(0, 0) = 0$ and that the function g may contain linear terms. The only equilibrium solution of the system is the origin and the Jacobian is

$$Df(0, 0) = \begin{pmatrix} -1 & 0 \\ g'(0) & 1 \end{pmatrix}.$$

Then the eigenvalues of $A = Df(0, 0)$ are $\lambda = \pm 1$, and the zero equilibrium is hyperbolic. If $\lambda = 1$, then its corresponding eigenvector is given by $v_1 = (0, 1)^T$. Similarly, $v_2 = (2, -g'(0))^T$ is the corresponding eigenvector for $\lambda = -1$. Thus

$$E^u = \{(x, y) \in \mathbb{R}^2 : x = 0\}$$

and

$$E^s = \{(x, y) \in \mathbb{R}^2 : y = -\frac{1}{2}g'(0)x\}.$$

To find the stable and unstable manifolds, we directly solve the system. Assume an initial point $x(0) = a$, $y(0) = b$ for $a, b \in \mathbb{R}$. The first equation in the system has the solution

$$x(t) = ae^{-t}.$$

Substituting x back into the second equation and using the variation of parameters formula, we arrive at

$$y(t) = be^t + e^t \int_0^t e^{-s} g(ae^{-s}) ds.$$

Making the substitution $u = e^{-s}$, we arrive at the flow

$$\phi_t(c) = \begin{pmatrix} ae^{-t} \\ be^t + e^t \int_{e^{-t}}^1 g(au) du \end{pmatrix}.$$

If $c = (a, b)^T \in \mathbb{R}^2$ is in the unstable manifold $W^u(0, 0)$, then $\lim_{t \downarrow -\infty} \phi_t(c) \to 0$. Then from the first component of ϕ_t it follows that $\lim_{t \downarrow -\infty} ae^{-t} = 0$ if and only if $a = 0$. Substituting $a = 0$ into the second component of ϕ_t and taking the limit, that is, $\lim_{t \uparrow \infty} (be^1 + e^t \int_{e^{-t}}^t g(0) ds) = \lim_{t \downarrow -\infty} be^t = 0$ (since $g(0) = 0$), it follows that

$$W^u(0, 0) = \{(x, y) \in \mathbb{R}^2 : x = 0\}.$$

Similarly, if $c = (a, b)^T \in \mathbb{R}^2$ is in the stable manifold $W^s(0, 0)$, then $\lim_{t \uparrow \infty} \phi_t(c) \to 0$. Then from the first component of ϕ_t it follows that $\lim_{t \uparrow \infty} ae^{-t} = 0$ for any a. From the second component of ϕ_t we have

$$\lim_{t \uparrow \infty} e^t \left(b + e^t \int_{e^{-t}}^t g(0) ds \right) = 0. \tag{6.7.4}$$

To that effect, we claim that

$$b = - \int_0^1 g(au) du.$$

To see that this satisfies (6.7.4), we first notice that

$$\int_0^1 g(au)du = \int_0^{e^{-t}} g(au)du + \int_{e^{-t}}^1 g(au)du.$$

Then it follows

$$\lim_{t\uparrow\infty} e^t\left(\int_{e^{-t}}^1 g(au)du - \int_0^1 g(au)du\right) = -\lim_{t\uparrow\infty} e^t\int_0^{e^{-t}} g(au)du.$$

We must show that the term $e^t\int_0^{e^{-t}} g(au)du$ is small for all $t > T$ for sufficiently large T. Since g is continuous with $g(0) = 0$, given $\varepsilon > 0$, there exists $\delta > 0$ such that $|g(au)| < \varepsilon$ for $|au| < \delta$. Moreover, as $0 \le u \le e^{-t}$, we have $|au| = |a||u| \le |a|e^{-t}$. Thus we can choose $T > 0$ sufficiently large so that $|a|e^{-t} < \delta$ for $t > T$. Therefore $\left|e^t\int_0^{e^{-t}} g(au)du\right| < \varepsilon$. As this can be done for any $\varepsilon > 0$, we conclude that (6.7.4) holds. It follows that the stable manifold is

$$W^s(0,0) = \{(x, y(x)) \in \mathbb{R}^2 : y(x) = -\int_0^1 g(xu)du\}.$$

Note that we could easily use L'Hospital's rule on $\lim_{t\uparrow\infty} e^t\int_0^{e^{-t}} g(xu)du$ to show that (6.7.4) holds. Next, we show that W^s is tangent to E^s:

$$\frac{dy}{dx}\Big|_{x=0} = -\int_0^1 g'(xu)udu\Big|_{x=0}.$$

Making the substitution $v = xu$, this expression becomes

$$\frac{dy}{dx}\Big|_{x=0} = -\frac{1}{x^2}\int_0^x g'(v)vdv\Big|_{x=0}.$$

Using L'Hospital's rule, we find the slope at the origin to be

$$-\frac{1}{x^2}\int_0^x g'(v)vdv\Big|_{x=0} = -\lim_{x\to0}\frac{1}{x^2}\int_0^x g'(v)vdv = -\lim_{x\to0}\frac{g'(x)x}{2x} = -\frac{1}{2}g'(0).$$

Thus W^s is tangent to E^s at the origin, and, moreover, $W^s \in C^1$.

Let $g(x) = xe^x$ in Example 6.20. Then $g(0) = 0$ and $g \in C^1$. Furthermore,

$$E^s = \{(x, y) \in \mathbb{R}^2 : y = -\frac{1}{2}g'(0)x\} = \{(x, y) \in \mathbb{R}^2 : y = -\frac{1}{2}x\},$$

and

$$y(x) = -\int_0^1 g(xu)du = \int_0^1 xue^{xu}du = -e^x + \frac{1}{x}e^x - \frac{1}{x}.$$

Substitute the Maclaurin series for e^x into the last expression and simplify to get

$$y = \frac{-x}{2}\left(1 + \frac{2x}{3} + \frac{x^2}{3} + \cdots\right).$$

Thus

$$W^s(0,0) = \left\{(x, y(x)) \in \mathbb{R}^2 : y(x) = \frac{-x}{2}\left(1 + \frac{2x}{3} + \frac{x^2}{3}\right)\right\}.$$

In Example 6.20, we were able to compute the solutions to find W^s and W^u. This is not the case for most nonlinear systems. The next theorem affirms the existence of such manifolds in the neighborhoods of hyperbolic equilibrium solution, regardless of whether solutions can be found or not. We have seen that the stable manifold of a hyperbolic critical point is the collection of all states that asymptotically approach that fixed point as time goes to infinity. Knowing the stable and unstable manifolds of a fixed point can be useful for simple purposes, for example, for knowing which initial conditions of a system of ordinary differential equations lead to which asymptotic behavior. It is also very useful for spotting "exotic" behavior in dynamical systems.

As mentioned before, if the equilibrium solution is not the origin, then we use transformation so that it coincides with the origin. Next, we define the differentiability of manifolds.

Definition 6.7.3. (Homeomorphism) Let K_1 and K_2 be two subsets of a metric space X. Then K_1 and K_2 are said to be *homeomorphic* if there is a continuous one-to-one map $g : K_1 \to K_2$ such that g^{-1} is also continuous.

Such two sets are called *homeomorphic* or topologically equivalent if there is a homeomorphism of K_1 onto K_2.

Definition 6.7.4. (Differentiable manifold) An n-dimensional differentiable manifold M is a connected metric space with an open covering $M = \cup_\alpha U_\alpha$ such that
(1) for all α, U_α is homeomorphic to the open unit ball $B = \{x \in \mathbb{R}^n : |x| < 1\}$ in $\mathbb{R}^n$ centered at the origin, that is, for all α, there exists a homeomorphism $g_\alpha : U_\alpha \to B$ of U_α onto B, and
(2) if $U_\alpha \cap U_\beta \neq \emptyset$ and $g_\alpha : U_\alpha \to B$, $g_\beta : U_\beta \to B$ are homeomorphisms, then $g_\alpha(U_\alpha \cap U_\beta)$ and $g_\beta(U_\alpha \cap U_\beta)$ are subsets of $\mathbb{R}^n$, the map

$$g = g_\alpha \circ g_\beta^{-1} : g_\beta(U_\alpha \cap U_\beta) \to g_\alpha(U_\alpha \cap U_\beta)$$

is differentiable (or of class C^n), and for all $x \in g_\beta(U_\alpha \cap U_\beta)$, the Jacobian determinant is not zero, that is, $\det\frac{\partial g}{\partial x} \neq 0$.

For complete proof of the next theorem, we refer to [36].

Theorem 6.7.2. *Suppose $f(0) = 0$ and let $E \subseteq \mathbb{R}^n$ be an open subset containing the origin with $f \in C^1(E)$. Let $\phi_t : E \to E$ be the flow of (6.7.1) defined for all $t \in \mathbb{R}$. Let*

$$x' = Ax := Df(0)x. \tag{6.7.5}$$

Suppose the linearized matrix $Df(0)$ has k eigenvalues with negative real parts and $n - k$ eigenvalues with positive real parts. Then there exists a k-dimensional differentiable manifold W^s that is tangent to the stable subspace E^s of (6.7.5), and there exists an $(n - k)$-dimensional differentiable manifold W^u that is tangent to the unstable space E^u of the linearized system (6.7.5). Moreover,

$$\lim_{t \uparrow \infty} \phi_t(p) = 0 \ \ for \ any \ \ p \in W^s,$$

and

$$\lim_{t \downarrow -\infty} \phi_t(p) = 0 \ \ for \ any \ \ p \in W^u.$$

Proof. (We only provide a sketch of the proof) This sketch of the proof is geared toward giving guidelines on how to find the stable and unstable manifolds.

Since $f \in C^1(E)$ and $f(0) = 0$, we may write system (6.7.1) as

$$x' = Ax + F(x), \tag{6.7.6}$$

where $A = Df(0)$ and $F(x) = f(x) - Ax$ with $F \in C^1(E)$, $F(0) = 0$ and $DF(0) = 0$. There is an invertible $n \times n$ matrix C such that

$$B = C^{-1}AC = \begin{pmatrix} P & 0 \\ 0 & Q \end{pmatrix},$$

where the $k \times k$ matrix P has the eigenvalues λ_i, $i = 1, 2, \ldots, k$, each with negative real part, and the $(n - k) \times (n - k)$ matrix Q has the eigenvalues λ_i, $i = k + 1, k + 2, \ldots, n$, each with a positive real part. Make the change of variables $y = C^{-1}x$. Then $x' = Ax + F(x)$ becomes $Cy' = ACy + F(Cy)$. Multiplying both sides from the left by C^{-1} gives $y' = C^{-1}ACy + C^{-1}F(Cy)$. Set $G(y) = C^{-1}F(Cy)$ to arrive at

$$y' = \begin{pmatrix} P & 0 \\ 0 & Q \end{pmatrix} y + G(y). \tag{6.7.7}$$

Since each step we have made is reversible, solving system (6.7.7) is equivalent to solving the original system (6.7.6). Next, we labor to find stable and unstable manifolds. Define

$$U(t) = \begin{pmatrix} e^{Pt} & 0 \\ 0 & 0 \end{pmatrix} \ \ and \ \ V(t) = \begin{pmatrix} 0 & 0 \\ 0 & e^{Qt} \end{pmatrix},$$

so that

$$e^{\begin{pmatrix} P & 0 \\ 0 & Q \end{pmatrix} t} = U(t) + V(t).$$

For $t \in \mathbb{R}$ and $a \in \mathbb{R}^n$, define the integral equation

$$u(t, a) = U(t)a + \int_0^t U(t - s)G(u(s, a))ds$$

$$-\int_t^\infty V(t-s)G(u(s,a))ds. \tag{6.7.8}$$

Perko showed the integral equation (6.7.8) satisfies (6.7.7) and $\lim\limits_{t\to\infty} u(t,a) = 0$. Define the iterative scheme for computing the solution:

$$u(t,a) = 0,$$

$$u^{(k+1)}(t,a) = U(t)a + \int_0^t U(t-s)G(u^{(k)}(s,a))ds$$

$$-\int_t^\infty V(t-s)G(u^{(k)}(s,a))ds.$$

Next, we give some insight on why this integral equation given by (6.7.8) is chosen. In general, using the variation of parameters formula, the solution of (6.7.7) is

$$u(t,a) = \begin{pmatrix} e^{Pt} & 0 \\ 0 & e^{Qt} \end{pmatrix} a + \int_0^t \begin{pmatrix} e^{P(t-s)} & 0 \\ 0 & e^{Q(t-s)} \end{pmatrix} G(u(s,a))ds.$$

Keep in mind that the goal is to remove the parts that blow up as $t \to \infty$. Thus we try to separate convergent and non-convergent parts from the above integral equation. Below we explain how this is done:

$$u(t,a) = \begin{pmatrix} e^{Pt} & 0 \\ 0 & e^{Qt} \end{pmatrix} a + \int_0^t \begin{pmatrix} e^{P(t-s)} & 0 \\ 0 & e^{Q(t-s)} \end{pmatrix} G(u(s,a))ds$$

$$= \begin{pmatrix} e^{Pt} & 0 \\ 0 & 0 \end{pmatrix} a + \begin{pmatrix} 0 & 0 \\ 0 & e^{Qt} \end{pmatrix} a + \int_0^t \begin{pmatrix} e^{P(t-s)} & 0 \\ 0 & 0 \end{pmatrix} G(u(s,a))ds$$

$$+ \int_0^t \begin{pmatrix} 0 & 0 \\ 0 & e^{Q(t-s)} \end{pmatrix} G(u(s,a))ds$$

$$= \begin{pmatrix} e^{Pt} & 0 \\ 0 & 0 \end{pmatrix} a + \begin{pmatrix} 0 & 0 \\ 0 & e^{Qt} \end{pmatrix} a + \int_0^t \begin{pmatrix} e^{P(t-s)} & 0 \\ 0 & 0 \end{pmatrix} G(u(s,a))ds$$

$$+ \int_0^\infty \begin{pmatrix} 0 & 0 \\ 0 & e^{Q(t-s)} \end{pmatrix} G(u(s,a))ds$$

$$- \int_t^\infty \begin{pmatrix} 0 & 0 \\ 0 & e^{Q(t-s)} \end{pmatrix} G(u(s,a))ds \quad \text{(remove terms that blow up to get)}$$

$$= \begin{pmatrix} e^{Pt} & 0 \\ 0 & 0 \end{pmatrix} a + \int_0^t \begin{pmatrix} e^{P(t-s)} & 0 \\ 0 & 0 \end{pmatrix} G(u(s,a))ds$$

$$- \int_t^\infty \begin{pmatrix} 0 & 0 \\ 0 & e^{Q(t-s)} \end{pmatrix} G(u(s,a))ds$$

$$= U(t)a + \int_0^t U(t-s)G(u(s,a))ds - \int_t^\infty V(t-s)G(u(s,a))ds.$$

It is evident that the last $n - k$ components of a do not enter the computation, and so we take them to be zero. As a consequence, let $u(t,a)$ be given by (6.7.8), and let us examine what that implies for the initial conditions $u(0,a)$. First, we notice that

$$u_j(0,a) = a_j, \quad j = 1, 2, \ldots, k,$$

$$u_j(0,a) = -\left(\int_0^\infty V(-s)G(u(s,a))ds \right)_j, \quad j = k+1, \ldots, n.$$

Hence the last $n - k$ components of the initial conditions satisfy

$$a_j = \psi_j(a_1, \ldots a_k) := u_j(0, a_1, \ldots, a_k, 0 \ldots, 0), \quad j = k+1, \ldots, n.$$

Finally, the stable manifold $W^s(0)$ is defined by

$$W^s = \{(y_1, \ldots, y_n) : y_j = \psi_j(y_1, \ldots, y_k), \quad j = k+1, \ldots, n\}.$$

Here is the iterative scheme for calculating an approximation to W^s:

(1) Calculate the approximate solution $u^{(m)}(t, a)$.
(2) For each $j = k+1, \ldots, n$, the function $\psi_j(a_1, \ldots, a_k)$ is given by the jth component of $u^{(m)}(0, a)$.

To find the unstable manifold W^u, do the same by replacing t by $-t$, that is,

$$y' = -By - G(y).$$

Then the stable manifold for this system will be the unstable manifold for (6.7.7). During the process, we must replace the vector y by the vector $(y_{k+1}, \ldots, y_n, y_1, \ldots, y_k)$ to determine the $(n - k)$-dimensional manifold W^u by the above process. $\square$

We end this section with the following popular example.

Example 6.21. Consider the planar nonlinear system

$$\begin{cases} x_1' = -x_1 - x_2^2, \\ y' = x_1^2 + x_2 \end{cases}$$

with the only equilibrium solution $(0,0)$. Then

$$A = Df(0) = \begin{pmatrix} -1 & 0 \\ 0 & 1 \end{pmatrix}.$$

The eigenvalues of A are $\lambda = 1, -1$ with corresponding eigenvectors $\begin{pmatrix} 0 \\ 1 \end{pmatrix}$ and $\begin{pmatrix} 1 \\ 0 \end{pmatrix}$, respectively. Thus the matrix C is given by $C = \begin{pmatrix} 1 & 0 \\ 0 & 1 \end{pmatrix}$. It follows that

$$A = B = \begin{pmatrix} -1 & 0 \\ 0 & 1 \end{pmatrix}, \quad F(x) = G(x) = \begin{pmatrix} -x_2^2 \\ x_1^2 \end{pmatrix},$$

and

$$U(t) = \begin{pmatrix} e^{-t} & 0 \\ 0 & 0 \end{pmatrix}, \quad V(t) = \begin{pmatrix} 0 & 0 \\ 0 & e^t \end{pmatrix}.$$

Since $n = 2$ and $k = 1$, the last components $n - k$ of the n vector a are zero. So we take $a = \begin{pmatrix} a_1 \\ 0 \end{pmatrix}$. Our goal is to find the first three successive approximations $u^{(1)}(t, a)$, $u^{(2)}(t, a)$, and $u^{(3)}(t, a)$ and use $u^{(3)}(t, a)$ to approximate the function ψ_2 describing the stable manifold $W^s : x_2 = \psi_2(x_1)$. Define the iterate of the function $u(t, a)$ using (6.7.8)

$$u^{(k+1)}(t, a) = \begin{pmatrix} e^{-t} & 0 \\ 0 & 0 \end{pmatrix} a + \int_0^t \begin{pmatrix} e^{-(t-s)} & 0 \\ 0 & 0 \end{pmatrix} G(u^{(k)}(s, a)) ds$$

$$- \int_t^\infty \begin{pmatrix} 0 & 0 \\ 0 & e^{(t-s)} \end{pmatrix} G(u^{(k)}(s, a)) ds,$$

where

$$u^{(0)}(t, a) = \begin{pmatrix} 0 \\ 0 \end{pmatrix},$$

$$u^{(1)}(t, a) = U(t)a = \begin{pmatrix} e^{-t} a_1 \\ 0 \end{pmatrix},$$

$$u^{(2)}(t, a) = \begin{pmatrix} e^{-t} & 0 \\ 0 & 0 \end{pmatrix} a + \int_0^t \begin{pmatrix} e^{-(t-s)} & 0 \\ 0 & 0 \end{pmatrix} G(u^{(1)}(s, a)) ds$$

$$- \int_t^\infty \begin{pmatrix} 0 & 0 \\ 0 & e^{(t-s)} \end{pmatrix} G(u^{(1)}(s, a)) ds$$

$$= \begin{pmatrix} e^{-t} a_1 \\ 0 \end{pmatrix} - \int_t^\infty \begin{pmatrix} 0 \\ e^{(t-s)} e^{-2s} a_1^2 \end{pmatrix} ds = \begin{pmatrix} e^{-t} a_1 \\ -\frac{1}{3} e^{-2t} a_1^2 \end{pmatrix},$$

$$u^{(3)}(t,a) = \begin{pmatrix} e^{-t} & 0 \\ 0 & 0 \end{pmatrix} a - \int_0^t \begin{pmatrix} e^{-(t-s)} & 0 \\ 0 & 0 \end{pmatrix} G(u^{(2)}(s,a))ds$$

$$- \int_t^\infty \begin{pmatrix} 0 & 0 \\ 0 & e^{(t-s)} \end{pmatrix} G(u^{(2)}(s,a))ds$$

$$= \begin{pmatrix} e^{-t}a_1 \\ 0 \end{pmatrix} - \int_0^t \begin{pmatrix} \frac{e^{-(t-s)}e^{42s}a_1^4}{9} \\ 0 \end{pmatrix} ds - \int_t^\infty \begin{pmatrix} 0 \\ e^{(t-s)}e^{-2s}a_1^2 \end{pmatrix} ds$$

$$= \begin{pmatrix} e^{-t}a_1 + \frac{1}{27}(e^{-4t} - e^{-t})a_1^4 \\ -\frac{1}{3}e^{-2t}a_1^2 \end{pmatrix}.$$

If we approximate $u(t,a)$ by $u^{(3)}(t,a)$, then

$$u(t,a) = \begin{pmatrix} e^{-t}a_1 + \frac{1}{27}(e^{-4t} - e^{-t})a_1^4 \\ -\frac{1}{3}e^{-2t}a_1^2 \end{pmatrix} = \begin{pmatrix} u_1(t,a_1,0) \\ u_2(t,a_1,0) \end{pmatrix},$$

and

$$\psi_1(a_1) = u_1(0,a_1,0), \quad \psi_2(a_1) = u_2(0,a_1,0)$$

are approximated by

$$\psi_1(a_1) = a_1, \quad \psi_2(a_1) = -\frac{1}{3}a_1^2.$$

Thus the stable manifold is approximated by

$$W^s = \{(y_1, y_2) \in \mathbb{R}^2 : y_2 = -\frac{1}{3}y_1^2\}.$$

However, since the matrix $C = I$ for this example, the x and y spaces are the same. It follows that the stable manifold is given by

$$W^s = \{(x_1, x_2) \in \mathbb{R}^2 : x_2 = -\frac{1}{3}x_1^2\}.$$

The unstable manifold W^u can be found by applying the same procedure to the system with $t \to -t$ and x_1 and x_2 interchanged. Thus the stable manifold for the resulting system will be the unstable manifold for the original system. With this in mind, we arrive at the unstable manifold

$$W^u = \{(x_1, x_2) \in \mathbb{R}^2 : x_1 = -\frac{1}{3}x_2^2\}.$$

Remark 6.5. We used $u^{(3)}(t,a)$ to approximate $u(t,a)$, and so

$$u(t,a) \approx \begin{pmatrix} e^{-t}a_1 + \frac{1}{27}(e^{-4t} - e^{-t})a_1^4 \\ -\frac{1}{3}e^{-2t}a_1^2 \end{pmatrix} \approx \begin{pmatrix} e^{-t}a_1 \\ -\frac{1}{3}e^{-2t}a_1^2 \end{pmatrix},$$

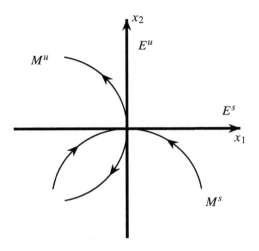

FIGURE 6.16

Local approximations of M^s and M^u of Example 6.21.

when $|a| = |(a_1, 0)^T|$ is small, or when a_1 is small. Thus the stable manifold can be approximated by

$$\psi_2(a_1) = -\left(\int_0^\infty V(-s)G(u(s, a))ds\right)_2$$

$$= -\left(\int_0^\infty \begin{pmatrix} 0 & 0 \\ 0 & e^{-s} \end{pmatrix} \begin{pmatrix} -\frac{1}{9}(e^{-4s} - e^{-t})a_1^4 \\ e^{-2s}a_1^2 \end{pmatrix} ds\right)_2$$

$$= -\int_0^\infty e^{-s}e^{-2s}a_1^2 ds = -\frac{1}{3}a_1^2.$$

Fig. 6.16 shows the stable and unstable manifolds in the neighborhood of $(0, 0)$.

6.7.2 Global manifolds

The stable and unstable manifolds of Section 6.7.1 are local since they are defined in a small neighborhood of the origin. For that reason, they are referred to as the *local* stable and unstable manifolds at the origin. In this section, we define the *global manifolds* around the origin.

Definition 6.7.5. Let ϕ_t be the flow of (6.7.1). The *global stable* and *unstable manifolds* of (6.7.1) at the origin are defined by

$$G^s(0) = \cup_{t \leq 0}\phi_t(W^s)$$

and

$$G^u(0) = \cup_{t \geq 0} \phi_t(W^s),$$

respectively.

The global stable and unstable manifolds are unique and invariant with respect to the flow. Furthermore, for all $x \in G^s(0)$, $\lim_{t \to \infty} \phi_t(x) = 0$, and for all $x \in G^u(0)$, $\lim_{t \to -\infty} \phi_t(x) = 0$.

For the next theorem, we define a neighborhood around zero. Let $\delta > 0$. Define the neighborhood of the origin by

$$N_\delta(0) = \{x \in \mathbb{R}^n : ||x|| < \delta\}.$$

Theorem 6.7.3. *Assume that the hypotheses of Theorem 6.7.2 hold. Suppose the linearized matrix $Df(0)$ has k eigenvalues with negative real parts and $n - k$ eigenvalues with positive real parts. Set*

$$\mathrm{Re}(\lambda_j) < -\alpha < 0 < \beta < \mathrm{Re}(\lambda_m)$$

for $j = 1, \ldots, k$ and $m = k + 1, \ldots, n$.
Then, given $\varepsilon > 0$, there exists $\delta > 0$ such that if $p \in N_\delta(0) \cap W^s$, then

$$|\phi_t(p)| \leq \varepsilon e^{-\alpha t}$$

for all $t \geq 0$. In a similar fashion, if $p \in N_\delta(0) \cap W^u$, then

$$|\phi_t(p)| \leq \varepsilon e^{\beta t}$$

for all $t \leq 0$.

Theorem 6.7.3 states that if solutions start in the stable manifold W^s, then they approach the origin exponentially. This is displayed in Example 6.19.

The center manifold theorem gives a detailed description of the dynamic of (6.7.1) in a neighborhood of a hyperbolic equilibrium once the center manifold W^c is determined.

6.7.3 Center manifold

Theorem 6.7.4. *Suppose $f(0) = 0$ and let $E \subseteq \mathbb{R}^n$ be an open subset containing the origin with $f \in C^1(E)$. Let $\phi_t : E \to E$ be the flow of (6.7.1) defined for all $t \in \mathbb{R}$. Let*

$$x' = Ax := Df(0)x. \qquad (6.7.9)$$

Suppose the linearized matrix $Df(0)$ has k eigenvalues with negative real parts, j eigenvalues with positive real parts, and $m = n - k - j$ eigenvalues with zero real parts. Then there exists an m-dimensional differentiable manifold $W^c(0)$ that is *tangent to the center subspace E^s of (6.7.9), a j-dimensional differentiable manifold $W^u(0)$ that is tangent to the unstable space E^u of the linearized system (6.7.9), and an m-dimensional differentiable manifold W^c that is tangent to the center space E^c of the linearized system (6.7.9). Moreover,*

$$\lim_{t\uparrow\infty} \phi_t(p) = 0 \text{ for any } p \in W^s,$$

$$\lim_{t\downarrow-\infty} \phi_t(p) = 0 \text{ for any } p \in W^u,$$

$$\lim_{t\downarrow-\infty} \phi_t(p) = 0 \text{ for any } p \in W^c.$$

In other words, the manifolds W^s, W^u, and W^c are invariant under the flow ϕ_t of (6.7.1).

Example 6.22. Consider the planar nonlinear system

$$\begin{cases} x_1' = x_1^2, \\ x_2' = -x_2 \end{cases}$$

with the only equilibrium solution $(0, 0)$. Linearization around the equilibrium solution gives

$$A = Df(0) = \begin{pmatrix} 0 & 0 \\ 0 & -1 \end{pmatrix}.$$

The eigenvalues of A are $\lambda = 0, -1$ with corresponding eigenvectors $\begin{pmatrix} 1 \\ 0 \end{pmatrix}$ and $\begin{pmatrix} 0 \\ 1 \end{pmatrix}$.
Thus the stable subspace is $E^s = \{(x_1, x_2) \in \mathbb{R}^2 : x_1 = 0\}$, and the center subspace is $E^c = \{(x_1, x_2) \in \mathbb{R}^2 : x_2 = 0\}$. For constants c_1 and c_2, the system has the solution

$$x_1(t) = \frac{c_1}{1 - c_1 t}, \quad x_2(t) = c_2 e^{-t}.$$

Note that if $c_1 = 0$, then $x_1 = 0$. Moreover, to get a better understanding of the dynamics of the solution, we eliminate t from both equations and arrive at the following relation between x_1 and x_2:

$$x_2 = c_2 e^{-\frac{1}{c_1} + \frac{1}{x_1}}.$$

Since $x_1 = 0$ for $c_1 = 0$, we have that the plane curves

$$x_2 = \Psi(x_1, c_1, c_2) = \begin{cases} c_2 e^{-\frac{1}{c_1} + \frac{1}{x_1}} & \text{for } x_1 < 0, \\ 0 & \text{for } x_1 = 0 \end{cases} \tag{6.7.10}$$

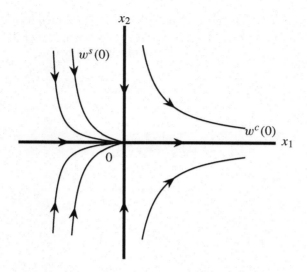

FIGURE 6.17

Center manifold.

satisfy

$$\Psi(0, c_1, c_2) = 0, \quad \frac{\partial \Psi}{\partial x_1}(0, c_1, c_2) = 0.$$

Now $c_1 < 0$ if and only if $x_1 < 0$ by $x_1(t) = \frac{c_1}{1-c_1t}$ for all $t \geq 0$. It follows that when $x_1 < 0$ is sufficiently small, (6.7.10) is a local center manifold that is tangent to E^c at the origin. However, since (6.7.10) holds for any $c_1 < 0$ and c_2, there are infinitely many center manifolds, and all have the same orientations as seen in Fig. 6.17.

6.7.4 Center manifold and reduced systems

Let $E \subseteq \mathbb{R}^n$ contain the origin, and let $f : E \to \mathbb{R}^n$ be C^2. Consider the autonomous nonlinear differential equation

$$x' = f(x). \tag{6.7.11}$$

Since $f \in C^1(E)$ and $f(0) = 0$, we may write system (6.7.11) as

$$x' = Ax + F(x), \tag{6.7.12}$$

where $A = Df(0)$ and $F(x) = f(x) - Ax$ with $F \in C^1(E)$, $F(0) = 0$, and $DF(0) = 0$. There is an $n \times n$ invertible matrix C such that

$$C^{-1}AC = \begin{pmatrix} P & 0 \\ 0 & Q \end{pmatrix},$$

where the $k \times k$ matrix P has eigenvalues λ_i, $i = 1, 2, \ldots, k$, each with zero real part, and the $(n - k) \times (n - k)$ matrix Q has the eigenvalues λ_i, $i = k + 1, k + 2, \ldots, n$, each with negative real part. Make the change of variables

$$\begin{pmatrix} y \\ z \end{pmatrix} = C^{-1}x, \quad y \in \mathbb{R}^k, \quad z \in \mathbb{R}^{n-k}$$

and transform (6.7.12) into the form

$$y' = Px + F_1(y, z),$$

$$z' = Qz + F_2(y, z), \tag{6.7.13}$$

where

$$F_i(0, 0) = 0, \quad \frac{\partial F_i}{\partial y}(0, 0) = \frac{\partial F_i}{\partial z}(0, 0), \quad i = 1, 2.$$

If $z = h(y)$ is an invariant manifold for (6.7.13) with

$$h(0) = 0 \quad \text{and} \quad \frac{\partial h}{\partial y}(0) = 0, \tag{6.7.14}$$

then we define the local *center manifold*

$$W^c = \{(y, z) \in \mathbb{R}^k \times \mathbb{R}^{n-k} : y = h(y), \ \|y\| < \delta\}$$

for small $\delta > 0$. This way, W^c is tangent to E^c at the origin. We have the following theorem.

Theorem 6.7.5. *For system (6.7.13), there exist $\delta > 0$ and $h \in C^1$ with $h(y)$ defined for all $\|y\| < \delta$ such that $z = h(y)$ is a center manifold.*

On the center manifold W^c, we get the *reduced system*

$$y' = Px + F_1(y, h(y)),$$

and z' can be calculated on the center manifold in two ways: directly from the z' equation above or by differentiating $z = h(y)$, that is,

$$z' = Qz + F_2(y, h(y)) \quad \text{and} \quad z' = \frac{d}{dt}h(y) = Dh(y)y' = Dh(y)[Px + F_1(y, h(y))],$$

where $Dh(y)$ is the matrix formed by partial derivatives of $h(y)$. Note that in one dimension, $Dh(y) = h'(y)$. Expanding h as a Taylor series (noting that the constant and linear terms vanish), the two equations for z' provide two different polynomials, and the coefficients of different monomials can be equated to determine the coefficients of the Taylor expansion. We end the section with the following two examples.

Example 6.23. Consider the planar nonlinear system

$$\begin{cases} x' = xy, \\ y' = -x^2 - y \end{cases}$$

with the only equilibrium solution $(0, 0)$. Linearization around the equilibrium solution gives

$$A = Df(0) = \begin{pmatrix} 0 & 0 \\ 0 & -1 \end{pmatrix}.$$

The eigenvalues of A are $\lambda = 0, -1$ with corresponding eigenvectors $\begin{pmatrix} 1 \\ 0 \end{pmatrix}$ and $\begin{pmatrix} 0 \\ 1 \end{pmatrix}$.

Thus the stable subspace is $E^s = \{(x, y) \in \mathbb{R}^2 : x = 0\}$, and the center subspace is $E^c = \{(x, y) \in \mathbb{R}^2 : y = 0\}$. Since the matrix A is already in normal form, no coordinate transformation is needed. We consider a center manifold of the form

$$y = h(x) = ax^2 + bx^3 + cx^4 + \mathcal{O}(x^5). \tag{6.7.15}$$

There is no constant term and no linear term in $h(x)$ since the center manifold passes through the origin and should be tangent to E^c. Now we calculate the constants. We have

$$y' = -x^2 - y = -x^2 - \left(ax^2 + bx^3 + cx^4\right) + \mathcal{O}(x^5). \tag{6.7.16}$$

Next, we differentiate (6.7.15) with respect to t. Using the chain rules, we get $y' = \dfrac{dh(x)}{dt} = x'h'(x)$ or

$$y' = x'h'(x) = xyh'(x) = x\left(ax^2 + bx^3 + cx^4 + \dots\right)\left(2ax + 3bx^2 + 4cx^3 + \dots\right). \tag{6.7.17}$$

Setting the right side of (6.7.17) equal to (6.7.16) and equating the coefficients at x^2, x^3, and x^4 give

$$a = -1, \quad b = 0, \quad \text{and} \quad c = -2.$$

Thus the center manifold is given by

$$W^c = \{(x, y) \in \mathbb{R}^2 : y = -x^2 - 2x^4\},$$

and the dynamics on W^c is

$$x' = xh(x) = -x^3 - 2x^5 + \mathcal{O}(x^7).$$

Now for $x > 0$, we have $x' < 0$ and $x' > 0$ if $x < 0$. Thus the origin is stable, and the solutions are stable nodes as shown in the graph below, but the motion onto the center manifold in the y-direction is much faster than the motion on the center manifold, leading to a phase portrait shown in Fig. 6.18.

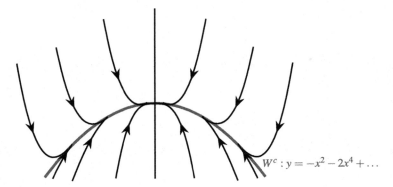

$$W^c : y = -x^2 - 2x^4 + \cdots$$

FIGURE 6.18

Exponential collapse of trajectories onto the center manifold and then slow convergence toward $(0, 0)$ on the center manifold.

The next example shows the need for a transformation.

Example 6.24. Consider the planar nonlinear system

$$\begin{cases} x' = y - x + xy, \\ y' = x - y - x^2 \end{cases} \tag{6.7.18}$$

with the only equilibrium solution $(0, 0)$. Linearization around the equilibrium solution gives

$$A = Df(0) = \begin{pmatrix} -1 & 1 \\ 1 & -1 \end{pmatrix}.$$

The eigenvalues of A are $\lambda = 0, -2$ with corresponding eigenvectors $\begin{pmatrix} 1 \\ 1 \end{pmatrix}$ and $\begin{pmatrix} -1 \\ 1 \end{pmatrix}$.

Thus the stable subspace is $E^s = \{(x, y) \in \mathbb{R}^2 : y = -x\}$, and the center subspace is $E^c = \{(x, y) \in \mathbb{R}^2 : y = x\}$.

Let $C = \begin{pmatrix} 1 & 1 \\ -1 & 1 \end{pmatrix}$. Then $C^{-1} = \frac{1}{2} \begin{pmatrix} 1 & 1 \\ -1 & 1 \end{pmatrix}$. Make the change of variable from x, y coordinates to u, v coordinates by

$$\begin{pmatrix} u \\ v \end{pmatrix} = C^{-1} \begin{pmatrix} x \\ y \end{pmatrix} = \frac{1}{2} \begin{pmatrix} x + y \\ y - x \end{pmatrix}$$

or

$$\begin{pmatrix} x \\ y \end{pmatrix} = C \begin{pmatrix} u \\ v \end{pmatrix} = \begin{pmatrix} u - v \\ u + v \end{pmatrix}.$$

Under these transformations, system (6.7.18) in the u, v coordinates becomes

$$u' = uv - v^2, \quad v' = -2v + uv - u^2.$$

The new system is now in the normal form since

$$Df(0) = \begin{pmatrix} 0 & 0 \\ 0 & -2 \end{pmatrix}.$$

For similar reasons as in the above example, we define the center manifold $h(u)$ by

$$v(u) = h(u) = au^2 + bu^3 + cu^4 + \cdots.$$

Then using the chain rule, we arrive at

$$
\begin{aligned}
v' &= \Big[2au + 3b^2 u^2 + 4cu^3 + \ldots\Big] u' \\
&= \Big[2au + 3b^2 u^2 + 4cu^3 + \ldots\Big] \\
&\quad \times \Big[u\big(au^2 + bu^3 + cu^4 + \ldots\big) - \big(au^2 + bu^3 + cu^4 + \ldots\big)^2\Big] \\
&= 2a^2 u^4 + (5ab - 2a^3)u^5 + \cdots.
\end{aligned}
\tag{6.7.19}
$$

On the other hand, using the second equation of the system with $v = h(u)$ gives

$$
\begin{aligned}
v' &= -2v + uv - u^2 \\
&= -2\big(au^2 + bu^3 + cu^4 + \ldots\big) + u\big(au^2 + bu^3 + cu^4 + \ldots\big) - u^2 \\
&= (2a + 1)u^2 + (2b - a)u^3 + (3c - b)u^4 + \cdots.
\end{aligned}
\tag{6.7.20}
$$

Setting (6.7.19) equal to (6.7.20) and then equating the coefficients at u^2, u^3, and u^4 give

$$a = -\frac{1}{2}, \quad b = -\frac{1}{4}, \quad c = -\frac{3}{8},$$

and it follows that the center manifold in uv coordinates is

$$W^c = \Big\{(u, v) \in \mathbb{R}^2 : v = -\frac{1}{2}u^2 - \frac{1}{4}u^3 - \frac{3}{8}u^4 + \ldots\Big\}.$$

The dynamics on the center manifold is

$$u' = uv - v^2 = -\frac{1}{2}u^3 - \frac{1}{4}u^4 - \frac{3}{8}u^5 + \cdots,$$

which is stable for small u. Finally, the center manifold W^c in the original coordinates x, y is approximated from replacing u with $x + y$ and v with $y - x$:

$$W^c = \Big\{(x, y) \in \mathbb{R}^2 : y - x = -\frac{1}{4}(x + y)^2 - \frac{1}{16}(x + y)^3 - \frac{3}{64}(x + y)^4 + \ldots\Big\}.$$

6.7.5 Hartman–Grobman theorem

Consider the nonlinear autonomous system

$$x' = f(x), \qquad (6.7.21)$$

where $E \subseteq \mathbb{R}^n$, and $f : E \to \mathbb{R}^n$ is a C^2 function. The *Hartman–Grobman theorem* states that near a hyperbolic equilibrium solution x_0, the nonlinear system given by (6.7.21) has the same qualitative structures as the linearized system

$$x' = Ax, \qquad (6.7.22)$$

where $A = Df(x_0)$. In this section, we assume that x_0 is the origin, $E \subseteq \mathbb{R}^n$ contains the origin, and $f(0) = 0$. Recall from Definition 6.7.3 that two sets K_1 and K_2 are said to be *homeomorphic* if there is a continuous one-to-one map $g : K_1 \to K_2$ such that g^{-1} is also continuous.

In our case the two autonomous systems (6.7.21) and (6.7.22) are said to be topologically equivalent in a neighborhood of the origin or to have the same qualitative structure near the origin if there is a homeomorphism H mapping an open set U containing the origin onto an open set V containing the origin that maps trajectories of (6.7.21) in U onto trajectories (6.7.22) in V and preserves the orientation. If the homeomorphism H preserves parameterization by time t, then (6.7.21) and (6.7.22) are said to be topologically conjugate in the neighborhood of the origin.

We have the following example.

Example 6.25. Consider the two linear systems $x' = Ax$ and $y' = By$ with

$$A = \begin{pmatrix} 2 & -1 \\ -1 & 2 \end{pmatrix} \text{ and } B = \begin{pmatrix} 3 & 0 \\ 0 & 1 \end{pmatrix}.$$

Define

$$R = \frac{1}{\sqrt{2}} \begin{pmatrix} 1 & -1 \\ 1 & 1 \end{pmatrix}; \text{ then } R^{-1} = \frac{1}{\sqrt{2}} \begin{pmatrix} 1 & 1 \\ -1 & 1 \end{pmatrix}.$$

It is easy to verify that $B = RAR^{-1}$. Let $y = H(x) = Rx$ or $x = R^{-1}y$. Then it follows that

$$y' = RAR^{-1}y = By.$$

Suppose $x(0) = x_0$. Then $x(t) = e^{At}x_0$ is the solution of $x' = Ax$, $x(0) = x_0$. Similarly, if $y(0) = y_0$, then $y = e^{Bt}y_0$ with $y_0 = Rx_0$.

Then by (3.4.3) it follows that

$$y(t) = H(x(t)) = Rx(t) = Re^{At}x_0 = Re^{At}R^{-1}y_0 = e^{Bt}y_0 = e^{Bt}Rx_0$$

is the solution of $x' = Bx$. In other words, H maps the trajectories of $x' = Ax$ onto trajectories of $x' = Bx$ and preserves the parameterization by t since

$$He^{At} = e^{Bt}H.$$

Thus H is a homeomorphism, and $x' = Ax$ and $y' = By$ are topologically conjugate.

Next, we introduce the Hartman–Grobman theorem, proved independently by Grobman in 1959 and Hartman in 1960. We only offer an outline of the proof, which should serve as guidelines on how to find the homeomorphism H. For complete proof, we refer the reader to [26] or [33]. In plain language the theorem says that if the linearization matrix has no zero or purely imaginary eigenvalues, then the phase portrait for the nonlinear system near an equilibrium point is similar to that of its linearization.

Theorem 6.7.6. *(Hartman–Grobman theorem) Let $f(0) = 0$, and let $E \subseteq \mathbb{R}^n$ be an open subset containing the origin with $f \in C^1(E)$. Let ϕ_t be the flow of (6.7.21). Suppose the linearized matrix $Df(0)$ has no eigenvalue with zero real part (the origin is hyperbolic). Then there is a homeomorphism H mapping an open set U containing the origin onto an open set V containing the origin such that for each $x_0 \in U$, there is an open interval $I_0 \subseteq \mathbb{R}$ containing zero such that for all $x_0 \in U$ and $t \in I_0$, we have*

$$H \circ \phi_t(x_0) = e^{At} H(x_0).$$

In other words, H maps trajectories of (6.7.21) in U onto trajectories (6.7.22) in V and preserves the orientation and parameterization by time t.

Proof. (Outline) Suppose that

$$A = Df(0) = \begin{pmatrix} P & 0 \\ 0 & Q \end{pmatrix},$$

where all the eigenvalues of the matrix P have negative real parts and all the eigenvalues of the matrix Q have positive real parts. Let

$$x_0 = \begin{pmatrix} y_0 \\ z_0 \end{pmatrix} \in \mathbb{R}^n,$$

let ϕ_t be the flow of (6.7.21) and write the solution

$$x(t, x_0) = \phi_t(x_0) = \begin{pmatrix} y(t, y_0, z_0) \\ z(t, y_0, z_0) \end{pmatrix}$$

for y_0 and z_0 in E^s and E^u, respectively.

Define locally ($t = 1$) the transformation

$$T^1 x_0 = \begin{pmatrix} e^P y_0 + Y(x_0) \\ e^Q z_0 + Z(x_0) \end{pmatrix}.$$

Now for $x \in \mathbb{R}^n$, define

$$H_0(x) = \begin{pmatrix} \Phi(x) \\ \Psi(x) \end{pmatrix} = \begin{pmatrix} \Phi(y, z) \\ \Psi(y, z) \end{pmatrix}.$$

Then

$$H_0 \circ T^1 = e^A \circ H_0,$$

which component-wise is equivalent to

$$e^P \Phi(x) = \Phi(e^P y + Y(x), e^Q z + Z(x)),$$
$$e^Q \Psi(x) = \Phi(e^P y + Y(x), e^Q z + Z(x)). \tag{6.7.23}$$

Next, we define a continuous successive approximation from the second equation by

$$\begin{cases} \Psi_0(x) = z, \\ \Psi_{k+1}(x) = e^{-Q} \Phi_k(e^P y + Y(x), e^Q z + Z(x)), \ k = 0, 1, \dots. \end{cases}$$

Then $\Psi_k(x)$ is a Cauchy sequence of continuous functions, which converges uniformly as $k \to \infty$ to a continuous function $\Psi(x)$. We write the first equation in (6.7.23) by

$$e^{-P} \Phi(x) = \Phi(e^{-P} y + Y_1(x), e^{-Q} z + Z_1(x)), \tag{6.7.24}$$

where the functions Y_1 and Z_1 are defined by the inverse of T,

$$T^{-1}(y, z) = \begin{pmatrix} e^{-P} y + Y_1(y, z) \\ e^{-Q} z + Z_1(y, z) \end{pmatrix}.$$

With this setup, Eq. (6.7.24) can be solved by the method of successive approximations exactly as above with $\Phi_0(x) = y$. Finally, define the homeomorphism

$$H = \int_0^1 e^{-As} H_0 T^s ds = \int_0^1 e^{-As} H_0 \circ \phi_s(x, y) ds,$$

which satisfies

$$H \circ \phi_t(x, y) = e^{At} H(x, y). \qquad \square$$

Remark 6.6. (1) From the computational aspect, Theorem 6.7.6 is difficult to use since it requires the computation of the flow, which is not possible in most nonlinear systems.

(2) However, conceptually, it suffices to know that such a homeomorphism exists. This allows us to understand the dynamics of the nonlinear system near a hyperbolic equilibrium solution by simply looking at the eigenvalues of the linear system $x' = Ax$, $A = Df(0)$, obtained from linearizing the nonlinear system near the origin.

In the next example, we consider a system for which the flows can be computed.

Example 6.26. Consider

$$\begin{cases} y' = -y, \\ z' = z + y^2 \end{cases}$$

with

$$x(0) = \begin{pmatrix} y_0 \\ z_0 \end{pmatrix},$$

$$Df(0) = \begin{pmatrix} -1 & 0 \\ 0 & 1 \end{pmatrix},$$

and hence $e^P = e^{-1}$, and $e^Q = e$. The equation $y' = -y$ has the solution $y = y_0 e^{-t}$. Substituting this solution into the second differential equation and using the variation of parameters formula, we arrive at

$$z = z_0 e^t + \frac{1}{3} y_0^2 (e^t - e^{-2t}).$$

Thus

$$\phi_t(x_0) = \begin{pmatrix} y_0 e^{-t} \\ z_0 e^t + \frac{1}{3} y_0^2 (e^t - e^{-2t}) \end{pmatrix},$$

and at $t = 1$, $Y(x) = 0$, $Z(x) = \frac{1}{3} y_0^2 (e - e^{-2})$, where $x = (y, z)$. As a consequence,

$$e^Q \Psi(x) = \Phi(e^P y + Y(x), e^Q z + Z(x)),$$

which results in

$$e\Psi(x) = \Phi(e^{-1} y, ez + \frac{1}{3} y_0^2 (e - e^{-2})).$$

Therefore the successive approximations are given by

$$\Psi_0(x) = z, \quad x = (y, z) \in \mathbb{R}^2,$$

$$\Psi_1(x) = e^{-1} \Psi_0(e^{-1} y, ez + \frac{1}{3} y^2 (e - e^{-2}))$$

$$= e^{-1} [ez + \frac{1}{3} y^2 (e - e^{-2})]$$

$$= z + \frac{1}{3}(1 - e^{-3}) y^2,$$

$$\Psi_2(x) = e^{-1} \Psi_1(e^{-1} y, ez + \frac{1}{3}(e - e^{-2}) y^2)$$

$$= e^{-1} [(ez + \frac{1}{3}(e - e^{-2}) y^2 + \frac{1}{3}(1 - e^{-3}) e^{-2} y^2]$$

$$= z + \frac{1}{3}(1 - e^{-3})(y^2 + e^{-3} y^2)$$

$$= z + \frac{1}{3}(1 - e^{-3})(1 + e^{-3}) y^2,$$

$$\Psi_3(x) = e^{-1}\Psi_2(e^{-1}y, ez + \frac{1}{3}(e - e^{-2})y^2)$$

$$= e^{-1}[(ez + \frac{1}{3}(e - e^{-2})y^2 + \frac{1}{3}(1 - e^{-3})(1 + e^{-3})e^{-2}y^2]$$

$$= z + \frac{1}{3}(1 - e^{-3})(y^2 + e^{-3}y^2)$$

$$= z + \frac{1}{3}(1 - e^{-3})(1 + e^{-3} + e^{-6})y^2,$$

$$\cdots$$

$$\Psi_m(x) = z + \frac{1}{3}(1 - e^{-3})(1 + e^{-3} + e^{-6} + \ldots + (e^{-3})^{m-1})y^2$$

$$= z + \frac{1}{3}(1 - e^{-3})y^2 \sum_{n=1}^{m}(\frac{1}{e^3})^{n-1} \to z + \frac{1}{3}(1 - e^{-3})\frac{1}{1 - e^{-3}}y^2$$

$$= z + \frac{1}{3}y^2 \text{ as } m \to \infty.$$

Thus as $m \to \infty$,

$$\Psi_m(y, z) \to \Psi(y, z) = z + \frac{1}{3}y^2$$

uniformly for all $(y, z) \in \mathbb{R}^2$. Now the functions Y_1 and Z_1 are defined by the inverse of T,

$$T^{-1}(y, z) = \begin{pmatrix} e^{-P}y + Y_1(y, z) \\ e^{-Q}z + Z_1(y, z) \end{pmatrix}.$$

However,

$$T^{-1}(y, z) = \begin{pmatrix} ey \\ e^{-1}z - e\frac{e^3-1}{3e^2}y^2 \end{pmatrix},$$

which implies that $Y_1(y, z) = 0$ and $Z_1(y, z) = -ery^2$, where $r = \frac{e^3-1}{3e^2}$. To fully determine $\Phi(y, z)$, we use (6.7.24):

$$e^{-P}\Phi(x) = \Phi(e^{-P}y + Y_1(x), e^{-Q}z + Z_1(x))$$

or

$$e\Phi(x) = \Phi(ey, e^{-1}z - ery^2).$$

We define the successive approximations by

$$\Phi_0(y, z) = y,$$

$$\Phi_{k+1}(y, z) = e^{-1}\Phi_k(ey, e^{-1}z - ery^2), \; k = 0, 1, \ldots,$$

and get

$$\Phi(y, z) = y, \quad (y, z) \in \mathbb{R}^2.$$

Thus

$$H_0(y, z) = \begin{pmatrix} y \\ z + \frac{1}{3}y^2 \end{pmatrix},$$

and the homeomorphism H is given by

$$
\begin{aligned}
H(y, z) &= \int_0^1 e^{-As} H_0 T^s \, ds \\
&= \int_0^1 \begin{pmatrix} e^s & 0 \\ 0 & e^{-s} \end{pmatrix} H_0[\begin{pmatrix} ye^{-s} \\ ze^s + \frac{1}{3}y^2(e^s - e^{-2s}) \end{pmatrix}] ds \\
&= \int_0^1 \begin{pmatrix} e^s & 0 \\ 0 & e^{-s} \end{pmatrix} \begin{pmatrix} ye^{-s} \\ ze^s + \frac{1}{3}y^2(e^s - e^{-2s}) + \frac{1}{3}y^2 e^{-2s} \end{pmatrix} ds \\
&= \begin{pmatrix} y \\ z + \frac{1}{3}y^2 \end{pmatrix}.
\end{aligned}
$$

Note that $H_0 T^s$ is taken as the composition or H_0 is acting on T^s. Let's dig deeper and see what subset gets mapped onto the stable subspace E^s. Looking at the corresponding eigenvector of $\lambda = -1$, we see that $E^s = \{(y, z) \in \mathbb{R}^2 : z = 0\}$. We are interested in finding the subset such that

$$H(y, z) = \begin{pmatrix} y \\ z + \frac{1}{3}y^2 \end{pmatrix} \in E^s.$$

It readily follows that the subset is

$$\{(y, z) \in \mathbb{R}^2 : z = -\frac{1}{3}y^2\},$$

which is the stable manifold W^s by simply taking $t \to \infty$ in the flow. In other words, the homeomorphism H maps W^s onto E^s. In addition, H maps $W^u = \{(y, z) \in \mathbb{R}^2 : y = 0\}$ onto $E^u = \{(y, z) \in \mathbb{R}^2 : y = 0\}$. Solving the linearized system $y' = -y$, $z' = z$, we get $y = c_1 e^{-t}$ and $z = c_2 e^t$ or $z = \frac{c}{y}$ for constants $c_1, c_2,$ and c. Thus

$$H(y, z) = H(y, \frac{c}{y}),$$

from which it follows that

$$(y, z + \frac{1}{3}y^2) = (y, \frac{c}{y}).$$

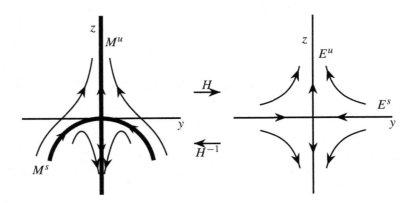

FIGURE 6.19

Under the homeomorphism H, the phase portrait for the nonlinear system near the origin is similar to that of its linearization.

Thus

$$z = \frac{c}{y} - \frac{1}{3}y^2.$$

We just showed that, under H, the curve $z = \frac{c}{y} - \frac{1}{3}y^2$ of the nonlinear system gets mapped onto $z = \frac{c}{y}$ of the linearized system. (See Fig. 6.19.)

6.8 Exercises

Exercise 6.1. For the following equations, find the equilibrium solutions, sketch the phase line, and determine the type of stability of all the equilibria.

(a) $x' = x^2(4 - x)$.

(b) $x' = 2x(1 - x) - \frac{1}{2}x$.

(c) $x' = (4 - x)(2 - x)^3$.

Exercise 6.2. Find the equilibrium solutions in terms of the parameter μ and determine their stability. Show that the origin is a turning point and sketch a bifurcation diagram and label all branches as stable or unstable. When possible, identify the type of bifurcation.

(a) $x' = \mu x(1 - x)$.

(b) $x' = (x^2 - \mu)(1 - x)$.

Exercise 6.3. Find the equilibrium solutions, determine their stability, and show that the origin is a turning point and sketch the bifurcation diagram.

$$x' = \mu - x^2.$$

Exercise 6.4. For the following equations, show that each of them undergoes a transcritical bifurcation and find the bifurcation value. Sketch the bifurcation diagram.
(a) $x' = \mu x + 2x^2$.
(b) $x' = (2 + \mu)x - 5x^2$.

Exercise 6.5. Find the equilibrium solution x_0 at μ_0 and use Theorem 6.1.2 to determine the type of bifurcation at (x_0, μ_0) for the scalar system

$$x' = \mu + x - e^x + 1.$$

Exercise 6.6. Show that

$$x' = \mu x + x^3$$

undergoes a pitchfork bifurcation at $\mu = 0$ and sketch the bifurcation diagram.

Exercise 6.7. Complete the proof of Theorem 6.2.1, that is, show that inequality (6.2.8) holds on $[0, \infty)$.

Exercise 6.8. Suppose the origin is uniformly stable for the linear system $x' = A(t)x$, where $A(t)$ is an $n \times n$ matrix with continuous entries. Suppose that $F : [0, \infty) \times \mathbb{R}^n \to \mathbb{R}^n$ is continuous with

$$|F(t, x)| \le \gamma(t)|x| \text{ and } \int_0^\infty \gamma(t)dt < \infty,$$

where the function $\gamma(t) > 0$ is continuous. Show that if $F(t, 0) = 0$, then the zero solution of

$$x' = A(t)x + F(t, x),$$

is uniformly stable.

Exercise 6.9. Investigate the stability of each of the equilibrium points for the given systems:
(a) $x' = x - 3y + 2xy$, $y' = 4x - 6y - xy$.
(b) $x' = 6x - 5y + x^2$, $y' = 2x - y + y^2$.
(c) $x' = x - 2y + 3xy$, $y' = 2x - 3y - x^2 - y^2$.
(d) $x' = x - y$, $y' = x^2 - y$.
(e) $x' = y^2 - 1$, $y' = x^3 - y$.
(f) $x' = 8x - y^2$, $y' = -y + x^2$.

Exercise 6.10. Consider the system

$$x' = -x + hy,$$
$$y' = x - y.$$

Show that

(1) $(0,0)$ is asymptotically stable if $h = 0$,
(2) $(0,0)$ is asymptotically stable spiral if $h < 0$, and
(3) $(0,0)$ asymptotically stable node if $0 < h < 1$.

Exercise 6.11. Study the stability of the origin for the system

$$x' = y + hx(x^2 + y^2),$$
$$y' = -x + hy(x^2 + y^2).$$

(Hint: use polar coordinates.)

Exercise 6.12. Study the stability of the origin for the system

$$x' = -2x - y + x^2 y,$$
$$y' = 4x + \mu y - y^2$$

for $\mu \neq 2$.

Exercise 6.13. Use polar coordinates to study the stability of the origin for the system

$$x' = -y - x\sqrt{x^2 + y^2}, \quad y' = x - y\sqrt{x^2 + y^2}.$$

Exercise 6.14. Show that the system

$$x' = y + \frac{x(1 - x^2 - y^2)}{\sqrt{x^2 + y^2}}, \quad y' = -x + \frac{y(1 - x^2 - y^2)}{\sqrt{x^2 + y^2}}$$

has a limit cycle.

Exercise 6.15. Show that the system

$$x' = y + x(1 - x^2 - y^2), \quad y' = -x + y(1 - x^2 - y^2)$$

has a limit cycle.

Exercise 6.16. Consider the system

$$x' = 4x + 4y - x(x^2 + y^2),$$
$$y' = -4x + 4y - y(x^2 + y^2).$$

(a) Show that there is a closed path in the region $1 \leq r \leq 3$, where $r^2 = x^2 + y^2$.
(b) Find the general solution.

Exercise 6.17. Show that $x'' + x' + f(x) = 0$ has no periodic solution.

Exercise 6.18. Determine the region on which there is no periodic solution for

$$x'' + \beta(x^2 - 1)x' + x = 0.$$

Exercise 6.19. Consider the predator–prey model

$$x' = x((1-x) - y), \quad y' = y(x - a), \ a > 0,$$

where $x(t)$ and $y(t)$ are the numbers of preys and predators at time t, respectively.
(a) Sketch the nullclines and indicate the direction field.
(b) Find and analyze the equilibrium points.
(c) What can you say for each of the cases $a > 1$, $a = 1/2$, and $0 < a < 1$?

Exercise 6.20. (a) Prove Theorem 6.4.2.
(b) Use (a) to show that the system

$$x' = x(A - ax + by), \quad y' = y(B - cy + dx),$$

where a, b, c, d, A, and B are constants with a, $c > 0$, has no closed trajectories or periodic orbits.
(Hint: let $h(x, y) = \frac{1}{xy}$.)

Exercise 6.21. Use polar coordinates to study the stability of the origin for the system

$$x' = -x - \frac{y}{\text{Ln}(\sqrt{x^2 + y^2})}, \quad y' = -y + \frac{x}{\text{Ln}(\sqrt{x^2 + y^2})}.$$

Exercise 6.22. Consider the nonlinear system

$$x' = y + \frac{1}{4}x[1 - 2(x^2 + y^2)], \quad y' = -x + \frac{1}{2}y[1 - (x^2 + y^2)].$$

(a) Show that $(0, 0)$ is the only critical point.
(b) Show, using polar coordinates (r, θ), that

$$rr' = \frac{1}{4}r^2[1 + \sin^2(\theta)] - \frac{1}{2}r^4.$$

(c) Find the region R of Theorem 6.4.4.

Exercise 6.23. Consider the nonlinear system

$$x' = x - y - x^3, \quad y' = x + y - y^3.$$

Given that $(0, 0)$ is the only critical point, use polar coordinates to show that there is a limit cycle in the region

$$R = \{(r, \theta) : 1 < r < \sqrt{2}\}.$$

Exercise 6.24. Use Theorem 6.6.1 to show that the system

$$x' = x\mu - y - x^3 - xy^2 - y, \quad y' = x + y\mu - y^3 - x^2y$$

undergoes a Hopf bifurcation, that is, the system has a stable periodic solution with radius $\sqrt{\mu}$.

Exercise 6.25. Decide if Theorem 6.6.1 applies to the van der Pol equation given in system form by

$$x' = y, \quad y' = -x - \mu(x^2 - 1)y.$$

Exercise 6.26. By solving each of the systems below find E^s, E^u, M^s, and M^u.
(a) $x' = -2x$, $y' = 3x + y$.
(b) $x' = 2x - y$, $y' = -y$.

Exercise 6.27. By solving the following system find E^s, E^u, M^s, and M^u:

$$x_1' = -x_1, \quad x_2' = x_2 + x_1^2, \quad x_3' = 0.$$

Exercise 6.28. Use the method of Example 6.21 to find M^s and M^u for

$$x_1' = -x_1, \quad x_2' = 2x_2 + x_1^2.$$

Exercise 6.29. Use the method of Example 6.21 to find M^s and M^u for

$$x_1' = -x_1 - x_2^3, \quad x_2' = x_2 + x_1^3.$$

(Hint: Find the first four successive approximations to $u^{(1)}(t, a)$, $u^{(2)}(t, a)$, $u^{(3)}(t, a)$, $u^{(4)}(t, a)$.)

Exercise 6.30. In Example 6.23, find the center manifold using

$$y = h(x) = ax^2 + bx^3 + cx^4 + dx^5 + ex^6 + \mathcal{O}(x^7).$$

Exercise 6.31. Find the center manifold W^c of the system

$$\begin{cases} x' = x^2 y, \\ y' = -x^2 - y. \end{cases}$$

Exercise 6.32. Find the center manifold W^c of the system

$$\begin{cases} x' = x + 2y - xy, \\ y' = 2x + 4y - y^2. \end{cases}$$

Exercise 6.33. For Example 6.26, verify that

$$e^{At} H(y, z) = H \circ \phi_t(x, y).$$

Exercise 6.34. Find the homeomorphism H for the nonlinear system

$$\begin{cases} y' = y + z^2, \\ z' = -z. \end{cases}$$

Exercise 6.35. Approximate the homeomorphism H, W^s, and W^u for the nonlinear system

$$\begin{cases} y_1' = -y_1, \\ y_2' = -y_2^2 + y_1^3, \\ z' = z + y_1^3. \end{cases}$$

Exercise 6.36. Consider the two linear systems $x' = Ax$ and $y' = By$ with

$$A = \begin{pmatrix} -3 & 0 \\ 0 & 1 \end{pmatrix} \text{ and } B = \begin{pmatrix} -1 & -2 \\ -2 & -1 \end{pmatrix}.$$

Find the homeomorphism H so that $e^{tA} H = He^{tB}$.

Lyapunov functions

Lyapunov functions are named after Aleksandr Mikhailovich Lyapunov, a Russian mathematician, who defended the thesis *The General Problem of Stability of Motion* at Kharkiv University in 1892. Lyapunov was a pioneer in creating the global approach to the analysis of the stability of nonlinear dynamical systems. His work, initially published in Russian and then translated to French, received little attention for many years. The current use of Lyapunov functions has proved that Lyapunov was ahead of his time in developing his theory that is being used in many different areas of sciences. In theory Lyapunov functions are scalar functions that may be used to prove the stability of equilibrium of a given dynamical system. Since the inception of Lyapunov functions, their successful usage has been extended to integral equations, integro-differential equations, and functional differential equations.

7.1 Lyapunov method

The existence of Lyapunov functions is a necessary and sufficient condition for stability in differential equations. There are no concrete procedures on how to find Lyapunov functions, but in some cases, the construction of Lyapunov functions is known. In the particular case of homogeneous autonomous systems with constant coefficients the Lyapunov function can be found as a quadratic form. As we will see later, Lyapunov functions allow us to study the stability of nonlinear dynamical systems when the equilibrium solution of an associated linear differential equation is a center.

Let D be an open set in $\mathbb{R}^n$ containing 0, and let t_0 be any initial time. Define

$$V : [t_0, \infty) \times D \to [0, \infty),$$

where V is any differentiable *scalar* function. If $\psi : [t_0, \infty) \times D \to D$ is any differentiable function, then $V(t) := V(t, \psi(t))$ is a scalar function of it, and using the chain rule, we can compute its derivative

$$V'(t) = \frac{\partial V}{\partial x_1} \psi_1'(t) + \ldots + \frac{\partial V}{\partial x_n} \psi_n'(t) + \frac{\partial V}{\partial t}.$$

Advanced Differential Equations. https://doi.org/10.1016/B978-0-32-399280-0.00013-9

Let $f : [t_0, \infty) \times D \to \mathbb{R}^n$ with $f(t, 0) = 0$. Assume the existence of the unknown solution $x : [t_0, \infty) \to D$ for all $(t^*, x_0) \in [t_0, \infty) \times D$ and consider the system

$$x' = f(t, x), \tag{7.1.1}$$

where

$$f(t, x) = \begin{pmatrix} f_1(t, x) \\ f_2(t, x) \\ \vdots \\ f_n(t, x) \end{pmatrix}.$$

Then it follows from the above argument that

$$V'(t, x(t)) = \frac{\partial V(t, x(t))}{\partial x_1} f_1(t, x(t)) + \dots$$
$$+ \frac{\partial V(t, x(t))}{\partial x_n} f_n(t, x(t)) + \frac{\partial V}{\partial t}. \tag{7.1.2}$$

Thus expression (7.1.2) defines the derivative of the function $V(t, x)$ along the unknown solutions of (7.1.1). Note that if $\nabla V = grad\, V = (\frac{\partial V}{\partial x_1}, \dots, \frac{\partial V}{\partial x_n})$, then expression (7.1.2) takes the form

$$V'(t, x(t)) = grad\, V \cdot f(t, x) + \frac{\partial V}{\partial t},$$

where "$\cdot$" represents the dot product.

Suppose we are able to find V such that for any $(t, x) \in [t_0, \infty) \times D$, we have

$$\frac{\partial V(t, x(t))}{\partial x_1} f_1(t, x(t)) + \dots + \frac{\partial V(t, x(t))}{\partial x_n} f_n(t, x(t)) + \frac{\partial V}{\partial t} \leq 0.$$

Then along the solutions of (7.1.1), we have $V'(t, \psi(t)) \leq 0$. Since V is nonincreasing along the solutions and V is increasing in $|x|$, with some work, we are able to get the boundedness of solutions and stability of equilibrium solutions.

The discussion that we gave above regarding the derivative of the function V is not rigor, and more precise definitions and conditions must be imposed. We start with the following definition.

Definition 7.1.1. Let $D \subset \mathbb{R}^n$ be an open set containing the origin, and let $V(t, x)$: $\mathbb{R}^+ \times D \to R$ be a given function. Then we define $W = \mathbb{R}^+ \times D$ and

$$V'_{(7.1.1)}(t, x(t)) = \limsup_{h \to 0^+} \frac{V(t + h, x + hf(t, x)) - V(t, x)}{h}, \quad (t, x) \in W,$$

where f is the right-hand side function of (7.1.1).

Let $x(t)$ be a solution of (7.1.1) and define

$$V'(t, x(t)) = \limsup_{h \to 0^+} \frac{V(t+h, x(t+h)) - V(t, x(t))}{h}.$$

Definition 7.1.2. Suppose $V(t, x) : \mathbb{R}^+ \times D \to \mathbb{R}^+$ is Lipschitz in x (uniformly in $t \in \mathbb{R}^+$), that is, there exists $L > 0$ such that for all $t \in \mathbb{R}^+$,

$$|V(t, x_1) - V(t, x_2)| \le L\|x_1 - x_2\|, \quad x_1, x_2 \in D.$$

We further assume that $V(t, x)$ is continuous in t and Lipschitz in x (uniformly in t) with Lipschitz constant $L > 0$. We have the following lemma.

Lemma 7.1. *Let $x(t)$ be a solution of (7.1.1). Then*

$$V'(t, x(t)) = V'_{(7.1.1)}(t, x(t)). \tag{7.1.3}$$

Moreover, if $Q : W \to [0, \infty)$ is continuous such that

$$V'(t, x(t)) \le -Q(t, x(t)), \tag{7.1.4}$$

then

$$V(t_2, x(t_2)) - V(t_1, x(t_1)) \le -\int_{t_1}^{t_2} Q(t, x(t))dt, \ 0 \le t_1 \le t_2. \tag{7.1.5}$$

Proof. First, we note that since V is a Lipschitz function, we have

$$-L\|x_1 - x_2\| \le V(s, x_1) - V(s, x_2) \le L\|x_1 - x_2\|, \ s \ge 0, \ x_1, x_2 \in D.$$

Let $x(t)$ be a solution of (7.1.1). Then

$$\begin{aligned}
V(t+h, &x(t+h)) - V(t, x(t)) \\
&= V(t+h, x(t+h)) - V(t+h, x+hf(t, x)) \\
&+ V(t+h, x+hf(t, x)) - V(t, x(t)).
\end{aligned}$$

Divide the above expression by h, take $\limsup_{h \to 0^+}$, and then apply the Lipschitz condition to get

$$\begin{aligned}
\limsup_{h \to 0^+} &\frac{V(t+h, x(t+h)) - V(t, x(t))}{h} \\
&\le \limsup_{h \to 0^+} \frac{V(t+h, x+hf(t, x)) - V(t, x(t))}{h} \\
&+ L \lim_{h \to 0^+} \frac{\|x(t+h) - x(t) - hf(t, x(t))\|}{h}
\end{aligned}$$

$$= \limsup_{h \to 0^+} \frac{V(t+h, x+hf(t,x)) - V(t,x(t))}{h}$$
$$+ L\|x' - f(t,x)\|,$$

which gives

$$V'(t, x(t)) \le \limsup_{h \to 0^+} \frac{V(t+h, x+hf(t,x)) - V(t,x)}{h} = V'_{(7.1.1)}(t, x(t)). \quad (7.1.6)$$

In a similar way, but reversing the use of Lipschitz inequality, we get

$$\limsup_{h \to 0^+} \frac{V(t+h, x(t+h)) - V(t,x(t))}{h}$$
$$\ge \limsup_{h \to 0^+} \frac{V(t+h, x+hf(t,x)) - V(t,x(t))}{h}$$
$$- L \lim_{h \to 0^+} \frac{\|x(t+h) - x(t) - hf(t,x(t))\|}{h}$$
$$= \limsup_{h \to 0^+} \frac{V(t+h, x+hf(t,x)) - V(t,x(t))}{h}$$
$$- L\|x' - f(t,x)\|,$$

which gives

$$V'(t, x(t)) \ge \limsup_{h \to 0^+} \frac{V(t+h, x+hf(t,x)) - V(t,x)}{h} = V'_{(7.1.1)}(t, x(t)). \quad (7.1.7)$$

This shows that

$$V'(t, x(t)) = V'_{(7.1.1)}(t, x(t)).$$

As for (7.1.5), we integrate both sides of (7.1.4) from t_1 to t_2. This completes the proof. $\square$

Remark 7.1. Lemma 7.1 states that the derivative of V with respect to (7.1.1) is the derivative of V along the solution of (7.1.1) and is given by (7.1.2).

The situation is considerably simplified when V is independent of t. In this case (7.1.2) simplifies to

$$V'(x(t)) = \frac{\partial V(x(t))}{\partial x_1} f_1(t, x(t)) + \cdots + \frac{\partial V(x(t))}{\partial x_n} f_n(t, x(t)). \quad (7.1.8)$$

Recall that if x^* is an equilibrium point of (7.1.1), then the change of variables $x = y + x^*$ translates every nonzero equilibrium solution to the origin.

Definition 7.1.3. Let D be an open set in $\mathbb{R}^n$ containing 0. A continuous autonomous function $V : D \to [0, \infty)$ is *positive definite* if

$$V(0) = 0 \text{ and } V(x) > 0 \text{ for } x \neq 0.$$

The function V is said to be *negative definite* if $-V$ is positive definite.

Definition 7.1.4. Let D be an open set in $\mathbb{R}^n$ containing 0. Let

$$V : D \to [0, \infty)$$

have continuous first partial derivatives. If V is positive definite and

$$V'(x(t)) = \frac{\partial V(x(t))}{\partial x_1} f_1(t, x(t)) + \cdots + \frac{\partial V(x(t))}{\partial x_n} f_n(t, x(t)) \leq 0$$

for $(t, x) \in [t_0, \infty) \times D$, and $x \neq 0$, then V is called a *Lyapunov function* for system (7.1.1).

If the inequality is strict, that is, $V'(x(t)) < 0$, then V is said to be a *strict Lyapunov function*.

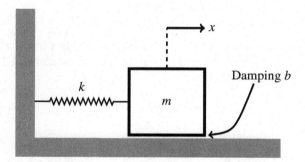

FIGURE 7.1

Damping harmonic motion.

Consider the damped harmonic motion depicted in Fig. 7.1, where m, k are positive constants and $a \geq 0$. If x is the displacement of the mass from its equilibrium position, then applying Newton's second law to all forces acting on the system, we arrive at the second-order differential equation

$$mx'' + bx' + kx = 0.$$

We introduce the transformation $x_1 = x$ and $x_2 = x'$ and arrive at the system

$$x_1' = x_2, \quad x_2' = -\frac{k}{m} x_1 - \frac{b}{m} x_2. \tag{7.1.9}$$

Clearly, if $b = 0$, then linearization gives no information regarding the stability of the origin. Next, we try to construct a Lyapunov function that overcomes such difficulties. Since we are dealing with a physical system, we expect that the system energy will decrease over time to zero. Thus a logical choice for a Lyapunov function would be the sum of *kinetic energy* and *spring potential energy*, that is,

$$V(x) = \frac{1}{2}m(x')^2 + \frac{1}{2}kx^2.$$

Clearly, V is positive definite, and

$$\begin{aligned} V' &= mx'x'' + kxx' \\ &= x'\left(mx'' + kx\right) \\ &= x'(-bx') \\ &= -b(x')^2. \end{aligned}$$

Later on, we show that this is enough to demonstrate that the origin is stable. We make the following definitions regarding stability.

Definition 7.1.5.

$$x' = f(t, x), \tag{7.1.10}$$

where $x(t) \in \mathbb{R}^n$. Assume that f is continuous and locally Lipschitz with respect to x. Let $t \rightarrow x(t, t_0, x_0)$ denote the maximal defined solution of (7.1.10) satisfying the initial condition $x(t_0) = x_0$. Let

$$\phi : [t_0, \infty) \rightarrow \mathbb{R}^n$$

be a solution of the differential equation (7.1.10).

(a) We say that the solution ϕ is *stable* (S) on $[t_0, \infty)$ if for each $\varepsilon > 0$, there is $\delta = \delta(t_0, \varepsilon) > 0$ such that whenever $|\phi(t_0) - x_0| < \delta$, the solution $x(t, t_0, x_0)$ is defined for all $t \in [t_0, \infty)$, and

$$|\phi(t) - x(t, t_0, x_0)| < \varepsilon \text{ for all } t \geq t_0.$$

(b) We say that the solution ϕ is *uniformly stable* (US) on $[t_0, \infty)$ if it is stable and given $\varepsilon > 0$ as before, δ in (a) is independent of t_0 and such that whenever $|\phi(t_0) - x_0| < \delta$, the solution $x(t, t_0, x_0)$ is defined for all $t \in [t_0, \infty)$, and

$$|\phi(t) - x(t, t_0, x_0)| < \varepsilon \text{ for all } t \geq t_0.$$

(c) We say that the solution ϕ is *asymptotically stable* (AS) on $[t_0, \infty)$ if it is stable and given $\varepsilon > 0$ as before, there is $\delta_1 < \delta$ such that whenever $|\phi(t_0) - x_0| < \delta_1$, we have $\lim_{t \to \infty} |\phi(t) - x(t, t_0, x_0)| = 0$.

(d) If ϕ is not stable, then we say that it is *unstable*, that is, there is some $\varepsilon > 0$ such that for every $\delta > 0$, there is some point x_0 with $|\phi(t_0) - x_0| < \delta$ such that $|\phi(t_1) - x(t_1, t_0, x_0)| \geq \varepsilon$ for some $t_1 \in [t_0, \infty)$.

(e) We say that the solution ϕ is *uniformly asymptotically stable* (UAS) on $[t_0, \infty)$ if it is US and there exists $\gamma > 0$ such that for each $\mu > 0$, there exists $T = T(\mu) > 0$ such that $|\phi(t_0) - x_0| < \gamma$, $t \geq t_0 + T$, implies $|x(t, t_0, x_0) - \phi(t)| < \mu$.

Since any equilibrium solution can be translated to the origin, we have the equivalent definitions regarding the zero solution $x = 0$.

Let $f : [t_0, \infty) \times D \to \mathbb{R}^n$ with $D \subset \mathbb{R}^n$ containing the origin. Assume the existence of the unknown solution $x : [t_0, \infty) \to D$ on each $(t^*, x_0) \in [t_0, \infty) \times D$ and consider system (7.1.1) with the initial condition $x(t_0) = x_0$ and $f(t, 0) = 0$.

Definition 7.1.6. The zero solution ($x = 0$) of (7.1.1):

(a) is *stable* (S) if for all $\varepsilon > 0$ and $t_0 \geq 0$, there is $\delta = \delta(t_0, \varepsilon) > 0$ such that $|x(t_0)| < \delta$ implies $|x(t, t_0, x_0)| < \varepsilon$,

(b) is *uniformly stable* (US) if δ in (a) is independent of t_0,

(c) is *unstable* if it is not stable,

(d) is *asymptotically stable* (AS) if it is stable and $\lim_{t \to \infty} |x(t, t_0, x_0)| = 0$, and

(e) is *uniformly asymptotically stable* (UAS) if it is US and there exists $\gamma > 0$ such that for each $\mu > 0$, there exists $T = T(\mu) > 0$ such that $|x(t_0)| < \gamma$, $t \geq t_0 + T$, implies $|x(t, t_0, x_0)| < \mu$.

Thus $x = 0$ is stable if solutions starting near $x = 0$ do not wander too far from $x = 0$ in future time. Now $x = 0$ is asymptotically stable if it is stable, and solutions near x_0 approach 0 as $t \to \infty$. The next result is useful in constructing positive definite or negative definite functions and is stated in the following theorem.

Theorem 7.1.1. *For constants a, b, and c, the function*

$$V(x, y) = ax^2 + bxy + cy^2$$

is positive definite if and only if

$$a > 0 \ and \ 4ac - b^2 > 0$$

and is negative definite if and only if

$$a < 0 \ and \ 4ac - b^2 > 0.$$

Example 7.1. Consider the planar system

$$x' = -x - xy^2,$$
$$y' = -y - yx^2.$$

Set

$$V(x, y) = ax^2 + bxy + cy^2.$$

Computing the derivative of V along the solutions of the system, we obtain

$$V'(x, y) = -2[2a(x^2 + x^2y^2) + b(2xy + xy^3 + yx^3) + 2c(y^2 + x^2y^2)].$$

If we choose $b = 0$, $a > 0$, and $c > 0$, then V' is negative definite, and V is positive definite by Theorem 7.1.1.

7.1.1 Stability of autonomous systems

Let D be an open subset of $\mathbb{R}^n$ containing 0, and let $f : D \to \mathbb{R}^n$ with $f(0) = 0$. Assume the existence of the unknown solution $x : [t_0, \infty) \to D$ for each $(t^*, x_0) \in [t_0, \infty) \times D$ and consider the system

$$x' = f(x), \; x(0) = x_0, \; t \geq 0, \tag{7.1.11}$$

where

$$f(x) = \begin{pmatrix} f_1(x) \\ f_2(x) \\ \vdots \\ f_n(x) \end{pmatrix}.$$

Note that Definitions 7.1.3, 7.1.4, and 7.1.6 carry over to the autonomous system (7.1.11). We make the following definitions regarding an open ball so that we may state the stability notion in terms of such balls.

Definition 7.1.7. Let $x^* \in \mathbb{R}^n$ and $r > 0$. Then $B_r(x^*)$ denotes the open ball centered at x^* with radius r, that is,

$$B_r(x^*) = \{x \in \mathbb{R}^n : ||x - x^*|| < r\}.$$

Definition 7.1.8. The equilibrium point x^* of (7.1.11):

(a) is *stable* (S) if for each ball $B_\varepsilon(x^*)$, there is a ball $B_\delta(x^*)$ ($\delta \leq \varepsilon$) such that if $x \in B_\delta(x^*)$, then $x(t, x_0)$ remains in the ball $B_\varepsilon(x^*)$ for $t \geq 0$,

(b) is *asymptotically stable* (AS) if it is *stable* and there is a ball $B_c(x^*)$ such that for each $x \in B_c(x^*)$, $x(t, x_0) \to 0$ as $t \to \infty$.

We warm up with the following example.

Example 7.2. The origin $(0, 0)$ is the only equilibrium solution and is a node for the system

$$x' = -x,$$
$$y' = -y.$$

We apply Definition 7.1.8. The solution is easily computed and is given by

$$\begin{pmatrix} x \\ y \end{pmatrix} = \begin{pmatrix} c_1 e^{-t} \\ c_2 e^{-t} \end{pmatrix}$$

for $x(0) = c_1$ and $y(0) = c_2$. For all $\varepsilon > 0$, let $\delta = \varepsilon > 0$. Then for all $c \in B_\delta(0) = \{(x, y) \in \mathbb{R}^2 : \sqrt{x^2 + y^2} < \delta\}$ and $t > 0$, we have

$$\|(x, y)\| = \sqrt{x^2(t) + y^2(t)} = \sqrt{(c_1^2 + c_2^2)e^{-2t}} \leq \delta = \varepsilon.$$

It follows that

$$(x(t, 0, c_1), y(t, 0, c_2)) \in B_\delta(0),$$

and hence $(0, 0)$ is stable. Moreover,

$$(x(t, 0, c_1), y(t, 0, c_2)) \to (0, 0) \text{ as } t \to \infty$$

shows that $(0, 0)$ is AS.

Theorem 7.1.2. *Let $V(x)$ be a Lyapunov function in D, that is, $V : D \to [0, \infty)$,*

$$V(0) = 0, \quad V(x) > 0, \quad \text{and } V' \leq 0 \text{ for } x \neq 0.$$

Then the zero solution of (7.1.11) is stable.

Proof. Pick $\varepsilon > 0$ such that the ball $B_\varepsilon(0)$ is in D. Let S_ε be the boundary of $B_\varepsilon(0)$. Due to the continuity of V at the origin and $V(0) = 0$, there must be a point on S_ε where V attains its minimum, which we denote by $m = \min_{x \in S_\varepsilon} V(x) > 0$. Let $B_\delta(0)$ be another ball in D. Since $V(0) = 0$ and V is continuous we can choose $\delta > 0$ such that $V(x) < m$ for all $x \in B_\delta(0)$. Our goal is to show that if $x_0 \in B_\delta(0)$, then $x(t, 0, x_0) \in B_\varepsilon(0)$ for all $t > 0$. Since $V' \leq 0$, then $V(x_0) < m$ implies that $V(x(t, 0, x_0)) < m$ for all $t > 0$ because the nonpositivity of the derivatives implies that V does not increase along the orbits with the initial condition x_0. This in turn implies that $x(t, 0, x_0)$ cannot cross S_ε since $V(x) \geq m$ for all $\in S_\varepsilon$. This implies that the origin is stable. This completes the proof. $\square$

Example 7.3. Consider the planar system

$$x' = -y + xy,$$
$$y' = x - x^2.$$

Clearly, $(0, 0)$ is an equilibrium point of the system. The linearization theorem, Theorem 6.2.1, is inconclusive since $(0, 0)$ is a center for the linear part

$$x' = -y,$$
$$y' = x.$$

Moreover, using the method of polar coordinates gives no information about the stability of the origin. Let us try to construct a Lyapunov function. Let a and b be positive constants and set

$$V(x, y) = ax^2 + by^2,$$

where a and b are to be determined. It is clear that $V(0,0) = 0$ and $V(x,y) > 0$ for $(x,y) \neq (0,0)$.

Computing the derivative of V along the solutions of the system using (7.1.8), we obtain

$$V'(x,y) = axx' + byy'$$
$$= ax(-y + xy) + by(x - x^2)$$
$$= 0$$

by setting $a = b > 0$. The simplest choice would be $a = \frac{1}{2} = b$. In this case,

$$V(x,y) = \frac{1}{2}(x^2 + y^2).$$

Thus by Theorem 7.1.2 the origin is stable.

Here is another example.

Example 7.4. Consider the three-dimensional system

$$x' = -2y + yz,$$
$$y' = x - xz,$$
$$z' = xy.$$

Clearly, $(0,0,0)$ is an equilibrium solution of the system, and

$$J(0,0,0) = \begin{pmatrix} 0 & -2 & 0 \\ 1 & 0 & 0 \\ 0 & 0 & 0 \end{pmatrix}.$$

Thus $J(0,0,0)$ has the eigenvalues $\lambda_1 = 0$ and $\lambda_2 = \lambda_3 = i\sqrt{2}$. The linearization theorem, Theorem 6.2.1, cannot be applied. Let us try to construct a Lyapunov function. Let a, b, and c be positive constants and set

$$V(x,y) = ax^2 + by^2 + cz^2,$$

where a and b are to be determined. It is clear that $V(0,0) = 0$ and $V(x,y,z) > 0$ for $(x,y,z) \neq (0,0,0)$. Computing the derivative of V along the solutions of the system using (7.1.8), we obtain

$$V'(x,y) = axx' + byy' + czz'$$
$$= ax(-2y + yz) + by(x - xz) + cz(xy)$$
$$= 2[(a - b + c)xyz + (-2a + c)xy].$$

Hence if $b = 2a$ and $c = a > 0$, then $V'(x, y, z) = 0$ for all $(x, y, z) \in \mathbb{R}^3$, and therefore by Theorem 7.1.2 the origin $(0, 0, 0)$ is stable. Furthermore, choosing $a = c = 1$ and $b = 2$, we see that the trajectories of this system lie on the ellipsoids

$$x^2 + 2y^2 + z^2 = k^2$$

for positive constants k.

Example 7.3 reveals the advantage of the use of Lyapunov functions over linearization since it showed that a center for a linear system is stable for the accompanying nonlinear system.

The next theorem provides criteria for asymptotic stability.

Theorem 7.1.3. *Let $V(x)$ be a strict Lyapunov function in D, that is, $V : D \to [0, \infty)$,*

$$V(0) = 0, \quad V(x) > 0, \text{ and } V' < 0 \text{ for } x \neq 0.$$

Then the zero solution of (7.1.11) is asymptotically stable.

Proof. Since the origin is stable and $V' < 0$ along the solutions of (7.1.11), it follows that $V(x)$ decreases to the value c as $t \to \infty$. The proof will be complete if we show that $c = 0$. Assume for contradiction that $c > 0$. Then there exists $\beta < \varepsilon$ (ε from the stability theorem) such that $V(x) < c$ for all $x \in B_\beta(0)$. However, then $x(t)$ cannot enter $B_\beta(0)$. Let

$$m = \min\{-V' : \beta \leq |x| \leq \varepsilon\}.$$

As $-V' > 0$, we have that $V' \leq -m$ for all $t \geq 0$. An integration of $V' \leq -m$ from 0 to t yields

$$V(x(t)) - V(x_0) \leq -mt.$$

Thus $\lim_{t \to \infty} V(x(t)) = -\infty$, which contradicts the assumption that V is positive definite in D and tends to c as $t \to \infty$. Hence c must vanish.

We next claim that $x(t) \to 0$ as $t \to \infty$. If this is not the case, then there are constants $\alpha > 0$ and a sequence $t_k \to \infty$ such that $|x(t)| \geq \alpha$ for all k. Since the closure $\bar{B}_\beta(0)$ is compact and all the points $x(t_k) \in \bar{B}_\beta(0)$, there is a subsequence that converges to a point $z \in B_\beta(0)$. We may assume that $x(t_k) \to z$, and we must have $0 \neq |z| \geq \alpha$. By the continuity of V we have that $V(x(t_k)) \to V(z) > 0$. However, since $t_k \to \infty$, we must have $V(x(t_k)) \to c = 0$, which is a contradiction. Thus the origin is asymptotically stable. This completes the proof. $\square$

Example 7.5. Consider the planar system

$$x' = -x - xy^2,$$
$$y' = -y + 3x^2y.$$

Clearly, $(0, 0)$ is an equilibrium point of the system. The linearization theorem, Theorem 6.2.1, tells us that $(0, 0)$ is AS. We verify the result by constructing a strict Lyapunov function. Let a and b be positive constants and set

$$V(x, y) = ax^2 + by^2,$$

where a and b are to be determined. Clearly, $V(0, 0) = 0$ and $V(x, y) > 0$ for $(x, y) \neq (0, 0)$. Computing the derivative of V along the solutions of the system using (7.1.8), we obtain

$$\begin{aligned}
V'(x, y) &= axx' + yy' \\
&= ax(-x - xy^2) + by(-y + 3x^2y) \\
&= -2ax^2 - 2ax^2y^2 - 2by^2 + 6bx^2y^2 \\
&= -(6x^2 + 2y^2) < 0
\end{aligned}$$

unless $x = y = 0$, where we choose $a = 3$, $b = 1$ to eliminate the mixed terms. Thus by Theorem 7.1.3 the origin is asymptotically stable.

Note that the origin $(0, 0)$ of the system in Example 7.1 is asymptotically stable by Theorem 7.1.3.

Let $V(x)$ be a strict Lyapunov function in D. Then *domain of attraction* of an equilibrium consists of all points such that a solution starting at them tends to the equilibrium. In engineering applications, it is often important to get an estimate of the size of the domain of attraction. Thus if V is a strict Lyapunov function on a set $D \in \mathbb{R}^n$, then the origin is asymptotically stable, and the *domain of attraction* is

$$\bar{N} = \{x \in D : V(x) \leq C\}$$

for some positive constant C.

Example 7.6. Consider the planar system

$$\begin{aligned}
x' &= -x - y, \\
y' &= 2x - y + y^3.
\end{aligned}$$

Clearly, $(0, 0)$ is an equilibrium point of the system. Set

$$V(x, y) = x^2 + \frac{y^2}{2}.$$

Clearly, $V(0, 0) = 0$, and $V(x, y) > 0$ for $(x, y) \neq (0, 0)$. Computing the derivative of V along the solutions of the system, we obtain

$$\begin{aligned}
V'(x, y) &= 2xx' + yy' \\
&= 2x(-x - y) + y(2x - y + y^3) \\
&= -2x^2 - y^2 + y^4.
\end{aligned}$$

To keep $V'(x, y) < 0$, we must restrict y so that $-y^2 + y^4 = y^2(-1 + y^2) < 0$, that is, $-1 < y < 1$. On the other hand, to keep the ellipses $V(x, y) = x^2 + \frac{y^2}{2} = C$ inside $D = \{(x, y) : x \in \mathbb{R}, \ -1 < y < 1\}$, we must take $0 < C < \frac{1}{2}$. Thus for any value of such C, the origin is asymptotically stable, and the domain of attraction is

$$\bar{N} = \{x \in D : V(x) \le C < \frac{1}{2}\}.$$

The next theorem only asks for the existence of a Lyapunov function to obtain the asymptotic stability but with a catch.

Theorem 7.1.4. *Let $V(x)$ be a Lyapunov function in D. Let*

$$Z = \{x : V'(x) = 0, x \in D\}.$$

Suppose that the only bounded solution of (7.1.11) that remains inside Z all the time is the equilibrium solution 0. Then the origin of (7.1.11) is asymptotically stable.

Proof. Theorem 7.1.4 is a particular case of Theorem 7.4.2, and we ask the reader to either wait or flip the pages.   □

As an application of Theorem 7.1.4, we provide the following example.

Example 7.7. Consider the planar system

$$x' = -x - 2y,$$
$$y' = 3x - 2y.$$

Clearly, $(0, 0)$ is an equilibrium point of the system. Set

$$V(x, y) = \frac{x^2}{2} + y^2.$$

Clearly, $V(0, 0) = 0$ and $V(x, y) > 0$ for $(x, y) \ne (0, 0)$. Computing the derivative of V along the solutions of the system, we obtain

$$\begin{aligned} V'(x, y) &= xx' + 2yy' \\ &= x(-x - 2y) + 2y(3x - 2y) \\ &= -(x - 2y)^2. \end{aligned}$$

It is clear that $V'(x, y) = 0$ if $x = 2y$. Thus $V(x, y)$ is a Lyapunov function (not strict) on the set

$$Z = \{(x, y) \in \mathbb{R}^2 : x = 2y\}.$$

Let (x, y) be a bounded solution (trajectory) in Z, that is, there exists a positive constant K such that $|x|, |y| \le K$. Substitute $x = 2y$ into the original system to get $x' = -2x$ and $y' = 4y$. By straightforward calculations we obtain the solutions

$$x(t) = x(0)e^{-2t} \quad \text{and} \quad y(t) = y(0)e^{4t},$$

respectively, for $t \in \mathbb{R}$. These functions cannot satisfy $|x|$, $|y| \leq K$, unless $x(0) = y(0) = 0$. To have you more convinced, the solutions must also satisfy $x = 2y$, or

$$x(0)e^{-2t} = 2y(0)e^{4t},$$

which can only hold if $x(0) = y(0) = 0$. Thus the only bounded solution in Z is the origin, and hence by Theorem 7.1.4 the origin is asymptotically stable.

The next theorem provides criteria for instability of the zero solution of (7.1.11).

Theorem 7.1.5. *Let D be an open set in $\mathbb{R}^n$ containing 0. Let $V : D \to [0, \infty)$ be continuous real-valued scalar differentiable function that is positive definite on D. If $V'(x) > 0$ for $x \neq 0$, then the origin of (7.1.11) is unstable.*

Proof. Assume that the zero solution is stable. Then, for $\varepsilon = 1$ and $t_0 = 0$, there exists $\delta > 0$ such that $|x_0| < \delta$ implies that $|x(t, 0, x_0)| \leq 1$. Hence for $x(t) = x(t, 0, x_0)$, $V(x(t))$ is bounded for $t \geq 0$ when $|x_0| \leq \delta$. Fix x_0 with $|x_0| = \frac{\delta}{2}$ such that $V(x_0) > 0$. Since $V' > 0$, it follows that V is increasing along the solutions. Therefore we have $V(x(t)) \geq V(x_0) > 0$, $t \geq 0$. Since V is continuous and $V(0) = 0$, there exists $p \in (0, 1)$ such that $p \leq |x(t)| \leq 1$, $t \geq 0$. Since $V' > 0$ along the solutions of (7.1.11) and positive on the closed and bounded set $p \leq |x(t)| \leq 1$, we have $l = \min_{p \leq |x| \leq 1} V'(x) > 0$. Thus

$$V(x(t)) - V(x_0) = \int_0^t \frac{d}{ds} V(x(s))ds \geq \int_0^t l\,ds = l(t - t_0) \to \infty, \ t \to \infty.$$

This contradicts the fact that $V(x(t))$ is bounded for $t \geq 0$ when $|x_0| < \delta$. It follows that the zero solution $x(t) = 0$ is unstable. $\square$

Example 7.8. Consider the planar system

$$x' = 2xy + x^3,$$
$$y' = -x^2 + y^5.$$

Clearly, $(0, 0)$ is an equilibrium point of the system. Linearization will not work here since there are no linear terms at $(0, 0)$. Set

$$V(x, y) = x^2 + 2y^2.$$

Clearly, $V(0, 0) = 0$ and $V(x, y) > 0$ for $(x, y) \neq (0, 0)$. Computing the derivative of V along the solutions of the system, we obtain

$$V'(x, y) = 2xx' + 4yy'$$
$$= 2x(2xy + x^3) + 4y(-x^2 + y^5)$$
$$= 2x^4 + 4y^6 > 0, \ (x, y) \neq (0, 0).$$

It follows from Theorem 7.1.5 that the origin is unstable.

Example 7.9. We examine the stability and asymptotic stability for different values of a for the planner system

$$x' = ax(x^2 + y^2) - xy^2,$$
$$y' = ay(x^2 + y^2) + x^2 y.$$

Clearly, $(0, 0)$ is an equilibrium point of the system. Linearization will not work here since there are no linear terms in the neighborhood of $(0, 0)$. Set

$$V(x, y) = x^2 + y^2.$$

Clearly, $V(0, 0) = 0$ and $V(x, y) > 0$ for $(x, y) \neq (0, 0)$. Set $V(t) := V(x(t), y(t))$. Then along the solutions of the system we have that

$$\begin{aligned} V'(t) &= 2xx' + 2yy' \\ &= 2x[ax(x^2 + y^2) - xy^2] + 2y[ay(x^2 + y^2) + x^2 y] \\ &= 2a(x^2 + y^2)^2. \end{aligned} \tag{7.1.12}$$

It is clear from relation (7.1.12) that the origin is stable if $a = 0$, asymptotically stable if $a < 0$, and unstable if $a > 0$. In the following, we will use expression (7.1.12) and try to understand the behavior of the Lyapunov function along the solutions.

We observe that (7.1.12) is equivalent to

$$V'(t) = 2aV^2(t),$$

which is a separable first-order differential equation in $V(t)$. Thus separating the variables and then integrating from 0 to t gives

$$V(t) = \frac{V(0)}{1 - 2aV(0)t}, \quad t \geq 0. \tag{7.1.13}$$

Case 1. If $a = 0$, then from (7.1.13) we have that $V(t) = V(0) = x(0)^2 + y(0)^2 > 0$. In this case, we have a periodic solution surrounding the origin

$$x^2 + y^2 = k^2$$

for nonzero constant k. Hence the origin is stable.

Case 2. If $a < 0$, then from (7.1.13) we have that $V(t) \to 0$ as $t \to \infty$, that is, $x^2 + y^2 \to 0$ as $t \to \infty$, which means that the solutions spiral toward the origin, and we conclude that the origin is asymptotically stable.

Case 3. If $a > 0$, then by (7.1.13) $V(t)$ is not defined for all $t \geq 0$. In fact, $V(t) \to \infty$ as $t \to \frac{1}{2aV(0)}$ from the left. We conclude that the origin is unstable.

The next theorem provides criteria of the instability of the zero solution, which is different from that given in Theorem 7.1.5.

Theorem 7.1.6. *Let x^* be an equilibrium solution of (7.1.11). Let U be a neighborhood of x^*. Let $V : U \to \mathbb{R}$ be a continuously differentiable function on U, except perhaps at x^*. If $V(x^*) = 0$, but there are points arbitrarily close to x^* where V is strictly positive and $V' > 0$ on $U \setminus \{x^*\}$, then x^* is unstable for (7.1.11).*

Proof. Without loss of generality, we take $x^* = 0$. Let $B_\beta(0) \subseteq U$, $\beta > 0$. We may choose a point $x_0 \in B_\beta(0)$ close to 0 such that $V(x_0) > 0$. Let (a, b) be the interval of existence of the solution $x(t) = x(t, x_0)$. The proof will be completed if we can show that $x(t)$ is outside the ball $\bar{B}_\beta(0)$. Assume the contrary, that is, solutions stay in $B_\beta(0)$ for all $t \geq 0$. Since $\bar{B}_\beta(0)$ is compact, V achieves its maximum, say $V_0 \in \bar{B}_\beta(0)$. Thus

$$V(x(t)) \leq V_0, \quad t \geq 0. \tag{7.1.14}$$

Since $V' > 0$ and V is increasing along the solutions and $V' \geq V(x_0) > 0$, we can find $\rho > 0$ such that $V(x) < V(x_0)$ for $x \in B_\rho(0)$. Then $x(t)$ is trapped in the compact annulus $\rho \leq |x| \leq \beta$. Let

$$m = \min\{V' : \rho \leq |x| \leq \beta\}.$$

An integration of $V' \geq m$ from 0 to t yields

$$V(x(t, x_0)) \geq V(x_0) + mt,$$

which contradicts (7.1.14) since the right side goes to infinity as t goes to infinity. Thus the solution $x(t)$ must escape $B_\beta(0)$, and the origin is unstable. This completes the proof. $\qquad\square$

Lyapunov functions can also be used to determine limit cycles, as the next example demonstrates.

Example 7.10. In Example 6.15 we considered the planar autonomous system

$$x' = -y - x(x^2 + y^2 - \mu), \quad y' = x - y(x^2 + y^2 - \mu)$$

with $\mu > 0$ and used polar coordinates to show the existence of a stable limit cycle. In this example, we use a Lyapunov function to arrive at the same result. Let

$$V(x, y) = x^2 + y^2.$$

Then along the solutions of the system we have the expression

$$\begin{aligned}
V'(t) &= 2xx' + 2yy' \\
&= 2x[-y - x(x^2 + y^2 - \mu)] + 2y[x - y(x^2 + y^2 - \mu)] \\
&= -2(x^2 + y^2 - \mu)(x^2 + y^2) \\
&= 2(\mu - V(x, y))V(x, y).
\end{aligned}$$

Now since $V > 0$, the sign of V' depends on the term $\mu - V(x, y)$. We have $V' = 0$ if $V = \mu$ or $x^2 + y^2 = \mu$, which is a circle centered at the origin with radius μ. If $0 < V < \mu$, then $V' > 0$. In this case the trajectories start inside the circle with radius μ and move upward toward it but do not cross it due to uniqueness. If $V > \mu$, then $V' < 0$. In this case the trajectories start outside the circle with radius μ and move toward it but do not cross it due to uniqueness. We conclude that $x^2 + y^2 = \mu$ is a stable limit cycle.

The next example shows the importance of Theorem 7.1.6 and illustrates its usage as Theorem 7.1.5 is not applicable.

Example 7.11. Considered the planar autonomous system

$$x' = 3x + y^2, \quad y' = -2y + x^3.$$

Clearly, the origin is the only equilibrium point of the system. We easily conclude that $(0, 0)$ is unstable using the linearization method. Let us see how we can implement Theorem 7.1.6. We consider the Lyapunov function

$$V(x, y) = \frac{x^2}{2} - \frac{y^2}{2}.$$

There are points in the neighborhood of the origin where V is positive, for example, the points on the x-axis. Since we are only interested what happens in the neighborhood of $(0, 0)$, we limit ourselves to $|x| < 1$ and $|y| < 1$. To better understand the derivative of V, we will utilize the inequality

$$0 < (a - b)^2 = a^2 - 2ab + b^2$$

for all real numbers a and b, from which it follows that

$$ab \leq \frac{a^2}{2} + \frac{b^2}{2}.$$

It readily follows that along the solutions of the system,

$$V' = (3x^2 + 2y^2) + (xy^2 - x^3y). \tag{7.1.15}$$

Next, we compare the term $(3x^2 + 2y^2)$ with $(xy^2 - x^3y)$:

$$\begin{aligned}
|xy^2 - x^3y| &\leq |xy||y - x^2| \\
&\leq \frac{1}{2}(x^2 + y^2)(x^2 + |y|) \\
&\leq \frac{1}{2}(x^2 + y^2)(|x| + |y|) \\
&\leq \frac{1}{2}(|x| + |y|)(|x| + |y|)
\end{aligned}$$

$$= \frac{1}{2}(|x|^2 + |y|^2) + |x||y|$$
$$\leq \frac{1}{2}(|x|^2 + |y|^2) + \frac{1}{2}(|x|^2 + |y|^2)$$
$$= x^2 + y^2$$
$$\leq 3x^2 + 2y^2.$$

Thus a substitution into (7.1.15) gives

$$V'(t) \geq \left| xy^2 - x^3y \right| + (xy^2 - x^3y) > 0$$

for $(x, y) \neq (0, 0)$. We conclude that $V' > 0$ on the set $|x|, |y| < 1$ minus the origin, and by Theorem 7.1.6 the origin is unstable. Note that the set U of Theorem 7.1.6 is defined by

$$U = \{x, y \in \mathbb{R} : |x| < 1, \ |y| < 1\}.$$

Theorem 7.1.6 also has its limitation since there may be some points near the equilibrium solution x^* that approach x^*, which makes it difficult to have $V' > 0$ on a whole neighborhood of x^*. The next generalization theorem is due to Četaev.

Theorem 7.1.7. *(Četaev) Let x^* be an equilibrium solution of (7.1.11) contained in the closure of an open set U, and let E be a neighborhood of x^*. Let $V : E \to \mathbb{R}$ be a continuously differentiable function on E such that*

(1) V and V' are both positive on $(U \cap E) \setminus \{x^*\}$.
(2) $V(x) = 0$ on the part of the boundary of U that is in E.

Then the equilibrium solution x^* is unstable.

Proof. Exercise. □

Next, we briefly discuss a special class of Lyapunov functions that includes the absolute value of the unknown solution. Let $x : \mathbb{R} \to \mathbb{R}$ be continuous. Observing that

$$|x| = \sqrt{x^2} = (x^2)^{\frac{1}{2}}$$

and using the chain rule, we arrive at

$$\frac{d}{dt}|x(t)| = \frac{1}{2}(x^2(t))^{-\frac{1}{2}}(2x(t)x'(t))$$
$$= \frac{x(t)}{(x^2(t))^{\frac{1}{2}}}x'(t)$$
$$= \frac{x(t)}{|x(t)|}x'(t).$$

We have the following lemma.

Lemma 7.2. *Let* $x : \mathbb{R} \to \mathbb{R}$ *be differentiable. Then*

$$\frac{d}{dt}|x(t)| = \frac{x(t)}{|x(t)|}x'(t).$$

Example 7.12. Consider the scalar nonlinear Volterra integro-dynamic equation

$$x'(t) = ax(t) + b\frac{x(t)}{1 + \int_0^t x^2(s)ds}, \quad t \in [0, \infty). \tag{7.1.16}$$

Suppose there exists a constant β such that

$$a + |b| = \beta.$$

Consider the function

$$V(t, x) = |x(t)|.$$

Then by Lemma 7.2, we have

$$
\begin{aligned}
V'(t, x) &= \frac{d}{dt}|x(t)| = \frac{x(t)}{|x(t)|}x'(t) \\
&= \frac{x(t)}{|x(t)|}\left(ax(t) + b\frac{x(t)}{1 + \int_0^t x^2(s)ds}\right) \\
&= ax^2\frac{1}{|x(t)|} + \frac{x}{|x|}b\frac{x(t)}{1 + \int_0^t x^2(s)ds} \\
&\leq a|x| + |b||x|\frac{1}{1 + \int_0^t x^2(s)ds} \\
&\leq a|x| + |b||x| \\
&= (a + |b|)|x|.
\end{aligned}
$$

Then it follows from Theorems 7.1.2–7.1.5 that the zero solution of (7.1.16) is stable if $\beta = 0$, asymptotically stable if $\beta < 0$, and unstable if $\beta > 0$.

7.1.2 Time-varying systems; non-autonomous

Now we turn our attention to systems that explicitly include the variable time t. Such systems are referred to as non-autonomous systems. Thus we consider the non-autonomous system

$$x' = f(t, x), \tag{7.1.17}$$

where $f \in C\big([0, \infty) \times D, \mathbb{R}^n\big)$, where $D \subset \mathbb{R}^n$ is open with $0 \in D$. We say that a vector $x^* \in \mathbb{R}^n$ is an *equilibrium solution*, or *constant solution*, or *equilibrium point* of (7.1.17) if

$$f(t, x^*) = 0.$$

This is the perfect place to make clear that the existence of equilibrium points for non-autonomous systems may depend on the initial time. To see this, we consider

$$x' = \begin{pmatrix} \frac{1}{t} & t \\ 1 & t \end{pmatrix} \begin{pmatrix} x_1 \\ x_2 \end{pmatrix}, \ t > 0, \ x(0) = x_0.$$

The system has a unique equilibrium solution $(0,0)$, provided that the matrix is non-singular, which is the case for $t \neq 1$. Thus, in this case, the system attains an equilibrium solution if and only if the initial time $t_0 > 1$.

Definition 7.1.6 of stability carries over to (7.1.17). In addition, the uniform asymptotic stability for (7.1.17) is equivalent to the asymptotic stability for autonomous system. We begin by considering the initial value time-varying scalar differential equation

$$x' = (6t \ \sin(t) - 2t)x, \ \ x(t_0) = x_0, \ \ t_0 \geq 0.$$

Its unique solution passing through (t_0, x_0) is given by

$$x(t) = x_0 \ \exp \left(\int_{t_0}^t (6s \sin(s) - 2s)ds \right)$$

$$= x_0 \ \exp \left(6\sin(t) - 6t\cos(t) - t^2 - 6\sin(t_0) + 6t_0 + 6t_0\cos(t_0) + t_0^2 \right).$$

Set

$$c(t_0) = \exp \left(6\sin(t) - 6t\cos(t) - t^2 - 6\sin(t_0) + 6t_0 + 6t_0\cos(t_0) + t_0^2 \right).$$

Then

$$|x(t)| \leq |x_0| c(t_0)|, \ \ t \geq t_0.$$

Thus for any $\varepsilon > 0$, choose $\delta(t_0, \varepsilon) = \frac{\varepsilon}{c(t_0)}$ so that

$$|x_0| < \delta \Longrightarrow |x(t)| < \varepsilon,$$

and hence the zero solution is stable but not uniformly stable. Note that $c(t_0)$ increases as t_0 increases, and hence δ shrinks as t_0 increases and t is fixed. To be more specific, suppose $t_0 = 2n\pi, n = 0, 1, 2, \ldots,$ and $x(t)$ is evaluated π seconds later in each case. Then

$$x(t_0 + \pi) = x(t_0)e^{(4n+1)(6-\pi)},$$

and if $x(t_0) \neq 0$, then we get

$$\frac{x(t_0 + \pi)}{x(t_0)} \to \infty \text{ as } t \to \infty.$$

Thus there is no δ independent of t_0 that would satisfy the requirement for uniform stability.

Definition 7.1.9. A strictly increasing continuous function $W : [0, \infty) \to [0, \infty)$ with $W(0) = 0$ is called a wedge. (In this book, wedges are always denoted by W or W_i, where i is a positive integer.)

Definition 7.1.10. Let D be an open set in $\mathbb{R}^n$ containing 0. A continuous autonomous function $V : [0, \infty) \times D \to [0, \infty)$ is *positive definite* if

$$V(t, 0) = 0 \text{ and } V(t, x) > 0 \text{ for all } x \neq 0 \text{ and all } t \in [0, \infty).$$

Thus $V(t, x)$ is positive definite if and only if there is a wedge independent of time t such that $V(t, x) \geq W(|x|)$ for all $x \in D$.

Definition 7.1.11. Let D be an open set in $\mathbb{R}^n$ containing 0. Let

$$V : [0, \infty) \times D \to [0, \infty)$$

have continuous first partial derivatives. If V is positive definite and

$$V'(t, x(t)) = \frac{\partial V(t, x(t))}{\partial x_1} f_1(t, x(t)) + \cdots + \frac{\partial V(t, x(t))}{\partial x_n} f_n(t, x(t)) + \frac{\partial V(t, x(t))}{\partial t} \leq 0$$

for $(t, x) \in [t_0, \infty) \times D$ with $x \neq 0$, then V is called *Lyapunov function* for system (7.1.17).

If the inequality is strict, that is, $V'(t, x(t)) < 0$ then V is said to be a *strict Lyapunov function*.

Definition 7.1.12. Let D be an open set in $\mathbb{R}^n$ containing 0. Let

$$V : [0, \infty) \times D \to [0, \infty).$$

We say $V(t, x)$ is:

(a) *decrescent* in D if there is a wedge W_1 such that

$$V(t, x) \leq W_1(|x|) \text{ for all } x \in D,$$

(b) *radially unbounded* if $V(t, x) \to \infty$ as $|x| \to \infty$, which means that for each $M > 0$, there is $N > 0$ such that $V(t, x) > M$ for all $t \in [0, \infty)$ and all x such that $|x| > N$,

(c) both positive definite and decrescent if

$$W_1(|x|) \leq V(t, x) \leq W_2(|x|).$$

Theorem 7.1.8. *Let $D \subset \mathbb{R}^n$ be an open set containing the origin, and let $V(t, x) : [0, \infty) \times D \to [0, \infty)$ be a given continuously differentiable function satisfying*

$$V(t, 0) = 0, \tag{7.1.18}$$

$$W_1(|x|) \le V(t, x), \qquad (7.1.19)$$

and

$$V'(t, x) \le 0 \text{ for all } t \ge 0 \text{ and } x \ne 0. \qquad (7.1.20)$$

Then the zero solution of (7.1.17) is stable.

Proof. By (7.1.20), we have $V'(t, x) \le 0$, V is continuous, $V(t, 0) = 0$, and $W_1(|x|) \le V(t, x)$. Let $\varepsilon > 0$ and $t_0 \ge 0$. We must find δ such that $|x_0| < \delta$ and $t \ge t_0$ imply $|x(t, t_0, x_0)| < \varepsilon$. (Throughout these proofs, we assume that ε is small enough so that $|x(t, t_0, x_0)| < \varepsilon$ implies that $x \in D$.) As V is continuous in t and x and $V(t, 0) = 0$, there is $\delta > 0$ such that $|x_0| < \delta$ implies $V(t_0, x_0) < W_1(\varepsilon)$. Thus if $t \ge t_0$, $|x_0| < \delta$, and $x = x(t, t_0, x_0)$, then we have

$$W_1(|x(t)|) \le V(t, x) \le V(t_0, x_0) < W_1(\varepsilon).$$

Applying W_1^{-1} to both sides of the above inequality gives

$$|x(t)| \le \varepsilon.$$

This completes the proof. $\qquad \square$

Theorem 7.1.9. *Let $D \subset \mathbb{R}^n$ be an open set containing the origin, and let $V(t, x) : [0, \infty) \times D \to [0, \infty)$ be a continuously differentiable function. Assume that (7.1.18), (7.1.19), and (7.1.20) hold. If, in addition, V satisfies*

$$V(t, x) \le W_2(|x|), \qquad (7.1.21)$$

then the zero solution of (7.1.17) is uniformly stable.

Proof. For a given ε, we choose $\delta > 0$ such that $W_2(\delta) < W_1(\varepsilon)$, where $W_1(|x|) \le V(t, x) \le W_2(|x|)$. If $t_0 \ge 0$, then we have

$$W_1(|x(t)|) \le V(t, x) \le V(t_0, x_0)$$
$$\le W_2(|x_0|) < W_2(\delta) < W_1(\varepsilon),$$

or $|x(t)| < \varepsilon$, as required. This completes the proof. $\qquad \square$

The next example shows that $V'(t, x) \le 0$ is not enough to drive solutions to zero.

Example 7.13. Let $g : [0, \infty) \to (0, \beta]$ with $g(0) = 1$ and $0 < \beta \le 1$. Consider the non-autonomous first-order differential equation

$$x'(t) = \left[g'(t)/g(t) \right] x(t). \qquad (7.1.22)$$

It is clear that $x(t) = g(t)$ is a solution of (7.1.22). Our goal is to construct a function

$$V(t, x) = a(t)x^2(t)$$

such that $V'(t,x) = -\alpha(t)x^2(t)$, where $a(t), \alpha(t) > 0$ for $t \in \mathbb{R}^+$, and $\int_0^\infty \alpha(s)ds < \infty$. Along the solutions of (7.1.22) we have

$$V'(t,x) = \left[a'(t) + 2a(t)\frac{g'(t)}{g(t)}\right]x^2(t).$$

By substituting $V'(t,x) = -\alpha(t)x^2(t)$ we arrive at

$$a'(t) + 2a(t)\frac{g'(t)}{g(t)} = -\alpha(t).$$

It readily follows that the solution is

$$a(t) = [a(0)g^2(0) - \int_0^t \alpha(s)g^2(s)ds]g^2(t)$$

$$\geq [a(0)g^2(0) - \beta^2\int_0^t \alpha(s)ds]/g^2(t).$$

Since $0 < g \leq 1$ and $\int_{s=0}^\infty \alpha(s) < \infty$, we may choose $a(0)$ large enough to imply that $a(t) > 1$ on $[0,\infty)$. Thus we have shown that $V \geq 0$ and $V' \leq 0$ do not imply that solutions tend to zero. As a matter of fact, there is no wedge W_2 with $V(t,x) \leq W_2(|x|)$. Notice that V is not decrescent.

Theorem 7.1.10. *Let $D \subset \mathbb{R}^n$ be an open set containing the origin, and let $V(t,x) : [0,\infty) \times D \to [0,\infty)$ be a continuously differentiable function satisfying (7.1.18) and (7.1.19). If, in addition,*

$$V'(t,x) \leq -W_3(|x|) \text{ for all } t \geq 0, \tag{7.1.23}$$

then the zero solution of (7.1.17) is asymptotically stable.

Proof. By Theorem 7.1.8, the zero solution is stable. Therefore given $\varepsilon > 0$, there are $\delta > 0$ and $\gamma > 0$ such that

$$\gamma \leq |x(t,t_0,x_0)| < \varepsilon, \quad t \geq t_0, \quad |x_0| < \delta.$$

Due to (7.1.23) and $|x(t,t_0,x_0)| \geq \gamma$ for $t \geq t_0$, there exists a constant $d > 0$ such that

$$V'(t,x) \leq -d < 0 \text{ for all } t \geq t_0.$$

An integration of $V' \leq -d$ from t_0 to t yields

$$V(t,x(t,t_0,x_0)) \leq V(t_0,x_0) - d(t - t_0).$$

Thus $\lim_{t\to\infty} V(x(t)) = -\infty$, which contradicts the assumption that V is positive definite. Hence no such γ exists, and since $V(t,x)$ is a positive decreasing function, it follows that $\lim_{t \to \infty} V(x(t)) = 0$. Therefore

$$\lim_{t \to \infty} x(t, t_0, x_0) = 0$$

which implies that the zero solution is asymptotically stable. This completes the proof. □

Theorem 7.1.11. *Let $D \subset \mathbb{R}^n$ be an open set containing the origin, and let $V(t, x) : [0, \infty) \times D \to [0, \infty)$ be a continuously differentiable function satisfying*

$$V(t, 0) = 0,$$

$$W_1(|x|) \leq V(t, x) \leq W_2(|x|),$$

and

$$V'(t, x) \leq -W_3(|x|) \text{ for all } t \geq 0.$$

Then the zero solution of (7.1.17) is uniformly asymptotically stable.

Proof. By Theorem 7.1.9, the zero solution is uniformly stable. Let $\varepsilon = 1$, and find $\delta > 0$ of uniform stability and call it η. Let $\gamma > 0$ be given. We must find $T > 0$ such that

$$|x_0| < \eta, \quad t_0 \geq 0, \quad \text{and } t \geq t_0 + T$$

imply $|x(t, t_0, x_0)| < \gamma$. Pick $\mu > 0$ with $W_2(\mu) < W_1(\gamma)$, so that there is $t_1 \geq t_0$ with $|x(t_1)| < \mu$. Then we have, for $t \geq t_1$,

$$W_1(|x(t)|) \leq V(t, x) \leq V(t_1, x_1)$$
$$\leq W_2(|x_1|) < W_2(\eta) < W_1(\gamma),$$

or $|x(t_1)| < \gamma$. Since $V'(t, x) \leq -W_3(|x|)$, and so as long as $|x(t)| > \mu$, it follows that $V'(t, x) \leq -W_3(\mu)$. Thus

$$V(t, x(t)) \leq V(t_0, x_0) - \int_{t_0}^{t} W_3(|x(s)|)ds$$
$$\leq W_2(|x_0|) - W_3(\mu)(t - t_0)$$
$$\leq W_2(\eta) - W_3(\mu)(t - t_0),$$

which vanishes at

$$t = t_0 + \frac{W_2(\eta)}{W_3(\mu)} \geq t_0 + T,$$

where $T \geq \frac{W_2(\eta)}{W_3(\mu)}$. Hence, if $T > \frac{W_2(\eta)}{W_3(\mu)}$, then $|x(t)| > \mu$ fails, and we have $|x(t)| < \gamma$ for all $t \geq t_0 + T$. This proves UAS. □

Example 7.14. Consider the second-order differential equation

$$u'' + e(t)u = 0,$$

where e is continuously differentiable with

$$e'(t) \leq 0 \text{ and } \lim_{t \to \infty} e(t) = e_0 > 0.$$

Making the substitution $x = u$, $y = u'$, we arrive at the equivalent system

$$x' = y, \quad y' = -e(t)x.$$

Let

$$V(t, \mathbf{x}) = e(t)x^2 + y^2,$$

where $\mathbf{x} = \begin{pmatrix} x \\ y \end{pmatrix}$. Then along the solutions we have

$$\begin{aligned} V'(t, \mathbf{x}) &= e'(t)x^2 + 2e(t)xy - 2e(t)xy \\ &= e'(t)x^2 \\ &\leq 0. \end{aligned}$$

It is clear that $V(t, 0) = 0$. We will show that the zero solution is uniformly stable using Theorem 7.1.9 if we show that $W_1(|\mathbf{x}|) \leq V(t, \mathbf{x}) \leq W_2(|\mathbf{x}|)$. Since e is decreasing and its limit at infinity exists and is equal to e_0, we may let

$$c = \min\{e_0, 1\} \text{ and } k = \max\{e(0), 1\}.$$

Then we can define

$$W_1(|\mathbf{x}|) = c(x^2 + y^2) = c|\mathbf{x}|, \text{ and } W_2(|\mathbf{x}|) = k(x^2 + y^2) = k|\mathbf{x}|,$$

so that we have $W_1(|\mathbf{x}|) \leq e(t)x^2 + y^2 \leq W_2(|\mathbf{x}|)$. This implies that the zero solution is uniformly stable.

Example 7.15. Consider the system

$$x_1' = -x_1 - e^{-2t}x_2, \quad x_2' = x_1 - x_2.$$

Here $(0, 0)$ is the only equilibrium solution. Let

$$V(x_1, x_2) = x_1^2 + (1 + e^{-2t})x_2^2.$$

Then along the solutions we have by Definition 7.1.11 that

$$V'(t, x_1, x_2) = -2\left[x_1^2 - x_1 x_2 + x_2^2(1 + 2e^{-2t})\right].$$

Using $x_1 x_2 \leq \frac{x_1^2}{2} + \frac{x_2^2}{2}$, we arrive at

$$V'(t, x_1, x_2) = -2\left(x_1^2 + x_2^2(1 + 2e^{-2t})\right) + 2x_1 x_2$$
$$\leq -(x_1^2 + x_2^2 + 4x_2^2 e^{-2t})$$
$$\leq -(x_1^2 + x_2^2) := -W_3(|x|).$$

In addition, if we set

$$W_1(|x|) = \frac{1}{2}(x_1^2 + x_2^2), \quad W_2(|x|) = x_1^2 + x_2^2,$$

then we have

$$W_1(|x|) \leq V(t, x) \leq W_2(|x|),$$

and by Theorem 7.1.11, the origin is uniformly asymptotically stable.

Remark 7.2. Theorems 7.1.8–7.1.10 hold for the autonomous system (7.1.11) with autonomous $V(x)$.

7.2 Global asymptotic stability

In this section we briefly discuss the notion of global asymptotic stability.

Definition 7.2.1. Assume that the zero solution of (7.1.11) is stable. If $|x(t, t_0, x_0)| \to 0$ as $t \to \infty$ for every $x(t_0) = x_0$, then we say that the zero solution is *globally asymptotically stable* (GAS).

Remark 7.3. Theorem 7.1.10 implies the zero solution is GAS if $D = \mathbb{R}^n$. In addition, for the next theorem we note that (7.1.19) implies

$$V(x) \to \infty \text{ as } \|x\| \to \infty,$$

since $W_1(|x|) \to \infty$ as $|x| \to \infty$.

Example 7.16. Consider the system

$$x_1' = -x_1(1 - 2x_1 x_2), \quad x_2' = -x_2.$$

It is clear that $(0, 0)$ is the only equilibrium solution. Let

$$V(x_1, x_2) = \frac{1}{2}x_1^2 + x_2^2.$$

Then along the solutions we have

$$V'(x_1, x_2) = -x_1^2(1 - 2x_1 x_2) - 2x_2^2.$$

Now $V'(x_1, x_2) < 0$ if $1 - 2x_1x_2 > 0$, that is, if $x_1x_2 < \frac{1}{2}$. Thus the origin is asymptotically stable but not GAS since we are confined to the bounded region that consists of all points (x_1, x_2) such that $x_1x_2 < \frac{1}{2}$. Therefore we cannot conclude that the zero solution is GAS.

We state the following theorem regarding GAS.

Theorem 7.2.1. *Suppose the hypothesis of Theorem 7.1.3 holds with $D = \mathbb{R}^n$. In addition, if $V(x)$ is radially unbounded, that is,*

$$V(x) \to \infty \ \ as \ \ ||x|| \to \infty, \tag{7.2.1}$$

then the zero solution of (7.1.11) is GAS.

Proof. By Theorem 7.1.2, the zero solution is stable. Suppose there is a solution or trajectory that does not converge to zero. Since V is decreasing and nonnegative, it converges to some γ as $t \to \infty$. Since x does not converge to zero, we have $\gamma \le V(x(t)) \le V(x_0)$. Let

$$L = \{z : \gamma \le V(z) \le V(x_0)\}.$$

Then L is closed and bounded and hence compact. Therefore V attains its supremum on L, that is, $\sup_{z \in L} V' = -d < 0$, or $V'(x(t)) \le -d$ for all $t \ge t_0$. An integration of $V' \le -d$ from 0 to T yields

$$V(x(T)) \le V(x_0) - dT.$$

This implies that $V(x_0) < 0$ for $T > V(x_0)/d$, which is a contradiction, and hence every trajectory converges to zero. This completes the proof. $\square$

Basically, the conditions on V imply that every trajectory $x(t)$ tends to the origin as $t \to \infty$. Since $x(t)$ is continuous, this requires that $x(t)$ remains bounded for all $t \ge t_0$ for arbitrary initial conditions $x(t_0)$. Hence the purpose of the radial unboundedness condition on V is to ensure that $x(t)$ remains at all times within the bounded region defined by $V(x) \le V(x_0)$. If V were not radially unbounded, then not all level curves given by $V(x) \le c$ for positive constant c would be closed curves, and it would be possible for $x(t)$ to drift away from the equilibrium even though $V' < 0$.

Example 7.17. The zero solution of the system in Example 7.5 is GAS.

Example 7.18. Consider the system

$$x_1' = (x_2 - 1)x_1^3, \ \ x_2' = -\frac{x_1^4}{(1 + x_1^2)^2} - \frac{x_2}{1 + x_2^2}.$$

It is clear that $(0, 0)$ is the only equilibrium solution. Let

$$V(x_1, x_2) = \frac{x_1^2}{1 + x_1^2} + x_2^2.$$

Then along the solutions we have

$$V'(x_1, x_2) = -2\frac{x_1^4}{(1+x_1^2)^2} - 2\frac{x_2}{1+x_2^2},$$

so that $V(x_1, x_2)$ is a strict Lyapunov function on the set $D = \mathbb{R}^n$, as required by Theorem 7.1.3, and the zero solution is AS. However, V is not radially unbounded since the level curves for $V = c$ are not closed curves for $c \geq 1$. Thus it is not possible to conclude that the zero solution is GAS. Neither is it possible to conclude that the zero solution is *not* GAS without further analysis, since we might find another function V that satisfies all the requirements for the GAS.

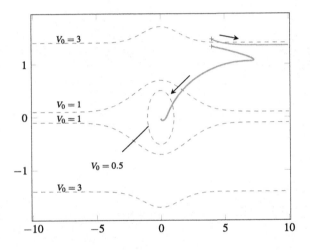

FIGURE 7.2

Level curves escaping to infinity.

As a matter of fact, according to Fig. 7.2, the zero solution is not GAS since solutions escape to infinity with the choice of the initial condition $(x_1(t_0), x_2(t_0)) = (3, 3/2)$.

Remark 7.4. We examine in detail why

$$V(x_1, x_2) = \frac{x_1^2}{1+x_1^2} + x_2^2$$

is not radially unbounded. As mentioned above, the condition that $V(x)$ is radially unbounded means that all level sets are closed. Letting $x_2 = 0$, we try to determine the level set $V(x) = 1$ or $V(x_1, 0) = \frac{x_1^2}{1+x_1^2} = 1$, which cannot hold for finite x_1. Thus V is not radially unbounded. Next, we examine the definition of being radially unbounded in the context of

$$V(x) \to \infty \text{ as } ||x|| \to \infty.$$

This means that any combinations of x_1, x_2 that make $||x|| \to \infty$ also have to make $V(x) \to \infty$. The Euclidean norm

$$||(x_1, 0)||_2 = \sqrt{x_1^2 + 0^2} = \sqrt{\infty^2 + 0^2} \to \infty.$$

However, in this case, we get for $V(x)$ that

$$\lim_{x_1 \to \infty} V(x_1, 0) = 1 \neq \infty.$$

It follows that V is not radially unbounded.

7.3 Instability

We consider once more the non-autonomous system

$$x' = f(t, x), \quad f(t, 0) = 0, \tag{7.3.1}$$

where $f \in C([0, \infty) \times D, \mathbb{R}^n)$, and $D \subset \mathbb{R}^n$ is open with $0 \in D$. We have discussed the instability for autonomous systems, but let us recall that if we negate the definition of stability, then there are $\varepsilon > 0$ and $t_0 \geq 0$ such that for any $\delta > 0$, there are x_0 with $|x_0| < \delta$ and $t_1 > t_0$ such that

$$|x(t_1, t_0, x_0)| \geq \varepsilon.$$

Then the zero solution is unstable. We make the following formal definition.

Definition 7.3.1. The zero solution of (7.3.1) is unstable if there are $\varepsilon > 0$ and $t_0 \geq 0$ such that for any $\delta > 0$, there are x_0 with $|x_0| < \delta$ and $t_1 > t_0$ such that $|x(t_1, t_0, x_0)| \geq \varepsilon$.

Theorem 7.3.1. *Suppose $V : [0, \infty) \times D \to [0, \infty)$ is a continuous function locally Lipschitz in x such that*

$$W_1(|x|) \leq V(t, x) \leq W_2(|x|), \tag{7.3.2}$$

and along the solutions of (7.3.1) we have

$$V'(t, x) \geq W_3(|x|). \tag{7.3.3}$$

Then the zero solution of (7.3.1) is unstable.

Proof. Suppose not. Then for $\varepsilon = \min\{1, d(0, \partial D)\}$, where ∂D denotes the boundary of the set D, we can find $\delta > 0$ such that $|x_0| < \delta$ and $t \geq 0$ imply that $|x(t, 0, x_0)| < \varepsilon$. We may pick x_0 such that $|x_0| = \delta/2$ and find $\gamma > 0$ with $W_2(\gamma) = W_1(\delta/2)$. Then for $x(t) = x(t, 0, x_0)$, we have $V'(t, x) \geq 0$ by (7.3.3), so that

$$W_2(|x(t)|) \geq V(t, x(t)) \geq V(0, x_0) \geq W_1(\delta/2) = W_2(\gamma),$$

from which we conclude that $\gamma \leq |x(t)|$ for $t \geq 0$. Thus

$$V'(t, x) \geq W_3(|x(t)|) \geq W_3(\gamma).$$

It follows that

$$W_2(|x(t)|) \geq V(t, x(t)) \geq V(0, x_0) + t W_3(\gamma),$$

from which we conclude that $|x(t)| \to \infty$, a contradiction. This completes the proof. $\square$

We ask the curious reader to compare condition (7.3.2) with the hypothesis of Theorem 7.1.6 and try to have an appreciation and some understanding of the differences between autonomous systems and non-autonomous systems.

In Theorem 7.3.1 we asked for $|x_0| = \delta/2$, which is too much to ask. This will be relaxed in the next theorem.

Theorem 7.3.2. *Suppose that $V : [0, \infty) \times D \to (-\infty, \infty)$ is a continuous function locally Lipschitz in x and that there is a sequence $x_n \in D$ starting at $t_0 \geq 0$ with $V(t_0, x_n) > 0$ and $x_n \to 0$ as $n \to \infty$. If*

$$W_1(|x|) \leq V(t, x) \leq W_2(|x|)$$

and if along the solutions of (7.3.1) we have

$$V'(t, x) \geq W_3(|x|),$$

then the zero solution of (7.3.1) is unstable.

Proof. Suppose not. Then there are ε and δ as in the proof of Theorem 7.3.1. For δ, we may find t_n and x_n with $|x_n| < \delta$ and $V(t_n, x_n) > 0$. The rest of the proof is identical to that of Theorem 7.3.1. This completes the proof. $\square$

7.4 ω-limit set

We discuss the Krasovskii–LaSalle invariance principle, which provides a way to prove the asymptotic stability when V' is negative *semi-definite*. It is usually much easier to construct such a function V. We begin with the following definition.

Definition 7.4.1. Let D be an open subset of $\mathbb{R}^n$ with $0 \in D$. A continuous autonomous function $V : D \to [0, \infty)$ is

(a) *positive definite* if

$$V(0) = 0 \text{ and } V(x) > 0 \text{ for } x \neq 0,$$

(b) *positive semi-definite* if

$$V(0) = 0 \text{ and } V(x) \geq 0 \text{ for } x \neq 0.$$

To illustrate the differences, consider the two functions

$$V_1(x) = x_1^2 \text{ and } V_2(x) = x_1^2 + x_2^2.$$

Clearly, both functions are nonnegative. However, it is possible for V_1 to be zero even when $x \neq 0$. Particularly, if we set $x = (0, d)$ where $d \in \mathbb{R}$ is any non-zero number, then $V_1(x) = V_1(0, d) = 0$, whereas $V_2(x) = 0$ if and only if $x = 0$. We conclude that V_1 is positive semi-definite and V_2 is positive definite.

Remark 7.5. If the signs on V' are reversed in Definition 7.4.1, then V is said to be *negative definite* and *negative semi-definite*, respectively.

Let $f : \mathbb{R}^n \to \mathbb{R}^n$ be continuous with continuous first-order partial derivatives with respect to $x_i, i = 1, 2, \ldots, n$. Consider the initial value autonomous system

$$x' = f(x), \; x(t_0) = x_0, \; t \geq t_0. \tag{7.4.1}$$

Let x^* be an equilibrium solution of (7.4.1).

Let $x(t, t_0, x_0)$ be a trajectory of (7.4.1) at time t starting from x_0 at the initial time t_0. Denote the set of all points lying along the trajectory by $x(\cdot, t_0, x_0)$

Definition 7.4.2. (ω-limit set) The ω-*limit set* of a trajectory $x(\cdot, t_0, x_0)$ is the set of all points $z \in \mathbb{R}^n$ such that

$$\lim_{n \to \infty} x(t_n, t_0, x_0) = z,$$

where the sequence t_n is strictly increasing.

Definition 7.4.3. (Invariant set) The set $M \in \mathbb{R}^n$ is said to be an *invariant set* if for all $y \in M$ and $t_0 \geq 0$, we have

$$x(t, t_0, y) \in M \quad \text{for all } t \geq t_0.$$

Theorem 7.4.1. *Let V be a Lyapunov function of (7.4.1) on a bounded open set $U \in \mathbb{R}^n$ with $x^* \in U$. Let c be a positive constant, and set*

$$S = \{x \in U : V(x) \leq c\},$$

which is closed in $\mathbb{R}^n$. Then S is invariant for $t \geq t_0 \geq 0$.

Proof. Let $x \in S$. Since V is a Lyapunov function on U, we have that $V(x(t, t_0, x_0)) \leq V(x_0) \leq c$ for $t \geq t_0 \geq 0$ as long as the solutions exist. Now since S is closed and bounded, the existence of solutions holds for all $t \geq t_0 \geq 0$. As a result, $x(t, t_0, x_0)$ remains in S for all $t \geq t_0 \geq 0$. This shows that the set S is invariant. This completes the proof. $\square$

In Example 7.3 we considered $V(x, y) = x^2 + y^2$ and obtained $V' = 0$. As a result of Theorem 7.4.1, we have that the set

$$S = \{(x, y) \in \mathbb{R}^2 : x^2 + y^2 \leq c\} \text{ for } c > 0$$

is invariant.

Theorem 7.4.2. *(Krasovskii–LaSalle invariance principle) Let D be an open subset of $\mathbb{R}^n$ containing x^*. Let $V : D \to \mathbb{R}$ be a positive definite function such that on the compact set*

$$\Omega_r = \{x \in D : V(x) \leq r, \ r > 0\},$$

we have $V'(x) \leq 0$. Define

$$S = \{x \in \Omega_r : V'(x) = 0\}.$$

Then, as $t \to \infty$, the trajectory tends to the largest invariant set inside S, that is, its ω-limit set is contained inside the largest invariant set in S. If S contains no invariant sets other than $x = x^*$, then the equilibrium solution x^* of (7.4.1) is asymptotically stable.

Proof. Let M be the largest invariant set in S, and let L^+ be the ω-limit set for the trajectory $x(t, t_0, x_0)$, which is invariant. Since $V'(x) \leq 0$ and is continuous in Ω_r, we have that $V(x) \geq b = \min_{x \in \Omega_r} V(x) > 0$. Thus there is a positive constant a such that

$$\lim_{t \to \infty} V(x(t, t_0, x_0)) = a.$$

Now $x(t, t_0, x_0) \in \Omega_r$ implies $x(t, t_0, x_0)$ is bounded, which implies that L^+ exists. Moreover, $L^+ \in \Omega_r$, and $x(t, t_0, x_0)$ approaches L^+ as $t \to \infty$. For any $z \in L^+$, there is an increasing sequence $\{t_n\}$ such that

$$\lim_{n \to \infty} x(t_n, t_0, x_0) = z.$$

As $V(x)$ is continuous, we have that

$$V(z) = \lim_{n \to \infty} V(x(t_n, t_0, x_0)) = a.$$

Since $V(x) = a$ on L^+ and L^+ is invariant, we get that $V'(x) = 0$ for all $x \in L^+$ and

$$L^+ \subset M \subset S \subset \Omega_r.$$

Thus $x(t, t_0, x_0) \to L^+$, and hence $x(t, t_0, x_0) \to M$ as $t \to \infty$. This completes the proof. $\qquad\qquad\qquad\qquad\qquad\qquad\qquad\qquad\qquad\qquad\qquad\qquad\qquad\square$

We provide a couple of examples.

Example 7.19. Consider the system

$$x' = y,$$
$$y' = -(x^3 - x^5) - y^3.$$

Let $D = \{(x, y) : x^2 + y^2 \le 1\}$ and define

$$V(x, y) = \frac{y^2}{2} + \frac{x^4}{4} - \frac{x^6}{6}.$$

Then V is positive definite on D. We can easily verify that $V' = -y^4$. Thus $V' = 0$ along the x-axis, and V' is negative semi-definite. Note that on D, we have $V(x, y) \le \frac{y^2}{2} + \frac{x^4}{4} \le \frac{1}{2}(\frac{1}{\sqrt{2}})^2 + \frac{1}{4}(\frac{1}{\sqrt{2}})^4 = \frac{5}{16}$. Thus $\Omega_r = \{(x, y) \in D : \frac{y^2}{2} + \frac{x^4}{4} \le \frac{5}{16}\}$. Moreover,

$$S = \{(x, y) \in \Omega_r : V'(x) = 0\} = \{(x, y) \in \Omega_r : y = 0\}.$$

Now V is constant only on solutions that lie on the x-axis, but by the uniqueness of solutions, the only such solution is the equilibrium $(0, 0)$. Thus $M = \{(x, y) \in S : x = 0, y = 0\}$ is the largest invariant subset of S. By Theorem 7.4.2, we have that all trajectories in S go to $(0, 0)$ as $t \to \infty$. Note that the other two equilibrium solutions, $(-1, 0)$ and $(1, 0)$, are outside the set Ω_r.

Corollary 7.1. *If the hypotheses of Theorem 7.4.2 hold with $D = \mathbb{R}^n$ and $V(x)$ is radially unbounded, then the origin is GAS.*

Example 7.20. Consider the system

$$x' = y,$$
$$y' = -h_1(x) - h_2(y),$$

where the functions h_1 and h_2 are continuous and satisfy

$$h_i(0) = 0, \quad uh_i(u) > 0, \quad i = 1, 2, \quad \text{for } 0 < |u| < a.$$

Let $D = \{(x, y) : |x| < a, \ |y| < a\}$ and set

$$V(x, y) = \frac{y^2}{2} + \int_0^x h_1(s)ds.$$

Then $V(0, 0) = 0$ and $V(x, y) > 0$ on D. In addition,

$$V'(x, y) = h_1(x)x' + yy'$$
$$= h_1(x)y + y\big[- h_1(x) - h_2(y)\big]$$
$$= -yh_2(y).$$

Now $V'(x, y) = 0$ implies $yh_2(y) = 0$, and hence $y = 0$. Thus $V' = 0$ along the x-axis, and V' is negative semi-definite. Note that on D, we have

$$V(x, y) \le \frac{a^2}{2} + \int_0^a ads = \frac{3}{2}a^2.$$

Thus $\Omega_a = \{(x, y) \in D : V(x, y) \le \frac{3}{2}a^2\}$. Moreover,

$$S = \{(x, y) \in \Omega_a : V'(x) = 0\} = \{(x, y) \in \Omega_a : y = 0\}.$$

As $y(t) = 0$, we have $y'(t) = 0$. The second equation of the system gives

$$h_1(x(t)) + h_2(0) = h_1(x(t)) = 0,$$

from which we have $x(t) = 0$. Hence the only solution that can stay identically in S is $(x(t), y(t)) = (0, 0)$, that is, $M = \{(x, y) \in S : x = 0, y = 0\}$ is the largest invariant subset of S. Thus the origin is asymptotically stable by Theorem 7.4.2.

Now we suppose $a = \infty$ and $\int_0^u h_1(s)ds \to \infty$ as $|u| \to \infty$. Then $D = \mathbb{R}^2$, and $V(x, y) = \frac{y^2}{2} + \int_0^x h_1(s)ds$ is radially unbounded. Moreover,

$$S = \{(x, y) \in \Omega_a : V'(x) = 0\} = \{(x, y) \in \Omega_a : y = 0\},$$

and the only solution that can stay identically in S is $(x(t), y(t)) = (0, 0)$. Thus the origin is GAS. Note that we did not have to worry about the set Ω_a since $a = \infty$ or $D = \mathbb{R}^2$.

We end this section by revisiting the pendulum with friction that we considered in Examples 3.5 and 6.9.

Example 7.21. In Example 3.5 we considered a pendulum with friction and arrived at the planar autonomous system

$$x_1' = x_2,$$

$$x_2' = -\frac{g}{L} \sin(x_1) - \frac{b}{m}x_2. \tag{7.4.2}$$

We are interested in examining the stability of the zero solution by constructing a suitable Lyapunov function. Let $D = \{(x_1, x_2) \in \mathbb{R}^2 : |x_1| < \pi\}$ and set

$$V(x_1, x_2) = \frac{g}{L}\big(1 - \cos(x_1)\big) + \frac{1}{2}x_2^2.$$

Then $V(0,0) = 0$, and $V(x_1, x_2) > 0$ on $D - (0,0)$. Moreover, along the solutions of (7.4.2) we have

$$V'(x_1, x_2) = \frac{g}{L} \sin(x_1) x_1' + x_2 x_2'$$

$$= \frac{g}{L} x_2 \sin(x_1) + x_2 \left(-\frac{g}{L} \sin(x_1) - \frac{b}{m} x_2 \right)$$

$$= -\frac{k}{m} x_2^2.$$

Now $V' = 0$ along the x_1-axis, and V' is negative semi-definite in D. Let

$$S = \{(x, y) \in D : V'(x) = 0\}.$$

As $x_2(t) = 0$, this implies that $x_1'(t) = 0$, and $x_1(t)$ is constant. Similarly, for $x_2(t) = 0$, this implies that $x_2'(t) = 0$, and hence $\sin(x_1) = 0$. However, the only point on the segment $-\pi < x_1 < \pi$ rendering $\sin(x_1) = 0$ is $x_1 = 0$. Hence the only solution that can forever stay in S is $(x_1(t), x_2(t)) = (0, 0)$. Thus $M = \{(x, y) \in S : x = 0, y = 0\}$ is the largest invariant subset of S, and the origin is asymptotically stable by Theorem 7.4.2.

Again, there was no need to construct the set Ω_a since $a = \infty$.

We end this section by giving an example in which we expose the difficulties of using Theorem 7.1.3, which requires V to be a strict Lyapunov function, and the advantage of using Theorem 7.4.2. First, we state the following lemma.

Lemma 7.3. *(Young's inequality) For any two nonnegative real numbers w and z, we have*

$$wz \leq \frac{w^e}{e} + \frac{z^f}{f} \quad \text{with} \quad \frac{1}{e} + \frac{1}{f} = 1, \ e, f \in (1, \infty).$$

Example 7.22. Consider the autonomous system

$$x' = -y - x^3, \quad y' = x^5.$$

We see that $(0, 0)$ is the only equilibrium solution. Let $D = \{(x, y) \in \mathbb{R}^2 : |y|^2 \leq 1/2\}$ and set

$$V(x, y) = x^6 + xy^3 + 3y^2.$$

We need to show V is a strict Lyapunov function on the set D. Clearly, $V(0, 0) = 0$. We need to worry about the mixed term xy^3. Let $e = 6$ and $f = 6/5$. Then by Lemma 7.3, for $|y| \leq 1$, we have

$$|xy^3| = |x||y^3| \leq \frac{|x|^6}{6} + \frac{5|y|^{18/5}}{6} \leq \frac{1}{6} x^6 + \frac{5}{6} y^2.$$

This implies that

$$3y^2 \geq \frac{18}{5}|xy^3| - \frac{3}{5}x^6.$$

Substituting $3y^2$ into V gives

$$V(x, y) \geq x^6 + xy^3 - \frac{3}{5}x^6 + \frac{18}{5}|xy^3|$$

$$= \frac{2}{5}x^6 + xy^3 + |xy^3| + \frac{13}{5}|xy^3|$$

$$\geq \frac{2}{5}x^6 + \frac{13}{5}|xy^3|.$$

Along the solutions of the system we have

$$V'(x, y) = -6x^8 - x^3y^3 + 3x^6y^2 - y^4.$$

Next, we apply Lemma 7.3 to reduce the two mixed terms. Let $(x, y) \in D$. Let $e = 8/3$ and $f = 8/5$. Then

$$|-x^3y^3| = |x^3||y^3| \leq \frac{3|x|^8}{8} + \frac{5|y|^{24/5}}{8} \leq \frac{3}{8}x^8 + \frac{5}{8}y^4,$$

and similarly, if we let $e = 8/6$ and $f = 8/2$, then

$$|3x^6y^2| = 3|x|^6|y|^2 \leq 3[\frac{3|x|^8}{4} + \frac{|y|^8}{4}] = \frac{9}{4}x^8 + \frac{3}{4}y^8.$$

Thus

$$V'(x, y) = -6x^8 + \frac{3}{8}x^8 + \frac{9}{4}x^8 - y^4 + \frac{5}{8}y^4 + \frac{3}{4}y^8$$

$$\leq -\frac{27}{8}x^8 - y^4 + \frac{5}{8}y^4 + \frac{3}{4}y^4(\frac{1}{4})$$

$$= -\frac{27}{8}x^8 - \frac{3}{16}y^4 < 0$$

on $D - \{(0, 0)\}$. By Theorem 7.1.3, the zero solution is asymptotically stable. Next, we use Theorem 7.4.2. Let

$$V(x, y) = x^6 + 3y^2.$$

Clearly, $V(0, 0) = 0$, and $V(x, y) > 0$ for $(x, y) \neq (0, 0)$. Moreover,

$$V'(x, y) = -6x^8 \leq 0.$$

Thus $V' = 0$ along the x-axis, and hence V is a Lyapunov function (semi-definite). Here $D = \mathbb{R}^2$, and

$$S = \{(x, y) \in D : V'(x) = 0\}.$$

As $y = 0$, this implies that $y' = 0$, from which we have $x^5 = 0$, that is, $x = 0$. Thus $M = \{(x, y) \in S : x = 0, y = 0\}$ is the largest invariant subset of S, and the origin is asymptotically stable by Theorem 7.4.2.

7.5 Connection between eigenvalues and Lyapunov functions

Consider the homogeneous system

$$x'(t) = Ax(t), \tag{7.5.1}$$

where A is an $n \times n$ constant matrix. We are interested in relations between the existence of a Lyapunov function for system (7.5.1) and its eigenvalues. It turns out that if all the eigenvalues of A have negative real parts, then there is a strict Lyapunov function. We begin with the following definition.

Definition 7.5.1. Let $x \in \mathbb{R}^n$, and let B be an $n \times n$ constant matrix. We say that B is *positive definite (negative definite)* if

$$x^T B x > 0 (< 0) \quad \text{for } x \neq 0.$$

Example 7.23. Let

$$B = I = \begin{pmatrix} 1 & 0 & \cdots & 0 \\ 0 & 1 & \ddots & \vdots \\ \vdots & \ddots & \ddots & \vdots \\ 0 & \cdots & 0 & 1 \end{pmatrix} \quad \text{and } x = \begin{pmatrix} x_1 \\ x_2 \\ \vdots \\ x_n \end{pmatrix}.$$

Then

$$x^T B x = (x_1, x_2, \ldots, x_n) \begin{pmatrix} 1 & 0 & \cdots & 0 \\ 0 & 1 & \ddots & \vdots \\ \vdots & \ddots & \ddots & \vdots \\ 0 & \cdots & 0 & 1 \end{pmatrix} \begin{pmatrix} x_1 \\ x_2 \\ \vdots \\ x_n \end{pmatrix}$$

$$= x_1^2 + x_2^2 + \ldots + x_n^2 > 0 \quad \text{for } x \neq 0.$$

Thus the $n \times n$ identity matrix I is positive definite.

To better understand where we are heading, we assume the existence of a positive definite symmetric $n \times n$ constant matrix B and define

$$V(x) = x^T B x. \tag{7.5.2}$$

Then $V(x)$ is positive definite, and, moreover, along the solutions of (7.5.1) we have that

$$
\begin{aligned}
V'(x) &= (x^T)'Bx + x^T Bx' \\
&= (x')^T Bx + x^T Bx' \\
&= (Ax)^T Bx + x^T B(Ax) \\
&= x^T A^T Bx + x^T BAx \\
&= x^T\left(A^T B + BA\right)x.
\end{aligned}
\tag{7.5.3}
$$

Notice that

$$
\begin{aligned}
\left(A^T B + BA\right)^T &= B^T A + A^T B^T \\
&= A^T B + BA.
\end{aligned}
$$

Thus $A^T B + BA$ is symmetric, and so is $A^T B + BA = -C$ for a symmetric matrix C; we wish $-C$ to be negative definite to obtain stability results concerning system (7.5.3).

Remark 7.6. If (7.5.2) holds, then there are positive constants α_1 and α_2 such that

$$
\alpha_1 x^T x \le x^T Bx \le \alpha_2 x^T x.
\tag{7.5.4}
$$

Then $W_1(|x|) = \alpha_1 x^T x$ and $W_2(|x|) = \alpha_2 x^T x$ could serve as two wedges such that

$$
\alpha_1 x^T x \le V(x) \le \alpha_2 x^T x.
$$

We have the following theorem.

Theorem 7.5.1. *Given an $n \times n$ constant matrix A and a positive definite symmetric matrix C, there exists a positive definite symmetric matrix B such that*

$$
A^T B + BA = -C
$$

if and only if the zero solution of (7.5.1) is asymptotically stable.

Proof. ($\Rightarrow$) Let V be given by (7.5.2) for a positive definite symmetric matrix B. Substituting $A^T B + BA = -C$ into (7.5.3) gives $V'(x) = -x^T Cx < 0$. Then by Theorem 7.1.3, the zero solution is AS.

($\Leftarrow$) Given A, suppose C is symmetric positive definite matrix. Assume that $x = 0$ is AS and let $\Phi(t)$ be the fundamental matrix of (7.5.1). Then by Theorem 4.2.3, there exist positive constants K and α such that

$$
\|\Phi(t)\Phi^{-1}(s)\| \le Ke^{-\alpha(t-s)}, \quad t > s.
$$

In particular, $||e^{At}e^{-As}|| \le Ke^{-\alpha(t-s)}, t > s$, or

$$||e^{A(t-s)}|| \le Ke^{-\alpha(t-s)}, \quad t > s.$$

Now $(e^A)^T = e^{A^T}$, and so the same above estimate holds for $e^{A^T(t-s)}$. Define

$$B(t) = \int_t^\infty e^{A^T(s-t)}Ce^{A(s-t)}ds,$$

which exists since $||e^{At}e^{-As}|| \le Ke^{-\alpha(t-s)}, t > s$. Differentiate both sides to get

$$\frac{d}{dt}B(t) = -C - A^T\int_t^\infty e^{A^T(s-t)}Ce^{A(s-t)}ds - \int_t^\infty e^{A^T(s-t)}Ce^{A(s-t)}ds\,A$$
$$= -C - A^T B(t) - B(t)A.$$

Let $u = s - t$ and rewrite the integral as

$$B(t) = \int_0^\infty e^{A^T u}Ce^{Au}du,$$

which does not depend on t at all, that is, $\frac{d}{dt}B(t) = 0$, or

$$-C - A^T B(t) - B(t)A = 0.$$

Finally, B is symmetric positive definite because C is. To see this,

$$B^T = \int_0^\infty \left(e^{A^T u}Ce^{Au}\right)^T du$$
$$= \int_0^\infty (e^{Au})^T C^T (e^{A^T u})^T du$$
$$= \int_0^\infty e^{A^T u}Ce^{Au}du = B.$$

Thus B is symmetric. It remains to show that B is positive definite. Let $x \in \mathbb{R}^n$ and consider the integrand. Then, for each t, we have

$$x^T e^{A^T t}Ce^{At}x = \left(e^{At}x\right)^T C\left(e^{At}x\right) := y^T Cy,$$

and if $x \ne 0$, then $y \ne 0$; hence, as C is positive definite, $y^T Cy > 0$ for $x \ne 0$. Thus B is positive definite. This completes the proof. $\qquad\square$

In Theorem 5.1.4 we showed if $Re(\sigma(A)) < 0$, then the zero solution of $x' = Ax$ is AS. Thus we may reformulate Theorem 7.5.1 in terms of the eigenvalues of the matrix A.

Theorem 7.5.2. *Given an $n \times n$ matrix A and a positive definite symmetric matrix C, there exists a positive definite symmetric matrix B such that*

$$A^T B + B A = -C$$

if and only if every eigenvalue of the matrix A has a negative real part.

Example 7.24. Let us find a strict Lyapunov function to show that the zero solution of

$$x' = \begin{pmatrix} -1 & 0 \\ 2 & -3 \end{pmatrix} \begin{pmatrix} x_1 \\ x_2 \end{pmatrix}, \ t > 0,$$

is asymptotically stable. Let $A = \begin{pmatrix} -1 & 0 \\ 2 & -3 \end{pmatrix}$. We will find a positive definite symmetric matrix B such that $A^T B + B A = -I$. If $B = \begin{pmatrix} a & b \\ b & c \end{pmatrix}$, then the relation $A^T B + B A = -I$ implies that

$$\begin{pmatrix} -2a + 4b & -4b + 2c \\ -4b + 2c & -6c \end{pmatrix} = \begin{pmatrix} -1 & 0 \\ 0 & -1 \end{pmatrix}.$$

The above equations give $a = \frac{2}{3}$, $b = \frac{1}{12}$, and $c = \frac{1}{6}$. Define

$$V(x_1, x_2) = x^T B x = (x_1 \ \ x_2) \begin{pmatrix} \frac{2}{3} & \frac{1}{12} \\ \frac{1}{12} & \frac{1}{6} \end{pmatrix} \begin{pmatrix} x_1 \\ x_2 \end{pmatrix}$$

$$= \frac{2}{3}x_1^2 + \frac{x_1 x_2}{6} + \frac{x_2^2}{6}$$

$$\geq \frac{1}{12}x_1^2 + \frac{x_1 x_2}{6} + \frac{x_2^2}{12}$$

$$= \frac{1}{12}(x_1 + x_2)^2.$$

Thus $V(x_1, x_2)$ is positive definite. On the other hand, after simple calculations, we arrive at

$$V'(x_1, x_2) = -x_1^2 - \frac{2}{3}x_2^2$$

$$\leq -\frac{2}{3}(x_1^2 + x_2^2).$$

Thus V is a strict Lyapunov function, and by Theorem 7.5.2, the zero solution of the system is AS. We could avoid verifying that V is a strict Lyapunov function and using Theorem 7.5.2 by finding a matrix B satisfying $A^T B + B A = -I$.

Remark 7.7. To ensure that the symmetric matrix B is positive definite, where $B = \begin{pmatrix} a & b \\ b & c \end{pmatrix}$, it is sufficient for its coefficients to satisfy

$$a > 0, \quad c > 0, \quad ac - b^2 > 0.$$

In the investigation of stability and boundedness for nonlinear non-autonomous systems, the effective approach is to study equations related in some way to the linear equations whose behavior is known to be covered by the known theory. We consider the perturbed nonlinear system

$$x'(t) = Ax(t) + g(t, x(t)), \; x(t_0) = x_0, \; t \geq t_0 \geq 0, \tag{7.5.5}$$

where A is an $n \times n$ constant matrix, and $g : [0, \infty) \times \mathbb{R}^n \to \mathbb{R}^n$ is a function continuous in t and x. Here $x = \begin{pmatrix} x_1 \\ x_2 \\ \vdots \\ x_n \end{pmatrix}$ and $g(t, x) = \begin{pmatrix} g_1(t, x) \\ g_2(t, x) \\ \vdots \\ g_n(t, x) \end{pmatrix}$. The perturbation term g is assumed to be small in some sense. The next theorem pertains to the stability of the perturbed nonlinear system (7.5.5).

By $|x|$ we denote the Euclidean norm, and for an $n \times n$ matrix C, by $|C|$ we denote a compatible norm such that $|Cx| \leq |C||x|$.

Theorem 7.5.3. *Suppose there exists a positive definite symmetric matrix B such that*

$$A^T B + BA = -I \tag{7.5.6}$$

and

$$\lim_{x \to 0} \frac{|g(t, x)|}{|x|} = 0 \; \text{uniformly in } t \in [0.\infty). \tag{7.5.7}$$

Then the zero solution of (7.5.5) is uniformly asymptotically stable.

Proof. Write x for $x(t)$ and define the function V by

$$V = x^T B x.$$

Using (7.5.6) and (7.5.7), along the solutions of (7.5.5) we have

$$\begin{aligned} V' &= (Ax + g(t, x))^T B + x^T B(Ax + g(t, x)) \\ &= x^T(-I)x + 2x^T Bg(t, x) \\ &\leq -|x|^2 + 2|x||B||g(t, x)|. \end{aligned}$$

Let $\varepsilon > 0$. Due to (7.5.7), we may choose ε sufficiently small so that for $|x| < \varepsilon$, we have

$$2|B||g(t, x)| < \frac{|x|}{2}.$$

Thus

$$V' \leq -|x|^2 + \frac{|x|^2}{2} = -\frac{|x|^2}{2}.$$

Since B is symmetric and positive definite, there are positive constants α and β such that

$$\alpha^2 x^T x \leq V(x) \leq \beta^2 x^T x, \tag{7.5.8}$$

from which we get

$$-|x|^2 \leq -\frac{1}{\beta^2} V(x) \leq -\frac{\alpha^2}{\beta^2}|x|^2.$$

As a consequence,

$$V' \leq -\frac{1}{2\beta^2} V(x).$$

An integration of the above inequality from t_0 to t yields

$$V(x(t)) \leq V(x_0) e^{-\frac{1}{2\beta^2}(t-t_0)}.$$

We have by (7.5.8)

$$\alpha^2 |x|^2 \leq V(x) \leq V(x_0) e^{-\frac{1}{2\beta^2}(t-t_0)}.$$

It follows that

$$\alpha^2 |x|^2 \leq x_0^T B x_0 e^{-\frac{1}{2\beta^2}(t-t_0)} \leq |x_0^T||B||x_0|e^{-\frac{1}{2\beta^2}(t-t_0)}.$$

For $\varepsilon > 0$ and $|x_0| < \delta$, we have

$$|x| \leq \frac{1}{\alpha}(\delta|B|\delta)^{\frac{1}{2}} = \frac{\delta}{\alpha}|B|^{\frac{1}{2}} < \varepsilon$$

for $\delta = \dfrac{\varepsilon\alpha}{|B|^{1/2}}$. Hence the zero solution is uniformly stable. Let $\delta = 1$ from the uniform stability, so that for $|x(t_0)| = |x_0| < \delta$, we need to show that $|x(t)| < \varepsilon$ for $t \geq t_0 + T(\varepsilon)$.

Set $T(\varepsilon) > -4\beta^2 \mathrm{Ln}\left(\frac{\alpha\varepsilon}{|B|^{1/2}}\right)$. Then

$$|x| \leq \frac{|B|^{1/2}}{\alpha} e^{-\frac{1}{4\beta^2}(t-t_0)}$$

$$\leq \frac{|B|^{1/2}}{\alpha} e^{-\frac{1}{4\beta^2}T(\varepsilon)}$$

$$< \varepsilon.$$

This completes the proof. $\square$

7.6 Exponential stability

We return to the non-autonomous system

$$x' = f(t, x), \quad f(t, 0) = 0, \tag{7.6.1}$$

where $f \in C\big([0, \infty) \times D, \mathbb{R}^n\big)$, and $D \subset \mathbb{R}^n$ is an open set with $0 \in D$.

Definition 7.6.1. We say that solutions of system (7.6.1) are *bounded* if any solution $x(t, t_0, x_0)$ of (7.6.1) satisfies

$$\|x(t, t_0, x_0)\| \leq C\big(|x_0|, t_0\big) \quad \text{for all } t \geq t_0,$$

where $C : \mathbb{R}^+ \times \mathbb{R}^+ \to \mathbb{R}^+$ is a constant that depends on t_0 and x_0. We say that solutions of system (7.6.1) are uniformly bounded if C is independent of t_0.

Definition 7.6.2. Suppose $f(t, 0) = 0$. We say the zero solution of (7.6.1) is α-*exponentially asymptotically stable* if there exists a continuous function $\alpha(t)$ such that $\int_{t_0}^{t} \alpha(s)ds \to \infty$ as $t \to \infty$ and constants d and $C \in \mathbb{R}^+$ such that for any solution $x(t, t_0, \varphi)$ of (7.6.1),

$$\|x(t, t_0, \varphi)\| \leq C\big(|\varphi|, t_0\big)(e^{-\int_{t_0}^{t} \alpha(s)ds})^d \quad \text{for all } t \in [t_0, \infty).$$

The zero solution of (7.6.1) is said to be α-uniformly exponentially asymptotically stable if C is independent of t_0.

Definition 7.6.3. (Exponential stability) The zero solution of system (7.6.1) said to be *exponentially stable* if any solution $x(t, t_0, x_0)$ of (7.6.1) satisfies

$$\|x(t, t_0, x_0)\| \leq C\big(|x_0|, t_0\big)e^{-d(t-t_0)} \quad \text{for all } t \geq t_0,$$

where d is a positive constant, and $C : [0, \infty) \times [0, \infty) \to [0, \infty)$. The zero solution of (7.6.1) said to be *uniformly exponentially stable* if C is independent of t_0.

Theorem 7.6.1. *Assume that $D \subset \mathbb{R}^n$ and $V : D \to [0, \infty)$ is a continuous function such that for all $(t, x) \in [0, \infty) \times D$,*

$$W_1(\|x\|) \leq V(t, x), \tag{7.6.2}$$

$$V'(t, x) \leq -\alpha(t)V^q(t, x) + \beta(t), \tag{7.6.3}$$

$$0 \leq V(t, x) - V^q(t, x) \leq \gamma(t), \tag{7.6.4}$$

where $\alpha(t)$, $\gamma(t)$, and $\beta(t)$ are nonnegative continuous functions, and p, q are positive constants. Then all solutions of (7.6.1) that start in D satisfy

$$\|x\| \leq W_1^{-1} \Big\{ V(t_0, x_0) e^{-\int_{t_0}^{t} \alpha(s)ds}$$

$$+ \int_{t_0}^{t} \Big(\gamma(u)\alpha(u) + \beta(u) \Big) e^{-\int_{u}^{t} \alpha(s)ds} \, du \Big\} \qquad (7.6.5)$$

for all $t \geq t_0$.

In case $q = 1$, the proof is a simple application of the variation of parameters formula. On the other hand, if $q \neq 1$, then the variation of parameters formula will not be of much help. Now we prove the theorem.

Proof. For any initial time $t_0 \geq 0$ and $x_0 \in D$, let $x(t, t_0, x_0)$ be a solution of (7.6.1) with $x_{t_0} = x_0$. Consider

$$\frac{d}{dt} \Big(e^{\int_{t_0}^{t} \alpha(s)ds} V(t, x) \Big) = \Big[V'(t, x) + \alpha(t)V(t, x) \Big] e^{\int_{t_0}^{t} \alpha(s)ds}$$

$$\leq \Big(-\alpha(t)V^q(t, x) + \beta(t) + \alpha(t)V(t, x) \Big) e^{\int_{t_0}^{t} \alpha(s)ds}$$

$$\leq \Big(\alpha(t)(V(t, x) - V^q(t, x)) + \beta(t) \Big) e^{\int_{t_0}^{t} \alpha(s)ds}$$

$$\leq \Big(\alpha(t)\gamma(t) + \beta(t) \Big) e^{\int_{t_0}^{t} \alpha(s)ds}.$$

Integrating both sides from t_0 to t, we obtain, for $t \in [t_0, \infty)$,

$$V(t, x) e^{\int_{t_0}^{t} \alpha(s)ds} \leq V(t_0, x_0)$$

$$+ \int_{t_0}^{t} \Big(\gamma(u)\alpha(u) + \beta(u) \Big) e^{\int_{t_0}^{u} \alpha(s)ds} \, du. \qquad (7.6.6)$$

It follows that

$$V(t, x) \leq V(t_0, x_0) e^{-\int_{t_0}^{t} \alpha(s)ds} + \int_{t_0}^{t} \Big(\gamma(u)\alpha(u) + \beta(u) \Big) e^{-\int_{u}^{t} \alpha(s)ds} \, du$$

for all $t \in [t_0, \infty)$. Inequality (7.6.2) implies that

$$\|x(t)\| \leq W^{-1} \Big\{ V(t_0, x_0) e^{-\int_{t_0}^{t} \alpha(s)ds} + \int_{t_0}^{t} \Big(\gamma\alpha(u) + \beta(u) \Big) e^{-\int_{u}^{t} \alpha(s)ds} \, du \Big\}$$

for all $t \geq t_0$. This concludes the proof. $\qquad \square$

It is easy to see that if $W_1(\|x\|) = \|x\|^p$ for some positive p, then inequality (7.6.5) becomes

$$\|x\| \leq \Big\{ V(t_0, x_0) e^{-\int_{t_0}^{t} \alpha(s)ds}$$

$$+ \int_{t_0}^{t} \Big(\gamma(u)\alpha(u) + \beta(u) \Big) e^{-\int_{u}^{t} \alpha(s)ds} \, du \Big\}^{1/p}. \qquad (7.6.7)$$

Corollary 7.2. *Assume that the hypotheses of Theorem 7.6.1 hold.*
(i) If

$$\int_{t_0}^{t} \Big(\gamma(u)\alpha(u) + \beta(u) \Big) e^{-\int_u^t \alpha(s)ds} \, du \le M, \forall t \ge t_0 \ge 0, \qquad (7.6.8)$$

for some positive constant M, then all solutions of (7.6.1) are uniformly bounded.
(ii) If

$$\int_{t_0}^{t} \Big(\gamma(u)\alpha(u) + \beta(u) \Big) e^{\int_{t_0}^u \alpha(s)ds} \, du \le M \qquad (7.6.9)$$

for some positive constant M, and

$$\int_{t_0}^{t} \alpha(s)ds \to \infty \text{ as } t \to \infty \text{ for all } t \ge t_0, \qquad (7.6.10)$$

then every solution of solution (7.6.1) decays α-exponentially asymptotically to zero with $d = 1/p$.
(iii) If

$$f(t,0) = 0$$

and (7.6.9) holds for q positive constant α, then the zero solution of (7.6.1) is exponentially stable.

Proof. Let x be a solution to (7.6.1) that starts in D for all $t \ge t_0 \ge 0$. Consequently, the proof of *(i)* is an immediate consequence of inequality (7.6.7). For the proof of *(ii)*, we consider the inequality from the proof of Theorem 7.6.1

$$V(t, x_t)e^{\int_{t_0}^t \alpha(s)ds} \le V(t_0, x_0) + \int_{t_0}^{t} \Big(\gamma(u)\alpha(u) + \beta(u) \Big) e^{\int_{t_0}^u \alpha(s)ds} \, du$$

for all $t \in [t_0, \infty)$. This yields

$$V(t, x_t) \le \left[V(t_0, x_0) + \int_{t_0}^{t} \Big(\gamma(u)\alpha(u) + \beta(u) \Big) e^{\int_{t_0}^u \alpha(s)ds} \, du \right] e^{-\int_{t_0}^t \alpha(s)ds}$$

for all $t \in [t_0, \infty)$. Using $W_1(||x||) = ||x||^p$, we have

$$|x| \le \left[V(t_0, x_0) + \int_{t_0}^{t} \Big(\gamma(u)\alpha(u) + \beta(u) \Big) e^{\int_{t_0}^u \alpha(s)ds} \, du \right]^{1/p} e^{-\frac{1}{p}\int_{t_0}^t \alpha(s)ds}. \qquad (7.6.11)$$

The proof of *(iii)* is an immediate consequence of inequality (7.6.11). This completes the proof. ☐

Remark 7.8. If $f(t,0) \ne 0$ and (7.6.9) and (7.6.10) hold, then all solutions of (7.6.1) decay exponentially to zero.

Example 7.25. For $a(t), b(t), h(t) \geq 0$, consider the scalar differential equation

$$x'(t) = -\left(a(t) + \frac{7}{6}\right)x(t) + b(t)x^{\frac{1}{3}}(t) + h(t) \quad t \geq t_0 \geq 0, \qquad (7.6.12)$$

with initial condition $x(t_0) = x_0 \in \mathbb{R}$. Let $V(t, x) = x^2$. Then along any solution $x := x(t)$ of (7.6.12), we have

$$V'(t, x) = 2xx'$$

$$= -2\left(a(t) + \frac{7}{6}\right)x^2 + 2b(t)x^{4/3} + 2xh(t)$$

$$\leq -2\left(a(t) + \frac{7}{6}\right)x^2 + 2b(t)x^{4/3} + x^2 + h^2(t).$$

To further simplify $V'(t, x)$, we use Young's inequality, which says for any two non-negative real numbers w and z, we have

$$wz \leq \frac{w^e}{e} + \frac{z^f}{f} \quad \text{with} \quad \frac{1}{e} + \frac{1}{f} = 1.$$

Thus for $e = 3/2$ and $f = 3$, we obtain

$$2|b(t)|x^{4/3} \leq 2\left[\frac{1}{3}|b(t)|^3 + \frac{(x^{4/3})^{3/2}}{3/2}\right] = \frac{4}{3}x^2 + \frac{2}{3}|b(t)|^3.$$

As a result, we have

$$V'(t, x) \leq -2a(t)x^2 + \frac{2}{3}b^3(t) + h^2(t). \qquad (7.6.13)$$

It follows that

$$V'(t, x) \leq -\alpha(t)x^2 + \beta(t), \qquad (7.6.14)$$

where $\alpha(t) = 2a(t)$ and $\beta(t) = \frac{2}{3}b^3(t) + h^2(t)$. We can easily check that conditions (7.6.2)–(7.6.4) of Theorem 7.6.1 are satisfied with $W_1(|x|) = x^2$, $q = 2$, and $\gamma = 0$. Next, we choose some convenient values for our variables.

Let $a(t) = \frac{t}{2}$, $b(t) = t^{\frac{1}{3}}$, and $h(t) = t^{1/2}$. Then $\alpha(t) = t$, $\beta(t) = \frac{2}{3}t$, and (7.6.14) becomes $V'(t, x) \leq -tx^2 + \frac{2}{3}t$. We note that

$$\int_{t_0}^t \left(\gamma\alpha(u) + \beta(u)\right)e^{-\int_u^t \alpha(s)ds} du = \frac{5}{3}\int_{t_0}^t ue^{-\int_u^t sds} du$$

$$\leq \frac{5}{3},$$

for all $t \geq t_0 \geq 0$. Thus condition (7.6.8) holds. By Corollary 7.2 all solutions of

$$x'(t) = -\left(\frac{t}{2} + \frac{7}{6}\right)x(t) + t^{\frac{1}{3}}x^{\frac{1}{3}}(t) + t^{1/2}, \quad x(t_0) = x_0, \ t \geq t_0 \geq 0, \qquad (7.6.15)$$

are uniformly bounded.

On the other hand, if we take $a(t) = 1/2$, $b(t) = e^{-\kappa_1 t/3}$, and $h(t) = 0$, where $\kappa_1 > 1$, then we have $\alpha(t) = 1$, $\beta(t) = \frac{2}{3}e^{-\kappa_1 t}$, and

$$\int_{t_0}^{t} (\gamma\alpha(u) + \beta(u))e^{\int_{t_0}^{u} \alpha(s)\,ds}\,du = \int_{t_0}^{t} \frac{2}{3}e^{-\kappa_1 u}e^{\int_{t_0}^{u} ds}\,du$$

$$\leq \frac{2}{3(\kappa_1 - 1)}e^{t_0(1-\kappa_1)}$$

for all $t \geq t_0 \geq 0$. Hence condition (7.6.9) is satisfied. Finally, (7.6.11) gives the inequality

$$|x(t)| \leq \left(x_0^2 + \frac{2}{3(\kappa_1 - 1)}e^{t_0(1-\kappa_1)}\right)e^{\kappa_1(t-t_0)}$$

$$\leq \left(x_0^2 + \frac{2}{3(\kappa_1 - 1)}\right)e^{\kappa_1(t-t_0)},$$

from which we arrive at the uniform exponential stability of the zero solution of

$$x'(t) = -\frac{5}{3}x(t) + e^{-\frac{\kappa_1}{3}t}x^{\frac{1}{3}}(t), \quad x(t_0) = x_0, \ t \geq t_0 \geq 0.$$

7.7 Exercises

Exercise 7.1. Use Lyapunov function to investigate the stability of the origin for each of the following systems and compare with the linearization method:
(a) $x' = y$, $y' = -x$.
(b) $x' = -xy^2$, $y' = 3yx^2$.
(c) $x' = y - 2x$, $y' = 2x - y - x^3$.
(d) $x' = x + 3y$, $y' = 2x$.
(e) $x' = x^3 + y$, $y' = x + y^3$.
(f) $x' = y$, $y' = -x - y - x^3$.

Exercise 7.2. Find the two equilibrium solutions and use Lyapunov functions to investigate their stability of

$$x' = y - x^2, \quad y' = x - y^2.$$

Exercise 7.3. Consider the planar system

$$x' = -xy^2,$$
$$y' = -y - yx^2.$$

Examine the stability of the origin using:
(a) Theorem 6.2.1,
(b) a Lyapunov function.

Exercise 7.4. Consider the planar system

$$x' = ax(x + y) - xy^3,$$
$$y' = ay(x + y) + x^2y^2.$$

Examine the stability of the origin for different values of a using:
(a) a Lyapunov function,
(b) the linearization method.

Exercise 7.5. Put the differential equation

$$x'' + (x')^2 + x' + x(x - 2) = 0$$

into a system and find its two equilibrium solutions and examine their stability using:
(a) a Lyapunov function,
(b) the linearization method.

Exercise 7.6. Use the indicated Lyapunov function V and Theorem 7.1.4 to show that the origin is asymptotically stable. Show that all solutions tend to the origin as $t \to \infty$.
(a) $x' = 2y$, $y' = -3x - 3y^3$, $V = 3x^2 + 2y^2$.
(b) $x' = -x + 6y$, $y' = -x - 2y$, $V = x^2 + 2y^2$.
(c) $x' = -2x - y$, $y' = 6x - y$, $V = 2x^2 + y^2$.
(d) $x' = y$, $y' = -x^2y - x^3$, $V = x^4 + 2y^2$.

Exercise 7.7. Study the stability of the origin for the planar system

$$x' = x(y - 1),$$
$$y' = -\frac{x^2}{1 + x^2},$$

by considering the Lyapunov function $V(x, y) = \mathrm{Ln}(1 + x^2) + y^2$.

Exercise 7.8. Use a Lyapunov function to show the existence of a stable limit cycle for the system

$$x' = y + x(\mu^2 - x^2 - y^2), \quad y' = -x + y(\mu^2 - x^2 - y^2)$$

for $\mu \neq 0$.

Exercise 7.9. Consider the planar system

$$x' = x + g_1(x, y), \quad y' = -y + g_2(x, y),$$

where g_1 and g_2 are continuous on a subset D of $\mathbb{R}^2$ that contains $(0, 0)$. Suppose that for a positive constant k,

$$||g_i|| \leq k(x^2 + y^2), i = 1, 2,$$

with $g_1(0, 0) = g_2(0, 0) = 0$. Use Theorem 7.1.6 to show that the origin is unstable.

Exercise 7.10. Consider the second-order differential equation

$$x'' + g(x) = 0,$$

where g is continuously differentiable, $g(0) = 0$, and $xg(x) > 0$. Write the equation as a system and use a Lyapunov function to show that its zero solution is stable.

Exercise 7.11. Consider the second-order differential equation

$$x'' + f(x)x' + g(x) = 0,$$

where $f(x)$ and $g(x)$ are even and odd polynomials, respectively.
(a) Show that the equation is equivalent to the system

$$x' = y - F(x), \quad y' = g(x),$$

where $F(x) = \int_0^x f(s)ds$.
(b) Let $G(x) = \int_0^x g(s)ds$ and suppose there are positive constants α and β such that $g(x)F(x) > 0$ for $0 \le |x| < \alpha$ and $G(x) < \beta$ for $|x| < \alpha$.
    Show that

$$V(x, y) = \frac{y^2}{2} + G(x)$$

is a Lyapunov function in the region $A = \{\alpha, \beta \in \mathbb{R} : |x| < \alpha, y^2 < \beta\}$ and that the origin is asymptotically stable.

Exercise 7.12. Consider the predator–prey model with carrying capacity

$$x' = x[a - dx - by], \quad y' = y[-c + kx],$$

where $x(t)$ and $y(t)$ are the numbers of preys and predators at time t, respectively, with positive constants a, b, c, d, and k.
(a) Find the equilibrium point (x^*, y^*) located in the first quadrant.
(b) Use the transformation

$$u = \mathrm{Ln}(\frac{x}{x^*}) \text{ and } v = \mathrm{Ln}(\frac{y}{y^*})$$

to obtain a system in (u, v).
(c) Show that

$$V(u, v) = kx^*(e^u - u) + by^*(e^v - v) - kx^* - by^*$$

is a Lyapunov function and that the equilibrium solution (x^*, y^*) is stable.

Exercise 7.13. Prove Theorem 7.1.7.

Exercise 7.14. Consider the second-order differential equation

$$x'' + (\mu + \nu x^2)x' + x + x^3 = 0$$

(a) If $\nu = 0$ and $\mu > 0$, then show that the zero solution is asymptotically stable.
(b) If $\mu = 0$ and $\nu > 0$, then show that the zero solution is stable using an appropriate Lyapunov function.

Exercise 7.15. Consider the planar system

$$x' = (x - by)(x^2 + y^2 - 1),$$
$$y' = (ax + y)(x^2 + y^2 - 1).$$

Show that the origin is stable if $a, b > 0$.

Exercise 7.16. Show that the origin is unstable for the planar system

$$x' = x + y,$$
$$y' = x - y + xy.$$

Exercise 7.17. Suppose f is continuously differentiable and consider the planar system

$$x' = y - xf(x, y),$$
$$y' = -x - yf(x, y).$$

Show that the origin is asymptotically stable if $f(x, y) > 0$ and unstable if $f(x, y) < 0$.

Exercise 7.18. Let $V = x^2 + y^2$ and consider the system

$$x' = (x - y)(x^2 + y^2 - 1),$$
$$y' = (x + y)(x^2 + y^2 - 1).$$

Show that the origin is asymptotically stable and find the region of attraction.

Exercise 7.19. Let $V = x^2 + y^2$ and consider the system

$$x' = -x + 2x^2 + y^2,$$
$$y' = -y + y^2.$$

Show that the origin is asymptotically stable and find the region of attraction.

Exercise 7.20. Let $V = x^2 + y^4$ and consider the system

$$x' = -x + 2y^3 - 2y^4,$$
$$y' = -x - y + xy.$$

Show that the origin is globally asymptotically stable.

Exercise 7.21. Let $V = ax^{2m} + by^{2m}$, $a, b > 0$ and consider the system

$$x' = -3x^3 - y,$$
$$y' = -x^5 - 2y^3.$$

Determine the values of a, b, m, and n, where m and n are positive integers, such that the origin is globally asymptotically stable.

Exercise 7.22. Decide if the functions given below are radially unbounded or not.
(a)

$$V(x_1, x_2) = (x_1 - x_2)^2.$$

(b)

$$V(x_1, x_2) = \frac{x_1^2 + x_2^2}{1 + x_1^2 + x_2^2} + (x_1 - x_2)^2.$$

Exercise 7.23. Consider the equation

$$y'' + By' + y + y^3 = 0, \quad B \geq 0,$$

which is a particular case of mass-spring nonlinear damper.
(a) Use the transformation $v = y'$ and obtain a corresponding system in y and v.
(b) Show that

$$V(y, v) = \frac{y^2 + y^4}{2} + \frac{v^2}{2}$$

is a Lyapunov function.
(c) Decide whether the origin is stable or asymptotically stable.

Exercise 7.24. Consider the non-autonomous system

$$x' = -x - g(t)y,$$
$$y' = x - y,$$

where $g(t)$ is continuously differentiable and, for positive constants k_1, k_2, satisfies

$$k_1 \leq g(t) \leq k_2 \quad \text{and} \quad g'(t) \leq g(t)$$

for all $t \geq 0$. Use a Lyapunov function to show that the zero solution is uniformly stable.

Exercise 7.25. Consider the scalar non-autonomous differential equation

$$x' = -\left(1 + g(t)\right)x^3,$$

where $g(t)$ is continuous, and

$$g(t) \geq 0 \quad \text{for all} \quad t \geq 0.$$

Use a Lyapunov function to show that the zero solution is uniformly asymptotically stable.

Exercise 7.26. Let $D = \{x \in \mathbb{R}^n : |x| \leq \delta\}$ for $\delta > 0$, and let $V(t, x) : [0, \infty) \times D \to [0, \infty)$ be a continuously differentiable function satisfying

$$V(t, 0) = 0,$$
$$W_1(|x|) \leq V(t, x) \leq W_2(|x|),$$

and

$$V'(t, x) \leq -W_3(V(t, x)) \quad \text{for all} \quad t \geq 0, \ |x| \leq \delta.$$

Show that the zero solution of (7.1.17) is uniformly asymptotically stable.

Exercise 7.27. Put the scalar equation

$$x'' - x^3$$

into a system and find a function that satisfies the requirements of Theorem 7.3.1.

Exercise 7.28. Write the scalar equation

$$x'' + f(x') + g(x) = 0$$

as a system. Assume that

$$xf(x') > 0, \ x' \neq 0, \ \text{and} \ xg(x) > 0, \ x \neq 0.$$

Use Theorem 7.3.1 to show that the origin is asymptotically stable.

Exercise 7.29. Use Exercise 7.28 to show that the zero solutions of given differential equations are asymptotically stable.
(a) $x'' + x'|x'| + x - \frac{x^3}{3} = 0$.
(b) $x'' + (x')^3 + x^5 - x^4 \sin^2(x) = 0$.

Exercise 7.30. Consider the system

$$x_1' = -x_1 + g(x_2),$$
$$x_2' = -x_2 + h(x_1),$$

where g and h are continuous and satisfy

$$|g(u)| \leq |u|/2, \ |h(u)| \leq |u|/2.$$

Use the Lyapunov function $V(x_1, x_2) = \frac{1}{2}(x_1^2 + x_2^2)$ to show that the origin is uniformly exponentially stable.

Exercise 7.31. Let $C = I$ and $A = \begin{pmatrix} 0 & 1 \\ -2 & -3 \end{pmatrix}$. Find a symmetric matrix B such that $A^T B + BA = -I$.

Exercise 7.32. Let B be chosen from Exercise 7.31 and set $V = x^T Bx$. Show that V is an autonomous Lyapunov function in some open set D containing $x = 0$ for the system

$$x_1' = x_2 + f_1(t, x),$$
$$x_2' = -2x_1 - 3x_2 + f(t, x),$$

where

$$\lim_{|x| \to 0} \frac{|f_i(t, x)|}{|x|} = 0, \ i = 1, 2,$$

uniformly for $t \in [0, \infty) \times B_1(0)$. Then find a wedge W such that $V'(x) \leq -W(|x|)$ on D.

Exercise 7.33. Consider the pendulum of Example 7.21 and define the function

$$V(x_1, x_2) = \frac{g}{L}(1 - \cos(x_1)) + \frac{1}{2}x^T Bx$$

with positive definite matrix B that is to be found.
Hint: See Remark 7.7 and then compute V' along the solutions. Pick the entries of B to ensure that V is a strict Lyapunov function on the set $D - \{(0, 0)\}$.

Exercise 7.34. Use Theorem 7.4.2 to show that the zero solution is asymptotically stable for

$$x' = -x^3 + 2y^3, \quad y' = -2xy^2.$$

Hint: See Example 7.22.

Exercise 7.35. (Harmonic oscillator) Let

$$x'' + \sin(x) = 0.$$

(a) Can you prove the stability of the origin using linearization? Use an appropriate Lyapunov function to show that the origin is a stable equilibrium solution.
(b) Add the damping term

$$x'' + \varepsilon x' + \sin(x) = 0.$$

Study the stability of the origin for $\varepsilon > 0$.

Exercise 7.36. Let $x(t) = (x_1(t), x_2(t))^T \in \mathbb{R}^2$ and consider

$$x_1'(t) = x_2(t) - x_1(t)|x_1(t)|,$$
$$x_2'(t) = -x_1(t) - x_2(t)|x_2(t)|$$

with initial condition $x(t_0) = x_0 \in \mathbb{R}^2$. Set $D = \{(x_1, x_2) \in \mathbb{R}^2 : x_1^2 + x_2^2 \leq \frac{8}{9}\}$. Let $V : [0, \infty) \times D \to [0, \infty)$ be given by $V(t, x) = \frac{1}{2}(x_1^2 + x_2^2)$. Show that:

(a)

$$V'(t, x) \leq -2V^{3/2}(t, x);$$

(b)

$$V(t, x) - V^q(t, x) = V(t, x) - V^{3/2}(t, x) \leq \frac{4}{27};$$

for $(x_1, x_2) \in D$.

(c) Refer to a specific theorem in the text to conclude that the zero solution is uniformly exponentially stable for any $x_0 \in D$.

(d) Explain why linearization around the origin cannot be used here.

Hint: Use the inequality

$$\left(\frac{a+b}{2}\right)^e \leq \frac{a^e}{2} + \frac{b^e}{2}, a, b > 0, e > 1.$$

Exercise 7.37. Let $x(t) = (x_1(t), x_2(t))^T \in \mathbb{R}^2$ and consider

$$x_1'(t) = x_2 - g(t)x_1(x_1^2 + x_2^2),$$

$$x_2'(t) = -x_1 - g(t)x_2(x_1^2 + x_2^2)$$

with initial condition $x(t_0) = x_0 \in \mathbb{R}^2$, where $g(t)$ is continuously differentiable and bounded, and $g(t) \geq k > 0$ for all $t \geq 0$. Set $D = \{(x_1, x_2) \in \mathbb{R}^2 : x_1^2 + x_2^2 \leq 1\}$. Let $V : [0, \infty) \times D \to [0, \infty)$ be given by $V(t, x) = \frac{1}{2}(x_1^2 + x_2^2)$. Show that the hypotheses of Theorem 7.6.1 are satisfied for $(x_1, x_2) \in D$ and conclude that the zero solution is uniformly exponentially stable for any $x_0 \in D$.

Exercise 7.38. Consider

$$x_1'(t) = x_2 + x_1 x_3,$$
$$x_2'(t) = -x_1 - x_2 + x_2 x_3,$$
$$x_3'(t) = -x_1^2 - x_2^2.$$

Use Theorem 7.4.2 to show that

$$x_1(t), \ x_2(t) \to 0 \text{ as } t \to \infty, \text{ and } x_3(t) \to c \text{ as } t \to \infty,$$

for some constant c.

Exercise 7.39. Repeat Example 7.24 for the system

$$x' = \begin{pmatrix} -1 & 0 \\ 1 & -2 \end{pmatrix} \begin{pmatrix} x_1 \\ x_2 \end{pmatrix}, \ t > 0.$$

Exercise 7.40. Consider the SIR epidemic model of Section 6.3 and let

$$T = \{(S, I) : S \geq 0, I \geq 0, \ S + I < 1\}.$$

Define the function V on T by

$$V(S, I) = S - S^* - S^* \text{Ln}(\frac{S}{S^*}) + I - I^* - I^* \text{Ln}(\frac{I}{I^*}).$$

(a) Show that along the solutions we have

$$V'(S, I) \leq -\mu \frac{(S - S^*)^2}{SS^*}.$$

(b) Show that V is positive definite on T.
Hint: Clearly, $V(S^*, I^*) = 0$. To show that $V(S, I) > 0$ for $(S, I) \neq (S^*, I^*)$, consider the function

$$f(x) = x - x^* - x^* \text{Ln}(\frac{x}{x^*}), \ 0 < x < 1, \ 0 < x^* < 1$$

and show it has a minimum of zero at x^* and $f(x) > 0$ for $x \neq x^*$.

(c) Use Theorem 7.4.2 to conclude that (S^*, I^*) is asymptotically stable.

Exercise 7.41. Consider the Lorenz system

$$x'(t) = \rho(y - x),$$
$$y'(t) = rx - y - xz,$$
$$z'(t) = xy - bz,$$

where ρ, r, and b are positive constants.
1) Show that, for $0 < r < 1$, the origin $(0, 0, 0)$ is the only equilibrium solution.
2) Let

$$V(x, y, z) = \frac{1}{2\rho}\left(x^2 + \rho y^2 + \rho z^2\right).$$

Show that the origin $(0, 0, 0)$ is asymptotically stable for $0 < r < 1$.

Delay differential equations

This chapter is introductory to the study of *delay differential equations*. We will touch on the existence and uniqueness of solutions and stability using the Lyapunov method. Toward the end of the chapter, we use the fixed point theory on Banach spaces developed in Chapter 2 to deduce boundedness and stability results for neutral delay differential equations. We end the chapter by utilizing Lyapunov functionals and obtain the exponential stability for totally nonlinear finite delay integro-differential equations.

8.1 Introduction

Delay differential equations are a type of equation in which the derivative of the unknown function at a certain time is given in terms of the values of the function at previous times. In the previous seven chapters, we considered finite-dimensional differential systems, where the state at any time can be obtained from a finite set of values. On the other hand, delay differential equations are *infinite-dimensional systems*. As a consequence, delay differential equations present us with some difficulties when we analyze them. Nevertheless, some known theories for differential equations can be extended to cover delay differential equations.

For the moment, we consider the scalar delay differential equation

$$x' = f(t, x(t), x(t - \tau)), \tag{8.1.1}$$

where τ is a positive constant known as the delay. If we set $t = 0$ in (8.1.1), then we must know the value of the dependent variable x at s, where $-\tau \le s \le 0$, that is, we need to define an initial function $\varphi : [-\tau, 0] \to \mathbb{R}$. In other words, the initial function φ gives the behavior of the system prior to time 0 (assuming that we start at time $t = 0$). In general, for system (8.1.1) to be *well posed*, we must specify an initial function and work with the delay system

$$x' = f(t, x(t), x(t - \tau)), \quad 0 \le t_0 \le t \le t_f, \tag{8.1.2}$$

$$x(s) = \varphi(s), \quad -\tau \le s \le t_0 \le 0,$$

where $f : [t_0, t_f] \times \mathbb{R}^2 \to \mathbb{R}$. Note that system (8.1.2) is referred to as a delay differential equation with finite delay.

Advanced Differential Equations. https://doi.org/10.1016/B978-0-32-399280-0.00014-0

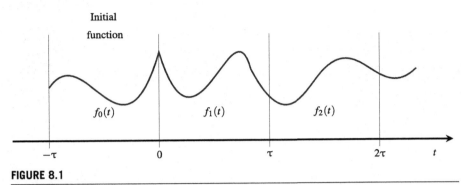

FIGURE 8.1

The effect of the initial function and the delay and the solution space.

Definition 8.1.1. We say that $x : [t_0 - \tau, t_f] \to \mathbb{R}$ is a solution of (8.1.2) if it is continuous and satisfies (8.1.2) for $t_0 \leq t \leq t_f$ and $x(s) = \varphi(s)$, $-\tau \leq s \leq t_0$. If every two solutions of (8.1.2) agree with each other as far as both are defined, then the solution is said to be *unique*.

We may think of the solution of (8.1.2) as a mapping from functions on the interval $[t - \tau, t]$ into functions on the interval $[t, t + \tau]$. In other words, the solutions can be thought of as a sequence of functions $f_0(t)$, $f_1(t)$, $f_2(t)$, ... defined over intervals of length τ. The points $t = j\tau$, $j = 0, 1, 2, \ldots$, where the solution segments meet are called knots. Fig. 8.1 illustrates this mapping.

It is to be understood that $x'(t_0)$ is the *right-hand derivative* of x at t_0. In most cases the solution is not differentiable at $t = t_0$ as seen in Fig. 8.1. In the case of differential equations, we only have to give an initial condition at a single point, known as the initial time.

For the delay system (8.1.2), we had to specify an initial function on the whole interval $[-\tau, t_0]$, $t_0 \leq 0$. The system inherits being infinite-dimensional from the fact that the initial function takes infinitely many values at infinitely many points in the interval $[-\tau, t_0]$.

There are different types of delay differential equations:

$$x' = f(t, x(t), x(t - \tau_1), \ldots, x(t - \tau_m)), \quad 0 \leq t_0 \leq t \leq t_f, \tag{8.1.3}$$

$$x(s) = \varphi(s), \quad -\tau \leq s \leq t_0 \leq 0,$$

where $\tau = \max\{\tau_j, 1 \leq j \leq m\}$. We can try to solve (8.1.3) for $0 \leq t_0 \leq t \leq t_0 + r$, where $r = \min\{\tau_j, 1 \leq j \leq m\}$. System (8.1.3) is known as a differential equation with *multiple delays*. For (8.1.3), the initial function $\varphi : [-\tau, t_0] \to \mathbb{R}^m$, $t_0 \leq 0$. On the other hand, if $a(t)$ is a continuous function with $a : [0, \infty) \to (-\infty, \infty)$, then the system

$$x' = a(t)x(t) + \int_{-\infty}^{t} f(s, x(s))ds, \quad 0 \le t_0 \le t \le t_f, \qquad (8.1.4)$$

$$x(s) = \varphi(s), \quad -\infty < s \le t_0 \le 0$$

is referred to as a *Volterra integro-differential equation with infinite delay*. System (8.1.4) is called a *Volterra integro-differential equation with finite delay* if the lower limit of integration is $t - \tau$, where $\tau > 0$, instead of $-\infty$. In this case the initial function is the same as in (8.1.2). Others may refer to it as a *Volterra integro-differential equation with distributed delay*. Finally, if the delay τ of system (8.1.2) depends on time, then the system is referred to as a *delay differential equation with time-varying delay or functional delay*.

8.2 Method of steps

The *Method of steps* can be used as a tool for simple delay differential equations to obtain a solution, given some initial data. The method of steps is best illustrated by looking at an example.

Example 8.1. (Method of steps) Consider the delay problem

$$x' = -x(t - 1), \ t > 0, \qquad (8.2.1)$$

$$x(s) = \varphi(s) = 10, \quad -1 \le s \le 0.$$

We will try to piece together the solution on $[0, 3]$.

   Step 1. In this step, we find the solution on $0 \le t \le 1$. For $0 \le t \le 1$, we have $-1 \le t - 1 \le 0$, and hence $x(t - 1) = 10$. Thus from (8.2.1) we obtain

$$\frac{dx}{dt} = -10,$$

which has the solution

$$x(t) = x(0) + \int_0^t (-10)du = 10 - 10t.$$

   Step 2. Now we find the solution on $1 \le t \le 2$. For $1 \le t \le 2$, we have $0 \le t - 1 \le 1$, and hence $x(t - 1) = 10 - 10(t - 1)$, where we have made use of the solution from Step 1. Thus from (8.2.1) we obtain

$$\frac{dx}{dt} = -10 + 10(t - 1),$$

which has the solution

$$x(t) = x(1) + \int_1^t [-10 + 10(u-1)]du$$
$$= 0 - 10u + 5(u-1)^2\big|_1^t$$
$$= -10(t-1) + 5(t-1)^2,$$

where we have used $x(1) = 0$ from the solution in Step 1.

Step 3. Now we find the solution on $2 \le t \le 3$. For $2 \le t \le 3$, we have $1 \le t-1 \le 2$, and hence $x(t-1) = -10(t-2) + 5(t-2)^2$, where we utilized the solution from Step 2. Thus from (8.2.1) we obtain

$$\frac{dx}{dt} = 10(t-2) - 5(t-2)^2,$$

which has the solution

$$x(t) = x(2) + \int_1^t [10(u-2) - 5(u-2)^2]du$$
$$= -5 + 5(t-2)^2 - \frac{5}{3}(t-2)^3,$$

where we have used $x(2) = -5$ from the solution in Step 2.

In Example 8.1

$$\lim_{t \to 0^-} x'(t) = \lim_{t \to 0^-} \varphi'(t) = 10,$$

and

$$\lim_{t \to 0^+} x'(t) = \lim_{t \to 0^+} (-10t + 10)' = -10.$$

Therefore the solution is not differentiable at the initial time $t_0 = 0$.

Now we consider the basic delay differential equation

$$x' = f(t, x(t-\tau)), \quad t \ge 0, \quad x(s) = \varphi(s), \quad -\tau \le s \le 0. \tag{8.2.2}$$

If f and φ are continuous on $\mathbb{R}^2$ and $\mathbb{R}$, respectively, then a unique solution defined by

$$x(t) = \varphi(0) + \int_0^t f(s, \varphi(s-\tau))ds \tag{8.2.3}$$

exists for $0 \le t \le \tau$. We may argue using the method of steps to show that a unique solution exists for all $t \ge 0$, which is valid regardless of how small the delay τ is.

Note that if $\tau = 0$, then our delay problem becomes an ordinary differential equation. Consider the delay problem

$$x' = x^2(t - \tau), \ t > 0, \ x(s) = 1, \ -\tau \le s \le 0. \tag{8.2.4}$$

Using the method of steps, we may find a unique solution. However, if we set $\tau = 0$ in (8.2.4) with $x(0) = 1$, then the solution of $x' = x^2(t), t > 0$, exists only for $t < 1$. This example should cement that the existence of solutions in finite delay differential equations is much easier to obtain, unlike ordinary differential equations.

Example 8.2. Consider the predator–prey model with carrying capacity

$$x' = ax - dx^2 - bxy, \quad y' = -cy + kxy,$$

where $x(t)$ and $y(t)$ are the numbers of preys and predators at time t, respectively, with positive constants a, b, c, d, and k. The term kxy represents the predator's utilization of the prey it captures. Hence there is certainly a delay, say $r > 0$. As a consequence, we have the predator–prey delay system

$$x' = ax - dx^2 - bxy, \quad y' = -cy + kx(t - r)y(t - r).$$

Suppose we are given an initial continuous function $(\phi_1(t), \phi_2(t))$ with $x(s) = \phi_1(s)$, $y(s) = \phi_2(s)$, $-r \le s \le 0$. Then by integrating the linear equation

$$y' = -cy + k\phi_1(t - r)\phi_2(t - r), \ 0 \le t \le r,$$

we arrive at the explicit solution of y given by $y(t) := \zeta(t)$. Substituting y back into the equation with x' gives the homogeneous Bernoulli differential equation

$$x' = ax - dx^2 - bx\zeta(t), \quad x(0) = \phi_1(0), \tag{8.2.5}$$

which can be easily solved using techniques from basic differential equations. To see this, we let $c(t) = a - b\zeta(t)$ and $v = x^{-1}$. Then we have the resulting differential equation in v,

$$v'(t) + c(t)v = b.$$

Multiplying by the integrating factor, integrating, and simplifying, we arrive at the solution

$$x(t) = \frac{\phi_1(0)}{e^{-\int_0^t c(s)ds} + b\phi_1(0) \int_0^t e^{-\int_u^t c(s)ds} du},$$

where we have used $v = x^{-1}$.

8.3 Existence and uniqueness

We examine the existence and uniqueness of systems with constant delays since they are involved in many applications. For simplicity of notation, we assume that $t_0 = 0$. This should not cause any difficulties or take away from the meaning of the problem.

Let x be an $n \times 1$ vector-valued function, let D be an open set in $\mathbb{R}^n$, and let f be defined on $[0, \infty) \times D^m$. Define an initial function $\varphi : [-\tau, 0] \to \mathbb{R}^m$, where $\tau = \max\{\tau_j, \ 1 \leq j \leq m\}$. Let $r = \min\{\tau_j, \ 1 \leq j \leq m\} > 0$ and consider the system with multiple constant delays

$$x' = f(t, x(t - \tau_1), \ldots, x(t - \tau_m)), \quad 0 \leq t \leq t_f, \tag{8.3.1}$$

$$x(s) = \varphi(s), \quad -\tau \leq s \leq 0. \tag{8.3.2}$$

We may rename the delays so that they are listed in an increasing order so that $\tau = \tau_m$ and $r = \tau_1$:

$$0 \leq \tau_1 < \tau_2 < \cdots < \tau_m.$$

Definition 8.3.1. We say that $x : [t_0 - \tau, t_\beta] \to D$ is a solution of (8.3.1)–(8.3.2), where $0 < t_\beta \leq t_f$, if it is continuous and satisfies (8.3.1) for $0 \leq t < t_\beta$, and $x(s) = \varphi(s)$, $-\tau \leq s \leq 0$. The solution of (8.3.1)–(8.3.2) is said to be unique if every two solutions agree with each other as far as both are defined.

Theorem 8.3.1. *Let f be continuous on $[0, t_f) \times D^m$, and let $\varphi : [-\tau, 0] \to D$ be continuous on $[-\tau, 0]$. Then system (8.3.1)–(8.3.2) has a unique solution x on $[-\tau, \beta_1)$, where $0 < \beta_1 \leq t_f$. Moreover, if $\beta_1 < t_f$, then $x(t)$ approaches the boundary of D as $t \to \beta_1$.*

Proof. Let $t \in [0, r]$. Then every solution resides over the domain of the initial function, that is,

$$x' = f(t, \varphi(t - \tau_1), \ldots, \varphi(t - \tau_m))$$

with $x(0) = \varphi(0)$. An integration from 0 to t gives

$$x(t) = \varphi(0) + \int_0^t f(u, \varphi(u - \tau_1), \ldots, \varphi(u - \tau_m)) du,$$

which uniquely defines $x(t)$ for $0 \leq t \leq r$, provided that $r < t_f$ and $x(t)$ remains in D. Now that we know $x(t)$ on $[-\tau, r]$, we may continue in this fashion on $[r, 2r]$ as long as $2r < t_f$ and $x(t)$ remains in D. We continue with this method of steps until we reach t_f or until $x(t)$ reaches the boundary of D. This completes the proof. $\square$

Remark 8.1. Suppose at least one of the $\tau_j = 0$, $j = 1, 2, \ldots, m$. To be specific, suppose only $\tau_1 = 0$ and the rest of the delays are distinct and greater than zero. Then system (8.3.1) will include a differential equation. In this case proving the existence and uniqueness requires a Lipschitz condition on f of the form

$$|f(t, y_{(1)}, y_{(2)}, \ldots, y_{(m)}) - f(t, z_{(1)}, z_{(2)}, \ldots, z_{(m)})| \leq K|y_{(1)} - z_{(1)}| \qquad (8.3.3)$$

for $y_{(j)}, z_{(j)} \in \mathbb{R}^n$ and positive constant K.

Corollary 8.1. *(Uniqueness) In addition to the hypotheses of Theorem 8.3.1, assume (8.3.3). Then (8.3.1)–(8.3.2) has at most one solution on any interval* $[-\tau, \beta_1)$*, where* $0 < \beta_1 \leq t_f$*.*

Proof. The case $r > 0$ is already covered by Theorem 8.3.1, so we let $r = 0$. Then for $t \in [0, \tau_2]$,

$$x' = f(t, x(t), \varphi(t - \tau_2), \ldots, \varphi(t - \tau_m))$$

with $x(0) = \varphi(0)$. Then by Theorem 2.2.1, it has at most one solution. We may repeat the above process for $t \in [\tau_2, 2\tau_2]$, for $\tau_2 < t_f$. We keep repeating the process until we reach t_f. This completes the proof. $\qquad \square$

Recall our remark in Chapter 3 that a multivariable function f satisfies a Lipschitz condition on some region if all its first partial derivatives are continuous on that region. We have the following example.

Example 8.3. Consider the sun flower model

$$u''(t) + bu'(t) + c \sin u(t - r) = 0, \ t \geq 0,$$

where b, c, and r are positive constants. Using the transformation $x_1 = u$, $x_2 = u'$, we arrive at the system

$$x_1'(t) = x_2,$$

$$x_2'(t) = -c \sin x_1(t - r) - bx_2.$$

Therefore

$$f(t, y_{(1)}, y_{(2)}) = \begin{pmatrix} y_{(1)2} \\ -c \sin y_{(2)1} - by_{(1)2} \end{pmatrix}.$$

Clearly, f is continuous, and so are the partial derivatives

$$\frac{\partial f_1}{\partial y_{(1)1}} = 0, \quad \frac{\partial f_1}{\partial y_{(1)2}} = 1, \quad \frac{\partial f_2}{\partial y_{(1)2}} = -b.$$

Thus by Corollary 8.1, given any continuous function $\varphi : [-r, 0] \to \mathbb{R}^2$, the delay system has at most one solution on $[-r, \beta_1)$ for any $\beta_1 > 0$.

8.4 Stability using Lyapunov functions

In this section, we consider general forms of delay differential equations that fall into the category of *functional delay differential equations*. A functional delay differential equation is a delay differential equation where the unknowns are functions. We will adopt the following notation. For any $r \geq 0$, any continuous function $x(u)$ defined on $-r \leq u \leq A$, $A > 0$, and any fixed t, $0 \leq t \leq A$, by x_t we denote the function $x_t(s) = x(t + s)$, $-r \leq s \leq 0$, that is, $x : [a, b] \to \mathbb{R}^n$ is continuous and is the segment of the function $x(s)$ defined by letting s range in the interval $t - r \leq s \leq t$. We consider the functional delay differential equation

$$x'(t) = F(t, x_t), \ t \geq t_0. \tag{8.4.1}$$

We assume that F is continuous in both arguments and that $F : \mathbb{R} \times C \to \mathbb{R}^n$, where C is the set of continuous functions $\phi : [-\alpha, 0] \to \mathbb{R}^n$, $\alpha > 0$. If $t_0 = -\infty$, then $t \geq t_0$ means that $t > t_0$. Let

$$C(t) = \{\phi : [t - \alpha, t] \to \mathbb{R}^n\}.$$

It is to be understood that $C(t)$ is C for $t = 0$. Also, ϕ_t denotes $\phi \in C(t)$, and $\|\phi_t\| = \max_{t-\alpha \leq s \leq t} |\phi(t)|$, where $|\cdot|$ is a convenient norm on $\mathbb{R}^n$. A solution is denoted by $x(t_0, \phi_{t_0})$, and its value by $x(t, t_0, \phi_{t_0})$ or, if there is no confusion, by $x(t_0, \phi)$ and $x(t, t_0, \phi)$.

Definition 8.4.1. Let $x(t) = 0$ be a solution of (8.4.1).

(a) The zero solution of (8.4.1) is stable if for all $\varepsilon > 0$ and $t_1 \geq t_0$, there exists $\delta > 0$ such that $[\phi \in C(t_1), \ \|\phi\| < \delta, t \geq t_1]$ imply that $|x(t, t_1, \phi)| < \varepsilon$.

(b) The zero solution of (8.4.1) is uniformly stable if it is stable and if δ is independent of $t_1 \geq t_0$.

(c) The zero solution of (8.4.1) is asymptotically stable if it is stable and if for each $t_1 \geq t_0$, there is $\eta > 0$ such that $[\phi \in C(t_1), \ \|\phi\| < \eta]$ imply that $|x(t, t_1, \phi)| \to 0$ as $t \to \infty$. Note that if this is true for every $\eta > 0$, then $x = 0$ is asymptotically stable in the large or globally asymptotically stable.

(d) The zero solution of (8.4.1) is uniformly asymptotically stable if it is uniformly stable and if there is $\eta > 0$ such that for each $\gamma > 0$, there exists $S > 0$ such that $[t_1 \geq t_0, \phi \in C(t_1), \ \|\phi\| < \eta, t \geq t_1 + S]$ imply that $|x(t, t_1, \phi)| < \gamma$. We also note that if this is true for every $\eta > 0$, then $x = 0$ is uniformly asymptotically stable in the large.

Let $W_i : [0, \infty) \to [0, \infty)$ be continuous with $W_i(0) = 0$, $W_i(r)$ strictly increasing, and $W_i(r) \to \infty$ as $r \to \infty$, $i = 1, 2, 3$.

Consider the totally nonlinear differential equation with bounded delay

$$x'(t) = -ax^3(t) + bx^3(t - r), \ t \geq t_0 \geq 0,$$

where a, b, and r are positive constants. Consider the Lyapunov-type functional

$$V(x_t) = |x(t)| + b \int_{t-r}^{t} |x^3(s)| ds.$$

Let $a \geq b$. Then along the solutions we have

$$
\begin{aligned}
V'(x_t) &= \frac{x}{|x|} x' - b|x^3(t-r)| + b|x^3(t)| \\
&= \frac{x}{|x|} \left(-ax^3(t) + bx^3(t-r) \right) - b|x^3(t-r)| + b|x^3(t)| \\
&\leq -a|x^3(t)| + b|x^3(t-r)| - b|x^3(t-r)| + b|x^3(t)| \\
&= [-a+b]|x^3| \leq 0.
\end{aligned}
\tag{8.4.2}
$$

Let $t_1 \in \mathbb{R}$, $\varphi \in C(t_1)$, and $x(t) = x(t, t_1, \varphi)$. Using the fact that $|x(t)| \leq V(x_t)$ and integrating (8.4.2) from $t_1 - r$ to t_1 yield

$$
\begin{aligned}
|x(t)| \leq V(x_t) &\leq V(\varphi_{t_1}) \\
&= |\varphi(t_1)|^3 + b \int_{t_1-r}^{t_1} |\varphi^3(s)| ds \\
&\leq (1 + br) \|\varphi_{t_1}\|^3.
\end{aligned}
\tag{8.4.3}
$$

Suppose $a > b$, let $\varepsilon > 0$, and choose $\delta = \left(\frac{\varepsilon}{1+br} \right)^{1/3}$. Then for $\|\varphi_{t_1}\| < \delta$, from (8.4.3) we have

$$|x(t)| \leq (1 + br) \left[\left(\frac{\varepsilon}{1+br} \right)^{1/3} \right]^3 = \varepsilon,$$

and the zero solution is uniformly stable. Note that stability is independent of the size of the delay r.

Integrate (8.4.2) for 0 to t and obtain $V(x_t) - V(\varphi_0) \leq -\int_0^t |x(s)|^3 ds$, which implies that

$$\int_0^t |x(s)|^3 ds \leq V(\varphi_0) - V(x_t) \leq V(\varphi_0) < \infty$$

for all $t \geq 0$. Hence $x^3(t) \in L^1[0, \infty)$.

Next, we prove parallel theorems to Theorems 7.1.8–7.1.11 concerning system (8.4.1).

Theorem 8.4.1. *Let $D > 0$. Suppose $V(t, \psi_t)$ is a scalar functional continuous in ψ and locally Lipschitz in ψ_t when $t \geq t_0$ and $\psi_t \in C(t)$ with $\|\psi_t\| < D$. Suppose also that $V(t, 0) = 0$ and*

$$W_1(|\psi(t)|) \leq V(t, \psi_t). \tag{8.4.4}$$

(a) *If*

$$V'(t, \psi_t) \leq 0 \text{ for } t_0 \leq t < \infty \text{ and } ||\psi_t|| \leq D, \tag{8.4.5}$$

then the zero solution of (8.4.1) is stable.
(b) *If in addition to (a),*

$$V(t, \psi_t) \leq W_2(||\psi_t||), \tag{8.4.6}$$

then the zero solution of (8.4.1) is uniformly stable.
(c) *If there is $M > 0$ such that $|F(t, \psi_t)| \leq M$ for $t_0 \leq t < \infty$ and $||\psi_t|| \leq D$, and if*

$$V'(t, \psi_t) \leq -W_2(|\psi(t)|), \tag{8.4.7}$$

then the zero solution of (8.4.1) is asymptotically stable.

Proof. Let $\varepsilon > 0$ be such that $\varepsilon < D$. Let $t_1 \geq t_0$. Since V is continuous and $V(t, 0) = 0$, there exists $\delta > 0$ such that $\phi \in C(t_1)$ with $||\phi_{t_1}|| < \delta$ implies that $V(t_1, \phi_{t_1}) < W_1(\varepsilon)$. By condition (8.4.5) we have

$$W_1(|x(t, t_1, \phi_{t_1})|) \leq V(t, x(t, t_1, \phi_{t_1})) \leq V(t_1, \phi_{t_1})$$
$$\leq W_1(\varepsilon),$$

from which it follows that

$$|x(t, t_1, \phi_{t_1})| \leq W_1^{-1}(W_1(\varepsilon)) = \varepsilon.$$

This concludes the proof of (a).

As for the proof of (b), we let $\varepsilon > 0$ be such that $\varepsilon < D$. We find $\delta > 0$ with $W_2(\delta) < W_1(\varepsilon)$. Let $t_1 \geq t_0$ and $\phi_{t_1} \in C(t_1)$ with $||\phi_{t_1}|| < \delta$. Then

$$W_1(|x(t, t_1, \phi_{t_1})|) \leq V(t, x_t(t_1, \phi_{t_1})) \leq V(t_1, \phi_{t_1})$$
$$\leq W_2(\delta) < W_1(\varepsilon),$$

from which it follows that

$$|x(t, t_1, \phi_{t_1})| \leq W_1^{-1}(W_1(\varepsilon)) = \varepsilon.$$

This concludes the proof of (b).

To prove (c), let $t_1 \geq t_0$, and let $0 < \varepsilon < D$. Find δ as in part (b) and take $\eta = \delta$. Let $\phi_{t_1} \in C(t_1)$ with $||\phi_{t_1}|| < \delta$. For simplicity, we write $x(t) = x(t, t_1, \phi_{t_1})$. We will establish the proof by contradiction. Assume that $x(t) \nrightarrow 0$ as $t \to \infty$. Then there are $\varepsilon_1 > 0$ and a sequence $\{t_n\} \to \infty$ with $|x(t_n)| \geq \varepsilon_1$. Since $|F(t, \psi_t)| \leq M$ for $t_0 \leq t < \infty$ and $||\psi_t|| \leq D$, there are $T > 0$ and $\varepsilon_2 < \varepsilon_1$ with $|x(t_n)| \geq \varepsilon_2$ for $t_n \leq t \leq t_n + T$.

From this and condition (8.4.7) it follows that

$$0 \leq V(t, x_t) \leq V(t_1, \phi_{t_1}) - \int_{t_1}^{t} W_2(|x(s)|)$$

$$\leq V(t_1, \phi_{t_1}) - \sum_{i=2}^{n} \int_{t_i}^{t_i + T} W_2(|x(s)|)$$

$$\leq V(t_1, \phi_{t_1}) - \sum_{i=2}^{n} \int_{t_i}^{t_i + T} W_2(\varepsilon_2)$$

$$= V(t_1, \phi_{t_1}) - (n-1)T W_2(\varepsilon_2) \to -\infty \text{ as } n \to \infty,$$

a contradiction. This concludes the proof of (c). $\square$

Theorem 8.4.2. *Let $D > 0$. Suppose $V(t, \psi_t)$ is a scalar functional continuous in ψ and locally Lipschitz in ψ_t when $t \geq t_0$ and $\psi_t \in C(t)$ with $||\psi_t|| < D$. In addition, we assume that if $x : [t_0 - \alpha, \infty) \to \mathbb{R}^n$ is bounded, then $F(t, x_t)$ is bounded on $[t_0, \infty)$. Suppose V satisfies $V(t, 0) = 0$,*

$$W_1(|\psi(t)|) \leq V(t, \psi_t) \leq W_2(||\psi_t||), \tag{8.4.8}$$

and

$$V'(t, \psi_t) \leq -W_3(|\psi(t)|). \tag{8.4.9}$$

Then the zero solution of (8.4.1) is uniformly asymptotically stable.

Proof. We refer to [15]. $\square$

In the next theorem, we use a Lyapunov function and obtain boundedness results concerning non-homogeneous delay differential equations.

Theorem 8.4.3. *Let $h > 0$ and consider the finite delay differential equation*

$$x'(t) = a(t)x(t) + b(t)x(t - h) + g(t), \ t \geq 0, \tag{8.4.10}$$

with $x(t_0) = \varphi(t_0)$, $-h \leq t_0 \leq 0$. Suppose $a(t)$, $b(t)$, and $g(t)$ are continuous for all $t \geq 0$.

Assume that

$$2a(t) + 3 + |b(t)| \leq 0, \tag{8.4.11}$$

$$|b(t)| - e^{-h} \leq 0, \tag{8.4.12}$$

and there exists a positive constant L such that

$$|g(t)| \leq L, \ t \geq 0.$$

Then all solutions of (8.4.10) are uniformly bounded. Moreover, if $g(t) = 0$ for all $t \geq 0$, then the zero solution is exponentially stable.

Proof. For $t \geq 0$, let

$$V(t, x_t) = x^2(t) + \int_{t-h}^{t} e^{-(t-s)} x^2(s) \, ds.$$

Then along solutions of (8.4.10) we have

$$V'(t, x) = 2a(t)x^2(t) + 2b(t)x(t)x(t-h) + 2x(t)g(t) + x^2(t)$$
$$- e^{-h}x^2(t-h) - \int_{t-h}^{t} e^{-(t-s)} x^2(s) \, ds$$
$$\leq (2a(t) + 2 + |b(t)|)x^2(t) + (|b(t)| - e^{-h})x^2(t-h)$$
$$- \int_{t-h}^{t} e^{-(t-s)} x^2(s) \, ds + g^2(t).$$

Using conditions (8.4.11) and (8.4.12), we have

$$V(t, x_t) + V'(t, x_t) \leq (2a(t) + 3 + |b(t)|)x^2(t) + (|b(t)| - e^{-h})x^2(t-h)$$
$$- \int_{t-h}^{t} e^{-(t-s)} x^2(s) \, ds + \int_{t-h}^{t} e^{-(t-s)} x^2(s) \, ds + g^2(t)$$
$$\leq g^2(t) \leq L^2.$$

So we have

$$V'(t, x_t) \leq -V(t, x_t) + L^2.$$

It follows from the variation of parameters formula that

$$|x| \leq \left[V(t_0, \varphi)e^{-(t-t_0)} + L^2 \right]^{1/2}$$
$$\leq \left[\left(1 + \int_{t_0-h}^{t_0} e^{-(t_0-s)} \, ds \right) ||\varphi||^2 e^{-(t-t_0)} + L^2 \right]^{1/2}$$
$$\leq \left[(2 - e^{-h}) ||\varphi||^2 e^{-(t-t_0)} + L^2 \right]^{1/2}.$$

This gives boundedness. Now if $g(t) = 0$ for all $t \geq 0$, then $L = 0$, and hence the above inequality gives the exponential stability. $\qquad \square$

We end this section by noting that despite the results of Exercises 8.8 and 8.9, researchers are still trying to substitute delay terms with terms that are independent of delays and arrive at ordinary differential equations that can be explicitly solved. This could be a good prospect for research using numerical analysis.

8.5 Stability using fixed point theory

Lyapunov functions and functionals have been successfully used to obtain boundedness, stability, and the existence of periodic solutions of differential equations, differential equations with functional delays, and functional differential equations. In the study of differential equations with functional delays by using Lyapunov functionals, many difficulties arise if the delay is unbounded or if the differential equation in question has unbounded terms. This section is mainly concerned with the asymptotic stability of the zero solution of the scalar neutral differential equation

$$x'(t) = -a(t)x(t) + c(t)x'(t - g(t)) + q(t, x(t), x(t - g(t))), \qquad (8.5.1)$$

where $a(t)$, $c(t)$, $g(t)$, and q are continuous in their respective arguments. What makes Eq. (8.5.1) interesting to study is the fact that it cannot be put in the form of

$$\frac{d}{dt}[x(t) + c(t)x(t - g(t))] = -a(t)x(t) + q(t, x(t), x(t - g(t))),$$

and hence a direct integration cannot be performed to arrive at the desired mapping. We could not construct a suitable Lyapunov functional that yielded any meaningful results about the asymptotic stability of the zero solution. Hence the construction of such a Lyapunov functional for Eq. (8.5.1) remains open. Moreover, in this study, we concentrate on the stability of the zero solution of nonlinear neutral differential equations with bounded and unbounded delays. To achieve our goal, we make use of the contraction mapping principle. To justify the need of using fixed point theory when studying stability, we consider the time-varying delay differential equation

$$x'(t) = -a(t)x(t) + b(t)x(t - g(t)), \qquad (8.5.2)$$

where a and b are bounded continuous functions, and g is continuously differentiable and nonnegative.

Theorem 8.5.1. *Suppose*

$$g'(t) \text{ is bounded, } 1 - g'(t) > 0, \qquad (8.5.3)$$

$$-2a(t) + 1 \leq -(1 - g'(t)), \qquad (8.5.4)$$

and

$$|b(t)| \leq \alpha(1 - g'(t)) \text{ for some } \alpha \in (0, 1). \qquad (8.5.5)$$

Then the zero solution of (8.5.2) is uniformly asymptotically stable.

Proof. Define the Lyapunov functional

$$V(t) = x^2(t) + \int_{t-g(t)}^{t} x^2(s)\, ds.$$

Then along solutions of (8.5.2) we have

$$
\begin{aligned}
V'(t) &= 2x(t)\Big(-a(t)x(t) + b(t)x(t - g(t))\Big) + x^2(t) - (1 - g'(t))x^2(t - g(t)) \\
&\leq -2a(t)x^2(t) + |b(t)|x^2(t) + |b(t)|x^2(t - g(t)) \\
&\quad + x^2(t) - (1 - g'(t))x^2(t - g(t)) \\
&= \Big(-2a(t) + |b(t)| + 1\Big)x^2(t) + \Big(|b(t)| - (1 - g'(t))\Big)x^2(t - g(t)) \\
&\leq (\alpha - 1)(1 - g'(t))\Big(x^2(t) + x^2(t - g(t))\Big) \\
&\leq (\alpha - 1)(1 - g'(t))x^2(t).
\end{aligned}
$$

The result follows from Theorem 8.4.2. $\qquad\qquad\square$

We remark that (8.5.4) and (8.5.5) imply that $|b(t)| \leq \alpha(2a(t) - 1)$. Thus we see that $|b(t)|$ must be bounded by $a(t)$. In addition, $a(t)$ must satisfy $a(t) \geq 1/2$.

8.5.1 Neutral differential equations

Conditions (8.5.3) and (8.5.4) are severe since a and b must be bounded and both a and b must satisfy an upper bound in terms of $1 - g'$. To relax the restrictions, we resort to the use of fixed point theory, namely, the contraction mapping principle. To better illustrate our procedure, we first consider the scalar linear neutral differential equation (neutral since there is a delay in the derivative) with unbounded delay

$$
x'(t) = -a(t)x(t) + b(t)x(t - g(t)) + c(t)x'(t - g(t)), \qquad (8.5.6)
$$

where $a(t)$ and $b(t)$ are continuous, $c(t)$ is continuously differentiable, and $g(t) > 0$ for all $t \in \mathbb{R}$ and is twice continuously differentiable. We note that Eq. (8.5.6) is more complicated than that we considered in Theorem 8.5.1 since the derivative on the right side has a delay. Hence the Lyapunov functional that was used in the proof of Theorem 8.5.1 will not work, and we have to construct a suitable one. However, the author believes that conditions similar to (8.5.3)–(8.5.5) have to be imposed. Using fixed point theory, a suitable mapping has to be formulated. To obtain the desired map, we invert Eq. (8.5.6). During the process of inverting (8.5.6), we have to integrate by parts the term involving $x'(t - g(t))$. We begin by requiring that

$$
g'(t) \neq 1, \quad \forall\, t \in \mathbb{R}. \qquad (8.5.7)
$$

Definition 8.3.1 naturally extends to suit Eq. (8.5.6).

Lemma 8.1. *Suppose (8.5.7) holds. Then $x(t)$ is a solution of Eq. (8.5.6) if and only if*

$$x(t) = \left(x(0) - \frac{c(0)}{1 - g'(0)}x(-g(0))\right)e^{-\int_0^t a(s)ds} + \frac{c(t)}{1 - g'(t)}x(t - g(t))$$

$$- \int_0^t (r(u) - b(u))x(u - g(u))e^{-\int_u^t a(s)ds}du, \tag{8.5.8}$$

where

$$r(u) = \frac{\left(c'(u) + c(u)a(u)\right)\left(1 - g'(u)\right) + g''(u)c(u)}{(1 - g'(u))^2}. \tag{8.5.9}$$

Proof. Multiplying both sides of (8.5.6) by $e^{\int_0^t a(s)ds}$ and then integrating from 0 to t, we obtain

$$\int_0^t \left[x(u)e^{\int_0^u a(s)ds}\right]'du = \int_0^t \left[b(u)x(u - g(u)) + c(u)x'(u - g(u))\right]e^{\int_0^u a(s)ds}du.$$

As a consequence, we arrive at

$$x(t)e^{\int_0^t a(s)ds} - x(0) = \int_0^t \left[b(u)x(u - g(u)) + c(u)x'(u - g(u))\right]e^{\int_0^u a(s)ds}du.$$

Dividing both sides of the above equation by $e^{\int_0^t a(s)ds}$, we obtain

$$x(t) = x(0)e^{-\int_0^t a(s)ds}$$

$$+ \int_0^t \left[b(u)x(u - g(u)) + c(u)x'(u - g(u))\right]e^{-\int_u^t a(s)ds}du. \tag{8.5.10}$$

Rewrite

$$\int_0^t c(u)x'(u - g(u))e^{-\int_u^t a(s)ds}du$$

$$= \int_0^t \frac{c(u)x'(u - g(u))(1 - g'(u))}{(1 - g'(u))}e^{-\int_u^t a(s)ds}du.$$

Integrating by parts the above integral with

$$U(u) = \frac{c(u)}{1 - g'(u)}e^{-\int_u^t a(s)ds}$$

and

$$dV = x'(u - g(u))(1 - g'(u))du,$$

we have

$$\int_0^t c(u)x'(u - g(u))e^{-\int_u^t a(s)ds}du = \frac{c(t)}{1 - g'(t)}x(t - g(t))$$

$$- \frac{c(0)}{1 - g'(0)}x(-g(0))e^{-\int_0^t a(s)ds} - \int_0^t r(u)e^{-\int_u^t a(s)ds}x(u - g(u))du, \tag{8.5.11}$$

where $r(u)$ is given by (8.5.9). Finally, substituting (8.5.11) into (8.5.10) completes the proof.   □

Next, let $\psi(t) : (-\infty, 0] \to \mathbb{R}$ be a continuous bounded initial function. We say that $x(t) := x(t, 0, \psi)$ is a solution of (8.5.6) if $x(t) = \psi(t)$ for $t \leq 0$ and satisfies (8.5.6) for $t \geq 0$.

We say that the zero solution of (8.5.6) is stable at t_0 if for each $\varepsilon > 0$, there is $\delta = \delta(\varepsilon) > 0$ such that $\left[\psi : [-\infty, t_0] \to \mathbb{R} \text{ with } |\psi(t)| < \delta \text{ on } (-\infty, t_0], \ t \geq t_0\right]$ implies $|x(t, t_0, \psi)| < \varepsilon$.

Let C be the space of all continuous functions from $\mathbb{R} \to \mathbb{R}$ and define the set

$$S = \left\{\varphi : \mathbb{R} \to \mathbb{R} \ \big| \ \varphi(t) = \psi(t) \text{ if } t \leq 0, \ \varphi(t) \to 0 \text{ as } t \to \infty, \ \varphi \in C,\right.$$

$$\left. \text{and } \varphi \text{ is bounded}\right\}.$$

Then $\left(S, || \cdot ||\right)$ is a complete metric space, where $|| \cdot ||$ is the supremum norm.

For the next theorem, we impose the following conditions:

$$e^{-\int_0^t a(s)ds} \to 0 \text{ as } t \to \infty, \tag{8.5.12}$$

there is $\alpha > 0$ such that

$$\left|\frac{c(t)}{1 - g'(t)}\right| + \int_0^t |r(u) - b(u)| e^{-\int_u^t a(s)ds} du \leq \alpha < 1, t \geq 0, \tag{8.5.13}$$

and

$$t - g(t) \to \infty \text{ as } t \to \infty. \tag{8.5.14}$$

Theorem 8.5.2. *If (8.5.7) and (8.5.12)–(8.5.14) hold, then every solution $x(t, 0, \psi)$ of (8.5.6) with small continuous initial function $\psi(t)$ is bounded and goes to zero as $t \to \infty$. Moreover, the zero solution is stable at $t_0 = 0$.*

Proof. Define the mapping $P : S \to S$ by

$$\left(P\varphi\right)(t) = \psi(t) \text{ if } t \leq 0$$

and

$$\left(P\varphi\right)(t) = \left(\psi(0) - \frac{c(0)}{1 - g'(0)}\psi(-g(0))\right)e^{-\int_0^t a(s)ds} + \frac{c(t)}{1 - g'(t)}\varphi(t - g(t))$$

$$- \int_0^t \left[r(u) - b(u)\right]\varphi(u - g(u))e^{-\int_u^t a(s)ds}du, \ t \geq 0.$$

It is clear that for $\varphi \in S$, $P\varphi$ is continuous. Let $\varphi \in S$ with $||\varphi|| \leq K$ for some positive constant K. Let $\psi(t)$ be a small continuous initial function with $|\psi| < \delta, \delta > 0$. Then using (8.5.13) in the definition of $(P\varphi)(t)$, we have

$$||(P\varphi)(t)|| \leq |(1 - \frac{c(0)}{1 - g'(0)})|K + |\frac{c(t)}{1 - g'(t)}|K$$

$$+ \int_0^t |r(u) - b(u)|e^{-\int_u^t a(s)ds} du\, K$$

$$\leq |(1 - \frac{c(0)}{1 - g'(0)})|\delta + \alpha K, \qquad (8.5.15)$$

which implies that $||(P\varphi)(t)|| \leq K$ for the right δ. Thus (8.5.15) implies that $(P\varphi)(t)$ is bounded. Next, we show that $(P\varphi)(t) \to 0$ as $t \to \infty$. The first term on the right side of $(P\varphi)(t)$ tends to zero by condition (8.5.12). Also, the second term on the right side tends to zero because of (8.5.14) and the fact that $\varphi \in S$. It remains to show that the integral term goes to zero as $t \to \infty$.

Let $\varepsilon > 0$, and let $\varphi \in S$ with $||\varphi|| \leq K$, $K > 0$. Then there exists $t_1 > 0$ such that $|\varphi(t - g(t))| < \varepsilon$ for $t > t_1$. Due to condition (8.5.12), there exists $t_2 > t_1$ such that $e^{-\int_{t_1}^t a(s)ds} < \frac{\varepsilon}{\alpha K}$ for $t > t_2$. Thus for $t > t_2$, we have

$$\left| \int_0^t (r(u) - b(u))\varphi(u - g(u))e^{-\int_u^t a(s)ds} du \right|$$

$$\leq K \int_0^{t_1} |r(u) - b(u)|e^{-\int_u^t a(s)ds} du + \varepsilon \int_{t_1}^t |r(u) - b(u)|e^{-\int_u^t a(s)ds} du$$

$$\leq K e^{-\int_{t_1}^t a(s)ds} \int_0^{t_1} |r(u) - b(u)|e^{-\int_u^{t_1} a(s)ds} du + \alpha\varepsilon$$

$$\leq \alpha K e^{-\int_{t_1}^t a(s)ds} + \alpha\varepsilon$$

$$\leq \varepsilon + \alpha\varepsilon.$$

Hence $(P\varphi)(t) \to 0$ as $t \to \infty$. It remains to show that $(P\varphi)(t)$ is a contraction under the supremum norm. Let $\zeta, \eta \in S$. Then

$$\left| (P\zeta)(t) - (P\eta)(t) \right| \leq \left\{ |\frac{c(t)}{1 - g'(t)}| + \int_0^t |r(u) - b(u)|e^{-\int_u^t a(s)ds} du \right\} ||\zeta - \eta||$$

$$\leq \alpha ||\zeta - \eta||.$$

Thus by the contraction mapping principle, P has a unique fixed point in S, which solves (8.5.6), is bounded, and tends to zero as t tends to infinity. The stability of the zero solution at $t_0 = 0$ follows from the above work by simply replacing K by ε. This completes the proof. $\qquad\qquad\square$

Example 8.4. Consider the linear neutral differential equation

$$x'(t) = -2x(t) + c_0\, x'(t - (\frac{t}{2})), \qquad (8.5.16)$$

where c_0 is a constant. Then $r(u) = -4c_0$, and condition (8.5.13) is satisfied for $|c_0| \leq \frac{\alpha}{4}$, $\alpha \in (0, 1)$. Let $\psi(t)$ be a given continuous initial function with $|\psi(t)| \leq \delta$. Let

$$S = \left\{ \varphi : \mathbb{R} \to \mathbb{R} \mid \varphi(t) = \psi(t) \text{ if } t \leq 0, \ \varphi(t) \to 0 \text{ as } t \to \infty, \ \varphi \in C, \right.$$
$$\left. \text{and } \varphi \text{ is bounded} \right\}.$$

Define

$$(P\varphi)(t) = \psi(t) \text{ if } t \leq 0$$

and

$$(P\varphi)(t) = \left((1 - 2c_0)\psi(0) \right) e^{-2t} + 2c_0\varphi(\frac{t}{2})$$
$$+ \int_0^t 4c_0\varphi(\frac{u}{2}) e^{-2(t-u)} du, \ t \geq 0.$$

Then, for $\varphi \in S$ with $\|\varphi\| \leq K$, where $K \geq \frac{(1-\alpha)}{|1-2c_0|\delta}$, we have $\|(P\varphi)(t)\| \leq K$. It is obvious that conditions (8.5.12) and (8.5.14) are satisfied. To see that P defines a contraction mapping, let $\zeta, \eta \in S$. Then

$$\left| (P\zeta)(t) - (P\eta)(t) \right| \leq 2|c_0| \|\zeta - \eta\| + 2|c_0|(1 - e^{-2t})\|\zeta - \eta\|$$
$$\leq 4|c_0| \|\zeta - \eta\| \leq \alpha \|\zeta - \eta\|.$$

Hence by Theorem 8.5.2 every solution $x(t, 0, \psi)$ of (8.5.16) with small continuous initial function $\psi(t) : (-\infty, 0] \to \mathbb{R}$ is in S, is bounded, and goes to zero as $t \to \infty$.

Next, we turn our attention to the nonlinear neutral differential equation with unbounded delay

$$x'(t) = -a(t)x(t) + c(t)x'(t - g(t)) + q\left(x(t), x(t - g(t)) \right), \tag{8.5.17}$$

where $a(t)$, $c(t)$, and $g(t)$ are defined as before. Here we assume that $q(0, 0) = 0$ and q is locally Lipschitz continuous in x and y in the sense that there is $K > 0$ such that if $|x|, |y|, |z|, |w| \leq K$, then

$$|q(x, y) - q(z, w)| \leq L|x - z| + E|y - w| \tag{8.5.18}$$

for some positive constants L and E.

Note that

$$|q(x, y)| = |q(x, y) - q(0, 0) + q(0, 0)|$$
$$\leq |q(x, y) - q(0, 0)| + |q(0, 0)|$$
$$\leq L|x| + E|y|.$$

Let

$$S = \left\{ \varphi : \mathbb{R} \to \mathbb{R} \,\middle|\, ||\varphi|| \leq K, \, \varphi(t) = \psi(t) \text{ if } t \leq 0, \, \varphi(t) \to 0 \text{ as } t \to \infty, \, \varphi \in C \right\}.$$

Define the map $P : S \to S$ by

$$(P\varphi)(t) = \psi(t) \quad \text{if } t \leq 0$$

and

$$(P\varphi)(t) = \left(\psi(0) - \frac{c(0)}{1 - g'(0)} \psi(-g(0)) \right) e^{-\int_0^t a(s)ds} + \frac{c(t)}{1 - g'(t)} \varphi(t - g(t))$$
$$+ \int_0^t \left[-r(u)\varphi(u - g(u)) + q(\varphi(u), \varphi(u - g(u))) \right] e^{-\int_u^t a(s)ds} du, \, t \geq 0.$$

It is clear that $P\varphi$ is continuous for $\varphi \in S$. If P has a fixed point, say φ, then φ is a solution of (8.5.17). For P to be a contraction, we assume that there is $\alpha > 0$ such that

$$\left| \frac{c(t)}{1 - g'(t)} \right| + \int_0^t (|r(u)| + L + E) e^{-\int_u^t a(s)ds} du \leq \alpha < 1, \, t \geq 0. \qquad (8.5.19)$$

Theorem 8.5.3. *If (8.5.7), (8.5.12), (8.5.14), and (8.5.19) hold, then every solution $x(t, 0, \psi)$ of (8.5.17) with small continuous initial function $\psi(t) : (-\infty, 0] \to \mathbb{R}$ goes to zero as $t \to \infty$. Moreover, the zero solution is stable at $t_0 = 0$.*

Proof. Let $\varphi \in S$, and let t_1 and t_2 be as in the proof of Theorem 8.5.2. Then for $t > t_2$,

$$\left| \int_0^t q(\varphi(u), \varphi(u - g(u))) e^{-\int_u^t a(s)ds} du \right|$$
$$\leq K(L + E) \int_0^{t_1} e^{-\int_u^t a(s)ds} du + \varepsilon(L + E) \int_{t_1}^t e^{-\int_u^t a(s)ds} du$$
$$\leq K(L + E) e^{-\int_{t_1}^t a(s)ds} \int_0^{t_1} e^{-\int_u^{t_1} a(s)ds} du + \varepsilon(L + E)\alpha$$
$$\leq \alpha K e^{-\int_{t_1}^t a(s)ds} + \varepsilon(L + E)\alpha$$
$$\leq \varepsilon + \varepsilon(L + E)\alpha.$$

This, along with the proof of Theorem 8.5.2, shows that $(P\varphi)(t) \to 0$ as $t \to \infty$. Let $\psi(t)$ be a small continuous initial function with $|\psi| < \delta$, $\delta > 0$. Then by using

(8.5.19) we arrive at

$$\|(P\varphi)(t)\| \leq |(1 - \frac{c(0)}{1 - g'(0)})|\delta + |\frac{c(t)}{1 - g'(t)}|K$$

$$+ \int_0^t (|r(u)| + L + E)e^{-\int_u^t a(s)ds}du\,K$$

$$\leq |(1 - \frac{c(0)}{1 - g'(0)})|\delta + \alpha K, \tag{8.5.20}$$

which implies that $\|(P\varphi)(t)\| \leq K$ with the right choice of δ and α. It remains to show that $(P\varphi)(t)$ is a contraction. Let $\zeta, \eta \in S$. Then

$$\left|(P\zeta)(t) - (P\eta)(t)\right|$$

$$\leq \left\{ |\frac{c(t)}{1 - g'(t)}|\,\|\zeta - \eta\| + \int_0^t (|r(u)||\zeta(u - g(u)) - \eta(u - g(u))|e^{-\int_u^t a(s)ds}du \right.$$

$$+ \int_0^t |q(\zeta(u), \zeta(u - g(u))) - q(\eta(u), \eta(u - g(u)))|e^{-\int_u^t a(s)ds}du \right\}$$

$$\leq \left\{ |\frac{c(t)}{1 - g'(t)}| + \int_0^t (|r(u)| + L + E)e^{-\int_u^t a(s)ds}du \right\}\|\zeta - \eta\|$$

$$\leq \alpha\|\zeta - \eta\|.$$

Thus by the contraction mapping principle, P has a unique fixed point in S, which solves (8.5.17) and tends to zero as t tends to infinity. This completes the proof. $\square$

8.5.2 Neutral Volterra integro-differential equations

Now we turn our attention to the scalar neutral Volterra integro-differential equation

$$x'(t) = -a(t)x(t) + c(t)x'(t - g(t)) + \int_{t-g(t)}^t k(t, s)h(x(s))ds, \tag{8.5.21}$$

where $0 \leq g(t) \leq g_0$ for some constant g_0. Here we assume that $g, a, c, h : \mathbb{R} \to \mathbb{R}$ and $k : \mathbb{R} \times \mathbb{R} \to \mathbb{R}$ are continuous. For the next theorem, we make the following assumptions:

$$\left|\frac{c(t)}{1 - g'(t)}\right| + \int_0^t \left[|r(s)| + \int_{s-g(s)}^s |k(s, u)|du\right]e^{-\int_s^t a(u)du}ds \leq \alpha < 1, t \geq 0; \tag{8.5.22}$$

for each $\varepsilon > 0$, there exist $t_1 > 0$ and $T > 0$ such that for $t_2 \geq t_1$ and $t \geq t_2 + T$, we have

$$e^{-\int_{t_2}^t a(s)ds} < \varepsilon \quad \text{and} \quad e^{-\int_0^t a(s)ds} \to 0 \text{ as } t \to \infty; \tag{8.5.23}$$

and there is $L > 0$ such that $|x|, |y| \leq L$ imply that

$$|h(x) - h(y)| \leq |x - y| \text{ and } h(0) = 0. \tag{8.5.24}$$

Theorem 8.5.4. *If (8.5.22)–(8.5.24) hold, then the zero solution of (8.5.21) is asymptotically stable at $t_0 = 0$.*

Proof. Let $\psi : [-g_0, 0] \to \mathbb{R}$ be a bounded initial function with $|\psi| < \delta$ for some positive constant δ. Define

$$S = \Big\{ \varphi : [-g_0, \infty) \to \mathbb{R} \mid \varphi(t) = \psi(t) \text{ if } -g_0 \leq t \leq 0, \; \|\varphi\| \leq L,$$
$$\varphi \in C, \; \varphi(t) \to 0 \text{ as } t \to \infty \Big\},$$

where $\| \cdot \|$ is the supremum norm.
Define the mapping $P : S \to S$ by

$$(P\varphi)(t) = \psi(t) \text{ if } -g_0 \leq t \leq 0$$

and

$$(P\varphi)(t) = \Big(\psi(0) - \frac{c(0)}{1 - g'(0)} \psi(-g(0)) \Big) e^{-\int_0^t a(s)ds} + \frac{c(t)}{1 - g'(t)} \varphi(t - g(t))$$
$$+ \int_0^t \Big[-r(s)\varphi(s - g(s)) + \int_{s-g(s)}^s k(s, u)h(x(u))du \Big] e^{-\int_s^t a(u)du} ds, \; t \geq 0,$$

where $r(u)$ is defined by (8.5.9). For $\varphi \in S$ with $\|\varphi\| \leq L$, we have that for any $\varepsilon > 0$, there exists $t_1 > 0$ such that $t \geq t_1 - g_0$ implies that $|\varphi(t)| < \varepsilon$. Similarly, $t_2 \geq t_1$ and $t \geq t_2 + T$ imply that $e^{-\int_{t_2}^t a(s)ds} < \varepsilon$. Thus $t \geq t_2 + T$ and t_1 large enough give

$$\|(P\varphi)(t)\| \leq |(1 - \frac{c(0)}{1 - g'(0)})|\delta\varepsilon + |\frac{c(t)}{1 - g'(t)}|\varepsilon$$
$$+ \int_0^{t_2} \Big[|r(s)|L + \int_{s-g(s)}^s k(s, u)Ldu \Big] e^{-\int_s^{t_2} a(u)du} e^{-\int_{t_2}^t a(u)du} ds$$
$$+ \int_{t_2}^t \Big[|r(s)|\varepsilon + \int_{s-g(s)}^s |k(s, u)|\varepsilon du \Big] e^{-\int_s^t a(u)du} ds$$
$$\leq |(1 - \frac{c(0)}{1 - g'(0)})|\delta\varepsilon + |\frac{c(t)}{1 - g'(t)}|\varepsilon + \varepsilon\alpha L + \varepsilon\alpha. \tag{8.5.25}$$

Thus $(P\varphi)(t) \to 0$ as $t \to \infty$. Also, as in the proof of (8.5.15), with the right choice of δ, we have that $\|P(\varphi)\| \leq L$ for $\|\varphi\| \leq L$. It remains to show that P is a contraction. For $\zeta, \eta \in S$, we have

$$\left|(P\zeta)(t) - (P\eta)(t)\right| \leq \left|\frac{c(t)}{1 - g'(t)}\right| \, ||\zeta - \eta||$$
$$+ \int_0^t \left[|r(s)| + \int_{s-g(s)}^s |k(s,u)| du\right] e^{-\int_s^t a(u) du} ds \, ||\zeta - \eta||$$
$$\leq \alpha ||\zeta - \eta||.$$

Hence P has a unique fixed point in S. This completes the proof. $\qquad\square$

We conclude this section by considering Eq. (8.5.21) when there is no delay in the derivative, that is, $c(t)$ is identically zero. We use the Lyapunov functional method to show that the zero solution is asymptotically stable. As we try to show that the derivative term of the Lyapunov functional is less than zero along the solutions of the desired equation, an unpleasant condition relating the size of $a(t)$ to the size of $g(t)$ arises, which limits the types of equations that can be discussed. In particular, we consider the scalar Volterra integro-differential delay equation

$$x'(t) = -a(t)x(t) + \int_{t-g(t)}^t k(t,s)h(x(s)) ds, \qquad (8.5.26)$$

where $0 \leq g(t) \leq g_0$ for some constant g_0.

Suppose there exists a positive constant β such that

$$|k(t,s)| \leq \beta \text{ for all } t \geq 0, \qquad (8.5.27)$$

$$g'(t) \leq 1 - \beta, \qquad (8.5.28)$$

and

$$a(t) \geq \frac{(\beta + 1)}{2} g(t) \text{ for all } t \geq 0. \qquad (8.5.29)$$

Theorem 8.5.5. *If* (8.5.24) *and* (8.5.27)–(8.5.29) *hold, then the zero solution of* (8.5.26) *is asymptotically stable.*

Proof. Define

$$V(t) = x^2(t) + \int_{t-g(t)}^t \int_s^t h^2(x(v)) dv ds.$$

It follows from (8.5.24) and (8.5.27) that along the solutions of (8.5.26)

$$V'(t) = 2x(t)\left[-a(t)x(t) + \int_{t-g(t)}^t k(t,s)h(x(s)) ds\right]$$
$$- (1 - g'(t)) \int_{t-g(t)}^t h^2(x(v)) dv + g(t)h^2(x(t))$$

$$\leq -2a(t)x^2(t) + 2\beta|x(t)| \int_{t-g(t)}^{t} |h(x(s))|ds$$

$$- (1 - g'(t)) \int_{t-g(t)}^{t} h^2(x(s))ds + g(t)h^2(x(t))$$

$$\leq -2a(t)x^2(t) + \beta \int_{t-g(t)}^{t} \left[x^2(t) + h^2(x(s))ds \right]$$

$$- (1 - g'(t)) \int_{t-g(t)}^{t} h^2(x(v))dv + g(t)h^2(x(t))$$

$$\leq x^2(t)\left[-2a(t) + \beta g(t) + g(t) \right] + (1 - g'(t) + \beta) \int_{t-g(t)}^{t} h^2(x(s))ds$$

$$\leq -\gamma x^2(t) \text{ for some positive constant } \gamma.$$

Then by (c) of Theorem 8.4.1, the zero solution is asymptotically stable. $\square$

8.6 Exponential stability

In this section we consider the scalar and totally nonlinear Volterra integro-differential equation with finite delay

$$x'(t) = - \int_{t-r}^{t} a(t, s)g(x(s))ds, \qquad (8.6.1)$$

where $r > 0$ is a constant, and $a : [0, \infty) \times [0, \infty] \to (-\infty, \infty)$. The function $g(x)$ is continuous in x.

The Lyapunov method allowed us to deduce inequalities that all solutions must satisfy and from which we deduce the exponential stability and instability.

Let $\psi : [-r, 0] \to (-\infty, \infty)$ be a continuous initial function with

$$||\psi|| = \max_{-r \leq s \leq 0} |\psi(s)|.$$

It should cause no confusion to denote the norm of a continuous function $\varphi : [-r, \infty) \to (-\infty, \infty)$ by

$$||\varphi|| = \sup_{-r \leq s < \infty} |\varphi(s)|.$$

The notation x_t means that $x_t(\tau) = x(t + \tau)$, $\tau \in [-r, 0]$, as long as $x(t + \tau)$ is defined. Thus x_t is a function mapping an interval $[-h, 0]$ into $\mathbb{R}$. We say that $x(t) \equiv x(t, t_0, \psi)$ is a solution of (8.6.1) if $x(t)$ satisfies (8.6.1) for $t \geq t_0$ and $x_{t_0} = x(t_0 + s) = \psi(s)$, $s \in [-r, 0]$.

Now we turn our attention to the totally nonlinear equation (8.6.1). We will construct a Lyapunov functional $V(t, x) := V(t)$ and show that for some positive α, under suitable conditions, $V'(t) \leq -\alpha V(t)$ along the solutions of (8.6.1). To rewrite (8.6.1) so that a suitable Lyapunov functional can be displayed, we let

$$A(t, s) := \int_{t-s}^{r} a(u + s, s)du, \ t, s \geq 0,$$

and assume that

$$A(t, t) = \int_{0}^{r} a(u + t, t)du \geq 0. \tag{8.6.2}$$

In preparation of the main results, we assume that

$$xg(x) \geq x^2 \text{ if } x \neq 0 \tag{8.6.3}$$

and there exists a positive constant λ such that

$$|g(x)| \leq \lambda |x|. \tag{8.6.4}$$

It is clear that conditions (8.6.3) and (8.6.4) imply that $g(0) = 0$. In addition to the above assumptions, we require that

$$A(t, s)\frac{\partial A(t, s)}{\partial t} \leq 0 \text{ for all } (t, s) \in [0, \infty) \times [t - r, t]. \tag{8.6.5}$$

Finally, we assume that for $1 < \alpha \leq 2$,

$$A^2\left(t - (\alpha - 1)r/\alpha, z\right) \geq A^2(t, z) \tag{8.6.6}$$

for all $t \in [0, \infty)$ and $z \in [t - r/\alpha, t - (\alpha - 1)r/\alpha]$. As a consequence of (8.6.5), we have

$$\int_{-r}^{0} \int_{t+s}^{t} A(t, z)\frac{\partial A(t, z)}{\partial t}g^2(x(z))dz \, ds$$

$$= \int_{t-r}^{t} \int_{-r}^{z-t} A(t, z)\frac{\partial A(t, z)}{\partial t}g^2(x(z))ds \, dz$$

$$= \int_{t-r}^{t} A(t, z)\frac{\partial A(t, z)}{\partial t}g^2(x(z))(z - t + r)dz$$

$$\leq 0, \tag{8.6.7}$$

which plays an essential role in the proof of the next lemma. We note that $g(x) = x(\sin^2(x) + 1)$ satisfies (8.6.3) and (8.6.4). To construct a suitable Lyapunov functional, we put (8.6.1) in the form

$$x'(t) = -A(t, t)g(x(t)) + \frac{d}{dt}\int_{t-r}^{t} A(t, s)g(x(s))ds. \tag{8.6.8}$$

Lemma 8.2. *Let (8.6.2)–(8.6.5) hold, and suppose that $0 < r \leq 1/2$ with*

$$[(r+1)\lambda^2 + 1]\left(\int_0^r a(u+t,t)du\right)^2 \leq \int_0^r a(u+t,t)du \leq \frac{1-2r}{2r}. \qquad (8.6.9)$$

If

$$V(t) = \left(x(t) - \int_{t-r}^t A(t,s)g(x(s))ds\right)^2$$

$$+ \int_{-r}^0 \int_{t+s}^t A^2(t,z)g^2(x(z))dz\,ds, \qquad (8.6.10)$$

then along the solutions of (8.6.1) we have

$$V'(t) \leq -A(t,t)V(t).$$

Proof. Let $x(t) = x(t,t_0,\psi)$ be a solution of (8.6.1) and define $V(t)$ by (8.6.10). Then along solutions of (8.6.1) we have

$$V'(t) = 2\left(x(t) - \int_{t-r}^t A(t,s)g(x(s))ds\right)[-A(t,t)g(x(t))]$$

$$+ rA^2(t,t)g^2(x(t)) - \int_{-r}^0 A^2(t,t+s)g^2(x(t+s))ds$$

$$+ \int_{-r}^0 \int_{t+s}^t 2A(t,z)\frac{\partial A(t,z)}{\partial t}g^2(x(z))dz\,ds$$

$$\leq -A(t,t)\left[x^2(t) - 2x(t)\int_{t-r}^t A(t,s)g(x(s))ds\right] + r\lambda^2 A^2(t,t)x^2(t)$$

$$- \int_{-r}^0 A^2(t,t+s)g^2(x(t+s))ds - A(t,t)x^2(t) \text{ by (8.5.6)}$$

$$- 2A(t,t)x(t)\int_{t-r}^t A(t,s)g(x(s))ds + 2A(t,t)g(x(t))\int_{t-r}^t A(t,s)g(x(s))ds$$

$$= -A(t,t)V(t) + A(t,t)\left(\int_{t-r}^t A(t,s)g(x(s))ds\right)^2$$

$$- 2A(t,t)x(t)\int_{t-r}^t A(t,s)g(x(s))ds + A(t,t)\int_{-r}^0 \int_{t+s}^t A^2(t,z)g^2(x(z))dz\,ds$$

$$+ 2A(t,t)g(x(t))\int_{t-r}^t A(t,s)g(x(s))ds$$

$$+ \left(r\lambda^2 A^2(t,t)x^2(t) - A(t,t)x^2(t)\right)$$

$$- \int_{-r}^0 A^2(t,t+s)g^2(x(t+s))ds. \qquad (8.6.11)$$

In what follows, we perform some calculations to simplify (8.6.11). First, if we let $u = t + s$, then

$$-\int_{-r}^{0} A^2(t, t+s)g^2(x(t+s))ds = -\int_{t-r}^{t} A^2(t, s)g^2(x(s))ds. \qquad (8.6.12)$$

Also, by Holder's inequality we have that

$$A(t, t)\left(\int_{t-r}^{t} A(t, s)g(x(s))ds\right)^2 \leq A(t, t)r \int_{t-r}^{t} A^2(t, s)g^2(x(s))ds. \qquad (8.6.13)$$

Finally, we easily observe that

$$A(t, t)\int_{-r}^{0}\int_{t+s}^{t} A^2(t, z)g^2(x(z))dz\,ds \leq A(t, t)r \int_{t-r}^{t} A^2(t, s)g^2(x(s))ds \qquad (8.6.14)$$

and

$$-2A(t, t)x(t)\int_{t-r}^{t} A(t, s)g(x(s))ds \leq A^2(t, t)x^2(t)$$
$$+ r\int_{t-r}^{t} A^2(t, s)g^2(x(s))ds. \qquad (8.6.15)$$

Similarly,

$$2A(t, t)g(x(t))\int_{t-r}^{t} A(t, s)g(x(s))ds \leq \lambda^2 A^2(t, t)x^2(t)$$
$$+ r\int_{t-r}^{t} A^2(t, s)g^2(x(s))ds. \qquad (8.6.16)$$

Invoking (8.6.12) and substituting expressions (8.6.12)–(8.6.16) into (8.6.11) yield

$$V'(t) \leq -A(t, t)V(t) + \left[(r+1)\lambda^2 A^2(t, t) + A^2(t, t) - A(t, t)\right]x^2(t)$$
$$+ \left[2r A(t, t) + 2r - 1\right]\int_{t-r}^{t} A^2(t, s)g^2(x(s))ds$$
$$\leq -A(t, t)V(t) \text{ by (8.6.9).} \qquad (8.6.17)$$

$\square$

Theorem 8.6.1. *Assume that the hypotheses of Lemma 8.2 and (8.6.6) hold, and let $1 < \alpha \leq 2$. Then any solution $x(t) = x(t, t_0, \psi)$ of (8.6.1) satisfies the exponential* *inequality*

$$|x(t)| \leq \sqrt{2 \frac{1 + \frac{\alpha-1}{\alpha}}{\frac{\alpha-1}{\alpha}} V(t_0)} e^{-\frac{1}{2} \int_{t_0}^{t-(\frac{\alpha-1}{\alpha})r} A(s,s)ds} \qquad (8.6.18)$$

for $t \geq t_0 + (\frac{\alpha - 1}{\alpha})r.$

Proof. By changing the order of integrations we have

$$\int_{-r}^{0} \int_{t+s}^{t} A^2(t,z)g^2(x(z))dz\,ds = \int_{t-r}^{t} \int_{-r}^{z-t} A^2(t,z)g^2(x(z))ds\,dz$$
$$= \int_{t-r}^{t} A^2(t,z)g^2(x(z))(z-t+r)dz. \qquad (8.6.19)$$

For $1 < \alpha \leq 2$, if $t - \dfrac{r}{\alpha} \leq z \leq t$, then $(\dfrac{\alpha-1}{\alpha})r \leq z - t + r \leq r$. Expression (8.6.19) yields

$$\int_{-r}^{0} \int_{t+s}^{t} A^2(t,z)g^2(x(z))dz\,ds = \int_{t-r}^{t} A^2(t,z)g^2(x(z))(z-t+r)dz$$
$$= \int_{t-r}^{t-\frac{r}{\alpha}} A^2(t,z)g^2(x(z))(z-t+r)dz + \int_{t-\frac{r}{\alpha}}^{t} A^2(t,z)g^2(x(z))(z-t+r)dz$$
$$\geq \int_{t-\frac{r}{\alpha}}^{t} A^2(t,z)g^2(x(z))(z-t+r)dz$$
$$\geq (\frac{\alpha-1}{\alpha})r \int_{t-\frac{r}{\alpha}}^{t} A^2(t,z)g^2(x(z))dz. \qquad (8.6.20)$$

Let $V(t)$ be given by (8.6.10). Then

$$V(t) \geq \int_{-r}^{0} \int_{t+s}^{t} A^2(t,z)g^2(x(z))dz\,ds$$
$$\geq (\frac{\alpha-1}{\alpha})r \int_{t-\frac{r}{\alpha}}^{t} A^2(t,z)g^2(x(z))dz. \qquad (8.6.21)$$

Using (8.6.6), this implies that, for $1 < \alpha \leq 2$,

$$V(t-(\alpha-1)r/\alpha) \geq (\alpha-1)r/\alpha \int_{t-r}^{t-r+r/\alpha} A^2(t,z)g^2(x(z))dz$$
$$\geq (\alpha-1)r/\alpha \int_{t-r}^{t-r/\alpha} A^2(t,z)g^2(x(z))dz. \qquad (8.6.22)$$

Note that since $V'(t) \leq 0$, for $t \geq t_0 + (\frac{\alpha-1}{\alpha})r$ we have that

$$0 \leq V(t) + V\left(t - (\frac{\alpha-1}{\alpha})r\right) \leq 2V\left(t - (\frac{\alpha-1}{\alpha})r\right).$$

In summary, (8.6.11), (8.6.21), and (8.6.22) imply that

$$V(t) + V\left(t - (\frac{\alpha-1}{\alpha})r\right) \geq \left(x(t) - \int_{t-r}^{t} A(t,s)g(x(s))ds\right)^2$$

$$+ \int_{-r}^{0}\int_{t+s}^{t} A^2(t,z)g^2(x(z))dz\,ds + (\frac{\alpha-1}{\alpha})r\int_{t-r}^{t-\frac{r}{\alpha}} A^2(t,z)g^2(x(z))dz$$

$$\geq \left(x(t) - \int_{t-r}^{t} A(t,s)g(x(s))ds\right)^2 + (\frac{\alpha-1}{\alpha})r\int_{t-\frac{r}{\alpha}}^{t} A^2(t,z)g^2(x(z))dz$$

$$+ (\frac{\alpha-1}{\alpha})r\int_{t-r}^{t-\frac{r}{\alpha}} A^2(t,z)g^2(x(z))dz$$

$$= \left(x(t) - \int_{t-r}^{t} A(t,s)g(x(s))ds\right)^2 + (\frac{\alpha-1}{\alpha})r\int_{t-r}^{t} A^2(t,z)g^2(x(z))dz$$

$$\geq \left(x(t) - \int_{t-r}^{t} A(t,s)g(x(s))ds\right)^2$$

$$+ (\frac{\alpha-1}{\alpha})\left(\int_{t-r}^{t} A(t,s)g(x(s))ds\right)^2 \text{ (by Hölder's inequality)}$$

$$= \frac{\frac{\alpha-1}{\alpha}}{1 + \frac{\alpha-1}{\alpha}}x^2(t) + \left[\frac{1}{\sqrt{1 + \frac{\alpha-1}{\alpha}}}x(t) - \sqrt{1 + \frac{\alpha-1}{\alpha}}\int_{t-r}^{t} A(t,s)g(x(s))ds\right]^2$$

$$\geq \frac{\frac{\alpha-1}{\alpha}}{1 + \frac{\alpha-1}{\alpha}}x^2(t). \tag{8.6.23}$$

Thus (8.6.23) shows that

$$\frac{\frac{\alpha-1}{\alpha}}{1 + \frac{\alpha-1}{\alpha}}x^2(t) \leq V(t) + V\left(t - (\frac{\alpha-1}{\alpha})r\right)$$

$$\leq 2V\left(t - (\frac{\alpha-1}{\alpha})r\right).$$

An integration of (8.6.17) from t_0 to t yields the inequality

$$V(t) \leq V(t_0)e^{-\int_{t_0}^{t} A(s,s)ds}.$$

As a consequence,

$$V\left(t - (\frac{\alpha-1}{\alpha})r\right) \leq V(t_0)e^{-\int_{t_0}^{t-(\frac{\alpha-1}{\alpha})r} A(s,s)ds},$$

and

$$|x(t)| \leq \sqrt{2\frac{1+\frac{\alpha-1}{\alpha}}{\frac{\alpha-1}{\alpha}}V(t_0)e^{-\frac{1}{2}\int_{t_0}^{t-(\frac{\alpha-1}{\alpha})r}A(s,s)ds}}$$

for $t \geq t_0 + (\frac{\alpha-1}{\alpha})r$. This completes the proof. □

Remark 8.2. It is clear that inequality (8.6.18) implies that the zero solution of (8.6.1) is asymptotically stable, provided that

$$\int^{\infty} A(t,t)dt \to \infty,$$

and exponentially stable if

$$\int_{t_0}^{t-(\frac{\alpha-1}{\alpha})r} A(s,s)ds \geq \beta(t-t_0)$$

for all $t \geq t_0 + (\frac{\alpha-1}{\alpha})r$ and a constant $\beta > 0$.

As an example, we take $g(x) = x(\frac{\sin^2(x)}{4}+1)$. Then $g(0) = 0$, $xg(x) > x^2$, and $|g(x)| \leq \frac{5}{4}|x|$. Choose $a(t,s) = \frac{1}{2}$ and $r = \frac{1}{3}$. Next, we make sure that (8.6.7) is satisfied:

$$\int_{-r}^{0}\int_{t+s}^{t} A(t,z)\frac{\partial A(t,z)}{\partial t}g^2(x(z))dz\,ds = -a^2\int_{t-r}^{t}(z-t+r)^2 g^2(x(z))dz \leq 0.$$

Thus we have shown that the zero solution of the nonlinear Volterra integro-differential equation

$$x'(t) = -\frac{1}{2}\int_{t-\frac{1}{3}}^{t} x(s)(\frac{\sin^2(x(s))}{4}+1)ds$$

is exponentially stable.

8.6.1 Instability

We turn our attention to the instability of the zero solution of (8.6.1). We start with the following lemma.

Lemma 8.3. *Suppose (8.6.3), (8.6.4), and (8.6.5) hold and there is a positive constant $D > r$ such that*

$$\frac{r}{r-D} \leq A(t,t) \leq -A^2(t,t)[(1+\lambda)^2 + D\lambda^2]. \tag{8.6.24}$$

If

$$V(t) = \left(x(t) - \int_{t-r}^{t} A(t,s)g(x(s))ds\right)^2$$
$$- D \int_{t-r}^{t} A^2(t,z)g^2(x(z))dz, \tag{8.6.25}$$

then along the solutions of (8.6.1) we have

$$V'(t) \geq -A(t,t)V(t).$$

Proof. Let $x(t) = x(t, t_0, \psi)$ be a solution of (8.6.1) and define $V(t)$ by (8.6.25). Then along solutions of (8.6.1) we have

$$V'(t) = 2\left(x(t) - \int_{t-r}^{t} A(t,s)g(x(s))ds\right)[-A(t,t)g(x(t))]$$
$$- DA^2(t,t)g^2(x(t)) - D \int_{t-r}^{t} 2A(t,z)\frac{\partial A(t,z)}{\partial t}g^2(x(z))dz$$
$$\geq 2\left(x(t) - \int_{t-r}^{t} A(t,s)g(x(s))ds\right)[-A(t,t)g(x(t))] - DA^2(t,t)g^2(x(t))$$
$$= -A(t,t)V(t)$$
$$- A(t,t)\left[-\left(\int_{t-r}^{t} A(t,s)g(x(s))ds\right)^2 + D \int_{t-r}^{t} A^2(t,z)g^2(x(z))dz\right]$$
$$- 2A(t,t)[x(t) - g(x(t))] \int_{t-r}^{t} A(t,s)g(x(s))ds$$
$$+ \left(-A(t,t) - \lambda^2 DA^2(t,t)\right)x^2(t). \tag{8.6.26}$$

First, we remark that, as a consequence of (8.6.5), we have

$$-D \int_{t-r}^{t} A(t,z)\frac{\partial A(t,z)}{\partial t}g^2(x(z))dz \geq 0.$$

We note that

$$2A(t,t)[g(x(t)) - x(t)] \int_{t-r}^{t} A(t,s)g(x(s))ds$$
$$\leq 2|A(t,t)|(|g(x)| + |x|)\left|\int_{t-r}^{t} A(t,s)g(x(s))ds\right|$$
$$\leq 2|A(t,t)||x|(1 + \lambda)\left|\int_{t-r}^{t} A(t,s)g(x(s))ds\right|$$

$$\leq A^2(t,t)(1+\lambda)^2 x^2(t) + \left(\int_{t-r}^{t} A(t,s)g(x(s))ds\right)^2$$

$$\leq A^2(t,t)(1+\lambda)^2 x^2(t) + r\int_{t-r}^{t} A^2(t,s)g^2(x(s))ds.$$

Similarly,

$$A(t,t)\left(\int_{t-r}^{t} A(t,s)g(x(s))ds\right)^2 \geq rA(t,t)\int_{t-r}^{t} A^2(t,s)g^2(x(s))ds.$$

Hence (8.6.26) reduces to

$$V'(t) \geq -A(t,t)V(t) - \{A(t,t) + (1+\lambda)^2 A^2(t,t) + D\lambda^2 A^2(t,t)\}x^2(t)$$

$$+ \left[A(t,t)(r-D) - r\right]\int_{t-r}^{t} A^2(t,s)g^2(x(s))ds$$

$$\geq -A(t,t)V(t). \tag{8.6.27}$$

This completes the proof.   □

Theorem 8.6.2. *Suppose hypotheses of Lemma 8.3 hold. Then the zero solution of (8.6.1) is unstable, provided that*

$$-\int_{t_0}^{\infty} A(s,s)\,ds = \infty.$$

Proof. An integration of (8.6.27) from t_0 to t yields

$$V(t) \geq V(t_0)e^{-\int_{t_0}^{t} A(s,s)\,ds}. \tag{8.6.28}$$

Let $V(t)$ be given by (8.6.25). Then

$$V(t) = x^2(t) - 2x(t)\int_{t-r}^{t} A(t,s)g(x(s))ds + \left[\int_{t-r}^{t} A(t,s)g(x(s))ds\right]^2$$

$$- D\int_{t-r}^{t} A^2(t,z)g^2(x(z))dz. \tag{8.6.29}$$

Let $\beta = D - r$. Then from

$$\left(\frac{\sqrt{r}}{\sqrt{\beta}}a - \frac{\sqrt{\beta}}{\sqrt{r}}b\right)^2 \geq 0$$

we have

$$2ab \leq \frac{r}{\beta}a^2 + \frac{\beta}{r}b^2.$$

With this in mind, we arrive at

$$-2x(t)\int_{t-r}^{t} A(t,s)g(x(s))ds \le 2|x(t)|\Big|\int_{t-r}^{t} A(t,s)g(x(s))ds\Big|$$

$$\le \frac{r}{\beta}x^2(t) + \frac{\beta}{r}\Big[\int_{t-r}^{t} A(t,s)g(x(s))ds\Big]^2$$

$$\le \frac{r}{\beta}x^2(t) + \frac{\beta}{r}r\int_{t-r}^{t} A^2(t,s)g^2(x(s))ds.$$

A substitution of the above inequality into (8.6.29) yields

$$V(t) \le x^2(t) + \frac{r}{\beta}x^2(t) + (\beta + r - D)\int_{t-r}^{t} A^2(t,s)g^2(x(s))ds$$

$$= \frac{\beta + r}{\beta}x^2(t)$$

$$= \frac{D}{D-r}x^2(t).$$

Using inequality (8.6.28), we get

$$|x(t)| \ge \sqrt{\frac{D-r}{D}}\, V^{1/2}(t)$$

$$= \sqrt{\frac{D-r}{D}}\, V^{1/2}(t_0)e^{-\frac{1}{2}\int_{t_0}^{t} A(s,s)\,ds}.$$

This completes the proof. □

As an example, we take $g(x) = x(\frac{\sin^2(x)}{100} + 1)$. Then $g(0) = 0$, $xg(x) > x^2$, and $|g(x)| \le \frac{101}{100}|x|$. Choose $a(t,s) = -\frac{1}{2}$, $r = \frac{1}{8}$, and $D = \frac{1}{4}$. Then (8.6.24) is satisfied. Also, it is clear that (8.6.7) holds.

Thus we have shown that the nonlinear Volterra integro-differential equation

$$x'(t) = \frac{1}{2}\int_{t-\frac{1}{8}}^{t} x(s)(\frac{\sin^2(x(s))}{100} + 1)ds$$

is unstable.

We end this section with the following open problem.

Open Problem. In light of this research, what can be said about the exponential stability and instability of the zero solution of the nonlinear Volterra integro-differential equation with infinite delay

$$x'(t) = -\int_{-\infty}^{t} a(t,s)g(x(s))ds?$$

8.7 **Existence of positive periodic solutions**

We consider the nonlinear neutral differential equation with functional delay

$$x'(t) = -a(t)x(t) + c(t)x'(t - g(t)) + q(t, x(t - g(t))), \qquad (8.7.1)$$

which arises in food-limited population models. For system (8.7.1), there may be a stable equilibrium point of the population. In the case the equilibrium point becomes unstable, there may exist a nontrivial periodic solution. Then the oscillation of solutions occurs. The existence of such a stable periodic solution is of quite fundamental importance biologically since it concerns the long-time survival of species. The study of such phenomena has become an essential part of the qualitative theory of differential equations. One of the most used models, a prototype of (8.7.1), is the system of Volterra integro-differential equations

$$\dot{N}(t) = -\gamma(t)N(t) + \alpha(t) \int_0^\infty B(s)e^{-\beta(t)N(t-s)}ds,$$

where $N(t)$ is the number of red blood cell at time t, $\alpha, \beta, \gamma \in C(\mathbb{R}, \mathbb{R})$ are T-periodic, and $B \in L^1(\mathbb{R}^+)$ is piecewise continuous. This is a generalized model of the red cell system introduced by Wazewska-Czyzewska and Lasota

$$\dot{n}(t) = -\gamma n(t) + \alpha e^{-\beta n(t-r)},$$

where α, β, γ, and r are constants with $r > 0$. The existence of positive periodic solutions was established for the neutral logistic equation with distributed delays

$$x'(t) = x(t)\left[a(t) - \sum_{i=1}^n a_i(t) \int_{-T_i}^0 x(t+\theta)\,d\mu_i(\theta) - \sum_{j=1}^m b_j(t) \int_{-\hat{T}_j}^0 x'(t+\theta)\,d\nu_j(\theta)\right],$$

$$(8.7.2)$$

where the coefficients a, a_i, and b_j are continuous periodic functions with the same period. The values $T_i, \hat{T}_j$ are positive, and the functions μ_i, ν_j are nondecreasing with $\int_{-T_i}^0 d\mu_i = 1$ and $\int_{-\hat{T}_j}^0 d\nu_j = 1$. Eq. (8.7.2) is of logistic form, and hence the method used to obtain the existence of positive periodic solutions will not work for our model (8.7.1). For example, in the above equation the transformation $x(t) = e^{N(t)}$ was used to put (8.7.2) in the form

$$N'(t) = a(t) - \sum_{i=1}^n a_i(t) \int_{-T_i}^0 e^{N(t+\theta)}\,d\mu_i(\theta)$$

$$- \sum_{j=1}^m b_j(t) \int_{-\hat{T}_j}^0 N'(t+\theta)e^{N(t+\theta)}\,d\nu_j(\theta). \qquad (8.7.3)$$

Eq. (8.7.1) represents a generalization of the hematopoiesis and blood cell production models.

The Krasnoselskii fixed point theorem has been extensively used in differential and functional differential equations by Burton [16] to prove the existence of periodic solutions. Also, Burton was the first to use the theorem to obtain stability results regarding solutions of integral equations and functional differential equations. For a collection of different type of results, we refer the reader to [16] and the references therein. The author is unaware of any results regarding the use of Krasnoselskii to prove the existence of a positive periodic solution.

Theorem 8.7.1. *(Krasnoselskii) Let* $\mathbb{M}$ *be a closed convex nonempty subset of a Banach space* $(\mathbb{B}, \|\cdot\|)$. *Suppose that A and B map* $\mathbb{M}$ *into* $\mathbb{B}$ *and satisfy the following conditions:*

 (i) *A is compact and continuous,*
 (ii) *B is a contraction mapping,*
 (iii) *If* $x, y \in \mathbb{M}$, *then* $Ax + By \in \mathbb{M}$.

Then there exists $z \in \mathbb{M}$ *such that* $z = Az + Bz$.

For $T > 0$, define $P_T = \{\phi \in C(\mathbb{R}, \mathbb{R}), \phi(t + T) = \phi(t)\}$, where $C(\mathbb{R}, \mathbb{R})$ is the space of all real-valued continuous functions. Then P_T is a Banach space when endowed with the supremum norm

$$\|x\| = \max_{t \in [0,T]} |x(t)| = \max_{t \in \mathbb{R}} |x(t)|.$$

We assume that

$$a(t + T) = a(t), \quad c(t + T) = c(t), \quad g(t + T) = g(t), \quad g(t) \geq g^* > 0 \qquad (8.7.4)$$

with continuously differentiable $c(t)$, twice continuously differentiable $g(t)$, and constant g^*. Usually, $a(t)$ is assumed to be positive, but here we only require that

$$\int_0^T a(s)ds > 0. \qquad (8.7.5)$$

It is interesting to note that Eq. (8.7.1) becomes of advanced type when $g(t) < 0$. Since we are searching for periodic solutions, it is natural to require that $q(t, x)$ is continuous in both arguments and periodic in t. Also, we assume that for all $0 \leq t \leq T$,

$$g'(t) \neq 1. \qquad (8.7.6)$$

Lemma 8.4. *Suppose (8.7.4)–(8.7.6) hold. If* $x(t) \in P_T$, *then* $x(t)$ *is a solution of Eq. (8.7.1) if and only if*

$$x(t) = \frac{c(t)}{1 - g'(t)} x(t - g(t))$$
$$+ \int_t^{t+T} \left[-r(u)x(u - g(u)) + q(u, x(u - g(u))) \right] \frac{e^{\int_t^u a(s)ds}}{e^{\int_0^T a(s)ds} - 1} du, \qquad (8.7.7)$$

where

$$r(t) = \frac{\left(c'(t) + c(t)a(t)\right)\left(1 - g'(t)\right) + g''(t)c(t)}{(1 - g'(t))^2}.$$ (8.7.8)

Proof. Let $x(t) \in P_T$ be a solution of (8.7.1). Multiply both sides of (8.7.1) with $e^{\int_0^t a(s)ds}$ and then integrate from t to $t + T$ to obtain

$$\int_t^{t+T} \left[x(u)e^{\int_0^u a(s)ds}\right]' du = \int_t^{t+T} \left[c(u)x'(u - g(u))\right.$$
$$\left. + q(u, x(u), x(u - g(u)))\right]e^{\int_0^u a(s)ds} du.$$

As a consequence, we arrive at

$$x(t + T)e^{\int_0^{t+T} a(s)ds} - x(t)e^{\int_0^t a(s)ds}$$
$$= \int_t^{t+T} \left[c(u)x'(u - g(u)) + q(u, x(u - g(u)))\right]e^{\int_0^u a(s)ds} du.$$

By dividing both sides of the above equation by $e^{\int_0^{t+T} a(s)ds}$ and the fact that $x(t + T) = x(t)$ we obtain

$$x(t) = \left(1 - e^{-\int_t^{t+T} a(s)ds}\right)^{-1} \int_t^{t+T} \left[c(u)x'(u - g(u))\right.$$
$$\left. + q(u, x(u), x(u - g(u)))\right]e^{-\int_u^{t+T} a(s)ds} du.$$ (8.7.9)

Rewrite

$$\int_t^{t+T} c(u)x'(u - g(u))e^{-\int_u^{t+T} a(s)ds} du$$
$$= \int_t^{t+T} \frac{c(u)x'(u - g(u))(1 - g'(u))}{(1 - g'(u))}e^{-\int_u^{t+T} a(s)ds} du.$$

Integrating by parts the above integral with

$$U = \frac{c(u)}{1 - g'(u)}e^{-\int_u^{t+T} a(s)ds}$$

and

$$dV = x'(u - g(u))(1 - g'(u))du,$$

we obtain

$$\int_t^{t+T} c(u)x'(u-g(u))e^{-\int_u^{t+T} a(s)ds}\,du = \frac{c(t)}{1-g'(t)}x(t-g(t))\left(1-e^{-\int_t^{t+T} a(s)ds}\right)$$
$$-\int_t^{t+T} r(u)e^{-\int_u^{t+T} a(s)ds}x(u-g(u))du,$$

where $r(u)$ is given by (8.7.8). Finally, due to the integration over one period and the periodicity of all functions, we have that

$$\frac{e^{-\int_u^{t+T} a(s)ds}}{1-e^{-\int_t^{t+T} a(s)ds}} = \frac{e^{-\int_u^{t+T} a(s)ds}}{e^{-\int_t^{t+T} a(s)ds}(e^{\int_t^{t+T} a(s)ds}-1)}$$
$$= \frac{e^{\int_t^u a(s)ds}}{e^{\int_t^{t+T} a(s)ds}-1} = \frac{e^{\int_t^u a(s)ds}}{e^{\int_0^T a(s)ds}-1}.$$

This completes the proof. □

To simplify notation, we let

$$M = \frac{e^{\int_0^{2T} |a(s)|ds}}{e^{\int_0^T a(s)ds}-1} \tag{8.7.10}$$

and

$$m = \frac{e^{-\int_0^{2T} |a(s)|ds}}{e^{\int_0^T a(s)ds}-1}. \tag{8.7.11}$$

Let

$$G(t,u) = \frac{e^{\int_u^t a(s)ds}}{e^{\int_0^T a(s)ds}-1}. \tag{8.7.12}$$

It is easy to see that for all $(t,u) \in [0,2T] \times [0,2T]$,

$$m \le G(t,u) \le M,$$

and for all $t, u \in \mathbb{R}$, we have

$$G(t+T,u+T) = G(t,u).$$

We obtain the existence of a positive periodic solution by considering two cases:
(a) $0 \le \dfrac{c(t)}{1-g'(t)} < 1,$
and
(b) $-1 \le \dfrac{c(t)}{1-g'(t)} \le 0.$

For some nonnegative constant L and positive constant K, we define the set

$$\mathbb{M} = \{\phi \in P_T : L \le ||\phi|| \le K\},$$

which is a closed convex bounded subset of the Banach space P_T. In addition, we assume that there are constants $0 \le \beta \le \alpha < 1$ such that

$$0 \le \beta \le \frac{c(t)}{1 - g'(t)} \le \alpha < 1, \tag{8.7.13}$$

and for all $u \in \mathbb{R}$ and $\rho \in \mathbb{M}$,

$$\frac{(1 - \beta)L}{mT} \le q(s, \rho) - r(s)\rho \le \frac{(1 - \alpha)K}{MT}, \tag{8.7.14}$$

where M and m are defined by (8.7.10) and (8.7.11), respectively. To apply Theorem 8.7.1, we need to construct two mappings: one is a contraction, and the other is compact. Thus we define the map $\mathbf{A} : \mathbb{M} \to P_T$ by

$$(\mathbf{A}\varphi)(t) = \int_t^{t+T} G(t, s)[q(s, \varphi(s - g(s))) - r(s)\varphi(s - g(s))]\, ds, t \in \mathbb{R}. \tag{8.7.15}$$

In a similar way, we set the map $\mathbf{B} : \mathbb{M} \to P_T$ by

$$(\mathbf{B}\varphi)(t) = \frac{c(t)}{1 - g'(t)}\varphi(t - g(t)), t \in \mathbb{R}. \tag{8.7.16}$$

It is clear from condition (8.7.13) that $\mathbf{B}$ defines a contraction mapping under the supremum norm.

Lemma 8.5. *If* (8.7.4)–(8.7.6), (8.7.13), *and* (8.7.14) *hold, then the operator* $\mathbf{A}$ *is completely continuous on* $\mathbb{M}$.

Proof. For $t \in [0, T]$, which implies that $u \in [t, t + T] \subseteq [0, 2T]$, and for $\varphi \in \mathbb{M}$, we have by (8.7.13) that

$$|(\mathbf{A}\varphi)(t)| \le ||\int_t^{t+T} G(t, s)[q(s, \varphi(s - g(s))) - r(s)\varphi(s - g(s))]\, ds||$$

$$\le TM\frac{(1 - \alpha)K}{MT}.$$

From the estimate of $|\mathbf{A}\varphi(t)|$ it follows that

$$||\mathbf{A}\varphi(t)|| \le (1 - \frac{c(t)}{1 - g'(t)})K \le Q_1$$

for some positive constant Q_1. This shows that $\mathbf{A}(\mathbb{M})$ is uniformly bounded. It remains to show that $\mathbf{A}(\mathbb{M})$ is equicontinuous. Let $\varphi \in \mathbb{M}$. Then a differentiation of

(8.7.13) with respect to t yields

$$(\mathbf{A}\varphi)'(t) = G(t, t+T)[q(t, \varphi(t-g(t))) - r(t)\varphi(t-g(t))] + a(t)(\mathbf{A}\varphi)(t).$$

Hence by taking the supremum norm in the above expression we have

$$\|(\mathbf{A}\varphi)'\| \le \frac{Q_1}{T} + \|a(t)\|Q_1.$$

Thus the estimation on $|(\mathbf{A}\varphi)'(t)|$ implies that $\mathbf{A}(\mathbb{M})$ is equicontinuous. Then by the Ascoli–Arzelà theorem, we obtain that A is a compact map. Due to the continuity of all terms in (8.7.13) for $t \in [0, T]$, we have that $\mathbf{A}$ is continuous. This completes the proof. $\qquad\square$

Theorem 8.7.2. *If (8.7.4)–(8.7.6), (8.7.13), and (8.7.14) hold, then Eq. (8.7.1) has a positive periodic solution z satisfying $L \le z \le K$.*

Proof. Let $\varphi, \psi \in \mathbb{M}$. Then by (8.7.13) and (8.7.16) we have that

$$
\begin{aligned}
(\mathbf{B}\varphi)(t) + (\mathbf{A}\psi)(t) &= \frac{c(t)}{1 - g'(t)}\varphi(t-g(t)) \\
&\quad + \int_t^{t+T} G(t, s)[q(s, \varphi(s-g(s))) - r(s)\varphi(s-g(s))]\, ds \\
&\le \alpha K + MT\frac{(1-\alpha)K}{MT} = K.
\end{aligned}
$$

On the other hand,

$$
\begin{aligned}
(\mathbf{B}\varphi)(t) + (\mathbf{A}\psi)(t) &= \frac{c(t)}{1 - g'(t)}\varphi(t-g(t)) \\
&\quad + \int_t^{t+T} G(t, s)[q(s, \varphi(s-g(s))) - r(s)\varphi(s-g(s))]\, ds \\
&\ge \beta L + m \int_t^{t+T} [q(s, \varphi(s-g(s))) - r(s)\varphi(s-g(s))]\, ds \\
&\ge \beta L + mT\frac{(1-\beta)L}{mT} = L.
\end{aligned}
$$

This shows that $\mathbf{B}\varphi + \mathbf{A}\psi \in \mathbb{M}$. All the hypotheses of Theorem 8.7.1 are satisfied, and therefore Eq. (8.7.1) has a periodic solution, say z, residing in $\mathbb{M}$. This completes the proof. $\qquad\square$

For the next theorem, we assume that there are constants $-1 < \beta \le \alpha \le 0$ such that

$$-1 < \beta \le \frac{c(t)}{1 - g'(t)} \le \alpha \le 0, \tag{8.7.17}$$

and for all $u \in \mathbb{R}$ and $\rho \in \mathbb{M}$,

$$\frac{(L - \beta K)}{mT} \leq q(s, \rho) - r(s)\rho \leq \frac{(K - \alpha L)}{MT}, \tag{8.7.18}$$

where M and m are defined by (8.7.10) and (8.7.11), respectively.

Theorem 8.7.3. *If (8.7.4)–(8.7.6), (8.7.17), and (8.7.18) hold, then Eq. (8.7.1) has a positive periodic solution z satisfying $L \leq z \leq K$.*

Proof. The proof follows along the lines of Theorem 8.7.2, and hence we omit it. $\square$

Example 8.5. The neutral differential equation

$$x'(t) = -\frac{1}{2} \sin^2(t)x(t) + \frac{1}{50}x'(t - \pi) + \frac{\cos^2(t)}{x^2(t - \pi) + 100} + \frac{1}{25} \tag{8.7.19}$$

has a positive π-periodic solution x satisfying

$$\frac{1}{10} \leq x \leq 2.$$

To see this, we let

$$q(s, \rho) = \frac{\cos^2(s)}{\rho^2 + 100} + \frac{1}{25}, \quad r(s) = \frac{1}{2} \sin^2(s), \quad \text{and} \quad T = g(t) = \pi.$$

Then

$$\frac{c(t)}{1 - g'(t)} = \frac{1}{50} < 1,$$

and

$$r(t) = \frac{1}{100} \sin^2(t).$$

A simple calculation yields

$$4.030 < M < 4.032 \quad \text{and} \quad 1.74 < m < 1.75.$$

Let $K = 2$ and $L = \frac{1}{10}$, and define the set $\mathbb{M} = \{\frac{1}{10} \leq \upsilon \leq 2\}$. Then for $\rho \in [\frac{1}{10}, 2]$, we have

$$q(s, \rho) - r(s)\rho = \frac{\cos^2(s)}{\sigma^2 + 100} + \frac{1}{100} \sin^2(s)\rho + \frac{1}{25}$$

$$\leq \frac{1}{100} + \frac{1}{50} + \frac{1}{25} = 0.07 < \frac{(1 - \frac{c(t)}{1 - g'(t)})K}{MT}.$$

On the other hand,

$$q(s, \rho) - r(s)\rho = \frac{\cos^2(u)}{\sigma^2 + 100} + \frac{1}{100}\sin^2(u)\rho + \frac{1}{25}$$

$$> \frac{1}{25} > \frac{(1 - \frac{c(t)}{1 - g'(t)})L}{mT}.$$

We see that all the conditions of Theorem 8.7.2 are satisfied, and hence (8.7.19) has a positive π-periodic solution x satisfying $\frac{1}{10} \leq x \leq 2$.

8.8 Exercises

Exercise 8.1. Use the method of steps to find the solution of the delay differential equation with delay 1,

$$x' = -tx(t-1), \ t > 0; \ x(s) = \varphi(s) = 12, \ -1 \leq s \leq 0,$$

on the interval $[0, 3]$ by considering the subintervals $[0, 1]$, $[1, 2]$, and $[2, 3]$.

Exercise 8.2. Consider the delay differential equation with delay 1,

$$x' = ax(t) + bx(t-1), \ t > 0,$$

with initial function

$$x(s) = \varphi(s) = 1 + t, \ -1 \leq s \leq 0,$$

where a and $b \neq 0$ are constants. Use the method of steps to find the solution on the interval $[0, 2]$ by considering the subintervals $[0, 1]$ and $[1, 2]$. Is the solution differentiable at the initial time $t_0 = 0$? What about the differentiability at $t = 1$?

Exercise 8.3. Consider the delay differential equation with delay 1,

$$x' = x(t) + x(t-1), \ t > 0,$$

with initial function

$$x(s) = \varphi(s) = t, \ -1 \leq s \leq 0.$$

Use the method of steps to find the solution on the interval $[0, 3]$. Is the solution differentiable at the initial time $t_0 = 0$? What about the differentiability at $t = 1$?

Exercise 8.4. Solve the Bernoulli equation (8.2.5).

Exercise 8.5. Repeat Example 8.3 for the delay second-order differential equation

$$mu''(t) + bu'(t) + qu'(t-r) + cu(t) = 0, \ t \geq 0,$$

where $m, b, q, c,$ and r are positive constants.

Exercise 8.6. Consider the delay second-order differential equation

$$u''(t) + au'(t) + bu'(t-r) = 0, \ t \geq 0,$$

where a, b, and r are positive constants. Write the delay equation into a system in x and y and use the Lyapunov-type functional

$$V(x_t, y_t) = y^2 + bx^2 + \gamma \int_{-r}^{0} \int_{t+s}^{t} y^2(u) du \, ds, \ \gamma > 0.$$

(i) Show that if

$$\gamma = b \ \text{and} \ -a + br < 0,$$

then the zero solution is uniformly stable.
(ii) Choose

$$\gamma > b \ \text{and} \ \frac{-2a}{r} + \gamma + b < 0$$

and show that the zero solution is asymptotically stable and uniformly asymptotically stable.

Exercise 8.7. Give an example of application of Theorem 8.7.3.

Exercise 8.8. Consider the delay differential equation

$$x'(t) = -ax(t) + bx(t-r), \ t \geq t_0 \geq 0,$$

where a, b, and r are positive constants. Consider the Lyapunov-type functional

$$V(x_t) = x^2(t) + b \int_{t-r}^{t} x^2(s) ds.$$

(i) Show that if $a \geq b$, then all solutions are bounded, and the zero solution is uniformly stable.
(ii) Show that if $a > b$, then the zero solution is uniformly asymptotically stable, and $x^2(t) \in L^1[0, \infty)$.
(iii) Use part (i) to show that all solutions of

$$x'(t) = -2x(t) + x(t-\tau), \ x(t) = \varphi(s), \ -\tau \leq s \leq 0, \ t \geq 0,$$

are bounded.
(iv) Replace the delay term $x(t-\tau)$ in the delay differential equation given in (iii) with

$$x(t) - \tau x'(t) + \frac{1}{2}\tau^2 x''$$

and show that the solutions of the resulting equation are unbounded regardless of the size of the delay τ.

Exercise 8.9. Consider the finite delay differential equation

$$x'(t) = -3x(t) + \frac{1}{e}x(t-1), \ t \geq 0, \tag{8.8.1}$$

with $x(t) = \varphi(s)$, $-1 \leq s \leq 0$.

(a) Use Theorem 8.4.3 to show that all solutions of (8.8.1) are bounded and its zero solution is exponentially stable.

(b) Replace the delay term $x(t-1)$ in (8.8.1) with $x(t) - x'(t)$. Are the solutions of the resulting equation bounded?

(c) Now replace the delay term $x(t-1)$ in (8.8.1) with $x(t) - x'(t) + \frac{1}{2}x''$ and show that the solutions of the resulting equation are unbounded.

Exercise 8.10. Construct an example that satisfies the hypotheses of Theorem 8.5.5.

Exercise 8.11. Let $h > 0$ and consider the finite delay differential equation

$$x'(t) = a(t)x(t) + b(t)x(t-h) + g(t), \ t \geq 0, \tag{8.8.2}$$

with $x(t_0) = \varphi$. Suppose $a(t)$, $b(t)$, and $g(t)$ are continuous for all $t \geq 0$. For some $\tau > 0$, define the continuous function

$$\xi(t) = \frac{e^{\int_0^t c(s)ds}}{1 + 2h \int_t^{t+\tau} e^{\int_0^u c(s)ds} du} \geq 0 \tag{8.8.3}$$

with

$$|b(t)| \leq h\xi(t), \tag{8.8.4}$$

where $c(t) := 2a(t) + |g(t)|$. Use the functional

$$V(t, x_t) = x^2(t) + h\xi(t) \int_{t-h}^t x^2(s) \, ds$$

to show that every solution of (8.8.2) with $x(t_0) = \varphi$ satisfies the inequality

$$\|x(t)\| \leq \left[V(t_0, \varphi) + \int_{t_0}^t |g(u)| e^{\int_{t_0}^u \alpha(s)ds} \, du \right]^{1/2} e^{-\frac{1}{2}\int_{t_0}^t \alpha(s)ds},$$

where $\alpha(t) = c(t) + 2h\xi(t)$ and $V(t_0, \varphi) = \varphi^2(0) + h\xi(t_0) \int_{t_0-h}^{t_0} \varphi^2(s)ds$.

Exercise 8.12. (Big project) Develop similar results as in Section 8.6 for the totally delayed differential equation

$$x' = ax(t-r), \ t > 0,$$

where $a \in \mathbb{R}$, and r is a positive constant.

New variation of parameters

In this chapter we introduce *new variation of parameters formula* and apply the idea to ordinary and delay differential equations. The results of this chapter are totally new and should serve as foundations for future research.

9.1 Applications to ordinary differential equations

It is customary in a nonlinear differential equation to add and subtract a convenient term that allows us to invert the equation in question and obtain a variation of parameters formula, which can be used to obtain different results on the solutions. However, the added term will cause restrictions on the coefficients, and as a result, limit the class of equations that can be considered. In this section we adopt the terminologies of Chapter 2 with respect to definitions of solutions and other important matters. We just try to avoid redundancies and repetitions. Consider the simplest ordinary differential equation

$$x'(t) = ax(t), \ \ x(0) = x_0, \tag{9.1.1}$$

which has the solution

$$x(t) = x_0 e^{at} \to 0 \text{ as } t \to \infty,$$

provided that

$$a < 0.$$

Suppose $a : \mathbb{R} \to \mathbb{R}$ is continuous and consider the ordinary differential equation with variable coefficient

$$x'(t) = a(t)x(t), \ \ x(0) = x_0, \tag{9.1.2}$$

which has the solution

$$x(t) = x_0 e^{\int_0^t a(s)ds} \to 0 \text{ as } t \to \infty,$$

provided that

$$\int_0^t a(s)ds \to -\infty. \tag{9.1.3}$$

Condition (9.1.3) implies that the function $a(t)$ can be positive or oscillates for a short time. Now let

$$v : [0, \infty) \to \mathbb{R}$$

be a continuous function. Multiply both sides of (9.1.1) by

$$e^{\int_0^t v(s)ds}$$

and then integrate from 0 to any $t \in [0, T)$:

$$\int_0^t e^{\int_0^u v(s)ds} x'(u)du = \int_0^t ax(u)e^{\int_0^u v(s)ds} du.$$

Integrating by parts the left side and simplifying, we get

$$x(t) = x_0 e^{-\int_0^t v(s)ds} + \int_0^t x(u)\big(v(u) + a\big)e^{-\int_u^t v(s)ds} du. \qquad (9.1.4)$$

Expression (9.1.4) is a *new variation of parameters formula for* (9.1.1) and of Volterra-type integral equation. Note that if

$$v(t) = -a,$$

then (9.1.4) becomes the regular solution $x(t) = x_0 e^{at}$ of (9.1.1). In a similar fashion,

$$x'(t) = a(t)x(t), \quad x(0) = x_0,$$

has the solution

$$x(t) = x_0 e^{-\int_0^t v(s)ds} + \int_0^t x(u)\big(v(u) + a(u)\big)e^{-\int_u^t v(s)ds} du. \qquad (9.1.5)$$

Again, letting

$$v(t) = -a(t),$$

we get the regular known solution $x(t) = x_0 e^{\int_0^t a(s)ds}$, and $x(t) \to 0$ as $t \to \infty$, provided that

$$\int_0^t a(s)ds \to -\infty.$$

Again, (9.1.5) is a new variation of parameters formula that we will analyze when considering nonlinear equations. In the mean time, for (9.1.5), by setting up the proper spaces and using the contraction mapping principle, we can show that $x(t) \to 0$ as $t \to \infty$, provided that

$$\int_0^t \big|v(u) + a(u)\big|e^{-\int_u^t v(s)ds} du \le \alpha, \ 0 < \alpha < 1,$$

and

$$\int_0^t v(s)ds \to \infty.$$

Suppose $f : \mathbb{R} \times \mathbb{R} \to \mathbb{R}$ is continuous and consider the nonlinear differential equation

$$x'(t) = f(t, x(t)), \ x(0) = x_0, \ \text{for a given constant } x_0. \qquad (9.1.6)$$

Then multiplying by a function $\int_0^t v(s)ds$ the solution of (9.1.6) is given by

$$x(t) = x_0 e^{-\int_0^t v(s)ds} + \int_0^t \big(x(u)v(u) + f(u, x(u))\big)e^{-\int_u^t v(s)ds}du. \qquad (9.1.7)$$

Next, we will use (9.1.7) to define a mapping on the proper space and show that the zero solution is AS. Let $\mathscr{C}$ be the set of all real-valued continuous functions. Define the space

$$\mathscr{S} = \{\phi : [0, \infty) \to \mathbb{R}/\phi \in \mathscr{C}, \ |\phi(t)| \leq L, \ \phi(t) \to 0 \text{ as } t \to \infty\}.$$

Then

$$(\mathscr{S}, ||\cdot||)$$

is a complete metric space under the uniform metric

$$\rho(\phi_1, \phi_2) = ||\phi_1 - \phi_2||,$$

where

$$||\phi|| = \sup\{|\phi(t)| : t \geq 0\}.$$

Let f be locally Lipschitz on the set $\mathscr{S}$, such that

$$f(t, 0) = 0, \qquad (9.1.8)$$

that is, for any ϕ_1 and $\phi_2 \in \mathscr{S}$, we have

$$|f(t, \phi_1) - f(t, \phi_2)| \leq \lambda(t)||\phi_1 - \phi_2|| \qquad (9.1.9)$$

with continuous $\lambda : [0, \infty) \to (0, \infty)$. Assume that for $\phi \in \mathscr{S}$, we have that

$$|x_0|e^{-\int_0^t v(s)ds} + L\int_0^t \big(|v(u)| + \lambda(u)\big)e^{-\int_u^t v(s)ds}du \leq L. \qquad (9.1.10)$$

Note that (9.1.10) implies that

$$\int_0^t \big(|v(u)| + \lambda(u)\big)e^{-\int_u^t v(s)ds}du \leq \alpha < 1.$$

Theorem 9.1.1. *Assume (9.1.8)–(9.1.10). Suppose there exists a positive constant k such that*

$$e^{-\int_0^t v(s)ds} \le k. \tag{9.1.11}$$

Then the unique solution of (9.1.6) is bounded, and its zero solution is stable.
If, in addition,

$$\int_0^t v(s)ds \to \infty, \tag{9.1.12}$$

then the zero solution of (8.5.2) is asymptotically stable.

Proof. For $\phi \in \mathscr{S}$, define the mapping $\mathfrak{P} : \mathscr{S} \to \mathscr{S}$ by

$$(\mathfrak{P}\phi)(t) = x_0 e^{-\int_0^t v(s)ds} + \int_0^t \big(\phi(u)v(u) + f(u, \phi(u))\big)e^{-\int_u^t v(s)ds}du. \tag{9.1.13}$$

It is clear that $(\mathfrak{P}\phi)(0) = x_0$. Now for $\phi \in \mathscr{S}$, we have that

$$|(\mathfrak{P}\phi)(t)| \le |x_0|k + \int_0^t \big(|\phi(u)||v(u)| + \lambda(u)|\phi(u)|\big)e^{-\int_u^t v(s)ds}du.$$

Consequently,

$$\|\mathfrak{P}\phi\| \le |x_0|k + \int_0^t \big(|v(u)| + \lambda(u)\big)e^{-\int_u^t v(s)ds}du\|\phi\|,$$

and

$$\|\mathfrak{P}\phi\| \le |x_0|k + \alpha\|\phi\| \le L, \quad \text{due to (9.1.10) and } \phi \in \mathscr{S}. \tag{9.1.14}$$

Since $\mathfrak{P}$ is continuous, we have that $\mathfrak{P} : \mathscr{S} \to \mathscr{S}$. Next, we show that $\mathfrak{P}$ is a contraction.

For $\phi_1, \phi_2 \in \mathscr{S}$, from (9.1.13) we have that

$$|(\mathfrak{P}\phi_1)(t) - (\mathfrak{P}\phi_2)(t)| \le \int_0^t \big(|v(u)| + \lambda(u)\big)e^{-\int_u^t v(s)ds}du\|\phi_1 - \phi_2\|$$
$$\le \alpha\|\phi_1 - \phi_2\|.$$

This shows that $\mathfrak{P}$ is a contraction. By Banach's contraction mapping principle, $\mathfrak{P}$ has a unique fixed point $x \in \mathscr{S}$, which is a bounded continuous function. Moreover, the unique fixed point is a solution of (9.1.6) on $[0, \infty)$. Let x be the unique solution. Let $\varepsilon > 0$ and choose $\delta = \varepsilon\frac{1-\alpha}{k}$. If $|x_0| < \delta$, then by (9.1.14) we have that

$$(1 - \alpha)\|x\| \le |x_0|k < \delta k$$

or

$$\|x\| \le \varepsilon.$$

The proof of

$$|x(t)| \to 0 \text{ as } t \to \infty$$

rests on condition (9.1.12), and we refer to the proof of Theorem 8.5.2. This completes the proof.   □

To see the benefits of our new inversion method, we consider a particular nonlinear equation and rewrite it so that we can invert the usual way. Consequently, the *contraction mapping principle* will no longer work. Let

$$f(t, x) = -x^3 + h(t, x),$$

where $h(t, x)$ satisfies a smallness condition. Thus we consider

$$x' = -x^3 + h(t, x). \tag{9.1.15}$$

Due to the absence of a linear term, we borrow one and try to invert the old way. We write (9.1.15) in the form

$$x' = -x + (x - x^3) + h(t, x) \tag{9.1.16}$$

and then use the variation of parameters formula to obtain

$$x(t) = x_0 e^{-t} + \int_0^t e^{-(t-s)} [x(s) - x^3(s) + h(s, x(s))] ds. \tag{9.1.17}$$

It is naive to believe that most maps can be defined so that they are contractions, even with the strictest conditions. To that effect, consider

$$g(x) = x - x^3.$$

Then for $x, y \in \mathbb{R}$ with $|x|, |y| \leq \frac{\sqrt{3}}{3}$, we have that

$$|g(x) - g(y)| = |x - x^3 - y + y^3| \leq |x - y| \left(1 - \frac{x^2 + y^2}{2} \right),$$

and the contraction constant tends to 1 as $x^2 + y^2 \to 0$. As a consequence, the regular contraction mapping principle failed to produce any results. Using (9.1.7), (9.1.15) has the solution

$$x(t) = x_0 e^{-\int_0^t v(s)ds} + \int_0^t (x(u)v(u) - x^3(u) + h(u, x(u))) e^{-\int_u^t v(s)ds} du. \tag{9.1.18}$$

Now using the contraction mapping principle, we would have to show that a function of the form

$$f(x) = v(r)x - x^3$$

is a contraction. This is true on some bounded and small set, provided that v is of sufficiently small magnitude.

Thus using (9.1.18) we can easily obtain stability and boundedness results regarding (9.1.15) under the right conditions.

9.1.1 Periodic solutions

Next, we apply our new method to linear or nonlinear differential equations to show the existence of periodic solutions without the requirement of some classic conditions. To better illustrate our approach, we let $f : \mathbb{R} \times \mathbb{R} \to \mathbb{R}$ and $a : \mathbb{R} \to \mathbb{R}$ be continuous and consider the nonlinear differential equation

$$x' = a(t)x(t) + f(t, x), \qquad (9.1.19)$$

where f is continuous in x. For $T \in R$, we assume the periodicity condition

$$a(t + T) = a(t) \quad \text{and} \quad f(t + T, \cdot) = f(t, \cdot). \qquad (9.1.20)$$

Let BC be the space of continuous bounded functions $\phi : \mathbb{R} \to \mathbb{R}$ with the maximum norm $|| \cdot ||$. Define

$$P_T = \{\phi \in BC, \phi(t + T) = \phi(t)\}.$$

Then P_T is a Banach space endowed with the maximum norm

$$\|x\| = \max_{t \in [0,T]} |x(t)|.$$

Also, we assume that

$$e^{\int_0^T a(s)ds} \neq 1. \qquad (9.1.21)$$

Throughout this section, we assume that $a(t) \neq 0$ for all $t \in [0, T]$. Now (9.1.19) is equivalent to

$$\left[x(t)e^{-\int_0^t a(s)ds}\right]' = f(t, x(t))e^{-\int_0^t a(s)ds}.$$

Integrating this expression from $t - T$ to t and using the fact that $x(t - T) = x(t)$ give

$$x(t) = \left(1 - e^{\int_0^T a(s)ds}\right)^{-1} \int_{t-T}^{t} f(u, x(u))e^{\int_u^t a(s)ds} du. \qquad (9.1.22)$$

Theorem 9.1.2. *Assume (9.1.20) and (9.1.21). Suppose the function f is Lipschitz continuous with Lipschitz constant k. If*

$$k\left|\left(1 - e^{\int_0^T a(s)ds}\right)^{-1}\right| \int_{t-T}^{t} e^{\int_u^t a(s)ds} du \leq \alpha$$

for $\alpha \in (0, 1)$, then (9.1.19) has a unique periodic solution.

Proof. Define $\mathfrak{P}$ by the right side of (9.1.22). It is easily verified that $(\mathfrak{P}\phi)(t+T) = (\mathfrak{P}\phi)(t)$, and hence $\mathfrak{P} : P_T \to P_T$. The rest of the proof is a direct application of the contraction mapping principle on the set P_T. $\qquad\square$

On the other hand, if we take a function $v(t)$ as before and assume that $v \in P_T$ with $v(t) \neq 0$ for all $t \in [0, T]$, then we obtain the following variation of parameters formula:

$$x(t) = \left(1 - e^{\int_0^T v(s)ds}\right)^{-1} \int_{t-T}^t \left[x(u)v(u) + x(u)a(u)\right.$$
$$\left. + f(u, x(u))e^{\int_u^t a(s)ds} du\right]. \qquad (9.1.23)$$

Theorem 9.1.3. *Suppose $v(t) \neq 0$ for all $t \in [0, T]$ and the function f is Lipschitz continuous with Lipschitz constant k. If*

$$\left|\left(1 - e^{\int_0^T v(s)ds}\right)^{-1}\right| \left|\int_{t-T}^t [|a(r)| + |v(r)| + k]\left|e^{\int_u^t a(s)ds} du\right|\right| \leq \alpha$$

for $\alpha \in (0, 1)$, then (9.1.19) has a unique periodic solution.

Proof. The proof is similar to that of Theorem 9.1.2. $\qquad\square$

Note that (9.1.23) can now handle equations of the form

$$x'(t) = \cos(t)x(t) + f(t, x(t)).$$

We easily see that

$$1 - e^{\int_0^T a(s)ds} = 1 - e^{\int_0^{2\pi} \cos(s)ds} = 0,$$

and hence (9.1.22) cannot be used.

9.2 Applications to delay differential equations

We begin by considering the following totally delayed nonlinear differential equation:

$$x'(t) = a(t)l(x_r), \qquad (9.2.1)$$

where $x_r(t) = x(t - r)$ for $r > 0$ constant, and $l = l(t)$ is a continuous function satisfying some conditions to be imposed later. Let $\psi : [-r, 0] \to \mathbb{R}$ be a continuous initial function. We rewrite (9.2.1) in the form

$$x'(t) = a(t + r)l(x) - \frac{d}{dt}\int_{t-r}^t a(s + r)l(x(s))ds$$

$$= a(t+r)x(t) - a(t+r)[x(t) - l(x(t))]$$

$$- \frac{d}{dt} \int_{t-r}^{t} a(s+r)l(x(s))ds. \tag{9.2.2}$$

Note that we added and subtracted $a(t+r)x$, so that the inversion is possible. Thus by the variation of parameters formula we obtain the integral

$$x(t) = \psi(0)e^{\int_0^t a(s+r)ds}$$

$$- \int_0^t e^{\int_s^t a(u+r)du} a(s+r)[x(s) - l(x(s))]ds$$

$$- \int_0^t e^{\int_s^t a(u+r)du} \frac{d}{ds} \int_{s-r}^{s} a(u+r)l(x(u))du\,ds. \tag{9.2.3}$$

The appearance of the term $x(s) - l(x(s))$ in (9.2.3) is a direct consequence of the borrowed term. In addition, to get any meaningful results, we would have to assume that $l(x(t))$ is odd. Otherwise, $x(s) - l(x(s))$ will not define a contraction. Next, we invert our way by assuming $v : [0, \infty) \to \mathbb{R}$ to be a nonnegative continuous function such that $0 < \int_0^\infty v(s)\,ds = m < \infty$. To solve (9.2.1), we apply an inversion technique that starts by multiplying both sides by $e^{\int_0^t v}$ and then integrates them. We simplify the notation here and drop the ds at the end of the exponent:

$$\int_0^t e^{\int_0^s v} x'(s)ds = x(t)e^{\int_0^t v} - x(0) - \int_0^t xv e^{\int_0^s v}\,ds.$$

Therefore (9.2.1) becomes

$$x(t)e^{\int_0^t v} = x(0) + \int_0^t xv e^{\int_0^s v}\,ds + \int_0^t e^{\int_0^s v} a(s)l(x_r)\,ds,$$

$$x(t) = x(0)e^{-\int_0^t v} + \int_0^t xv e^{-\int_s^t v}\,ds + \int_0^t e^{-\int_s^t v} a(s)l(x_r)\,ds. \tag{9.2.4}$$

By imposing adequate conditions on $v(t)$, $a(t)$, and $l(t)$, we will prove the existence of bounded solutions to (9.2.1). To this end, we will use the right-hand side to define a contraction map on a complete metric space $\mathscr{S}$. The resulting unique fixed point will be the solution we are looking for. Let $K > 0$ be a constant and fix an initial continuous function $\Psi : [-r, 0] \to \mathbb{R}$ with $|\Psi(t)| < K$ for $t \in [-r, 0]$ and $|\Psi(0)| > 0$. Define the space

$$\mathscr{S} := \{x : [-r, \infty) \to \mathbb{R} \mid x \in \mathscr{C}^1[-r, \infty), \|x\|_\infty \le K, x \equiv \Psi \text{ on } [-r, 0]\}, \tag{9.2.5}$$

where $\mathscr{C}^1[-r, \infty)$ denotes the space of continuously differentiable functions, and $\| \cdot \|_\infty$ is the supremum norm. By general principles, it follows that $\mathscr{S}$ is a complete metric space. Define the map $\mathfrak{P}$ on $\mathscr{S}$ by

$$\mathfrak{P}(x) := x(0)e^{-\int_0^t \nu} + \int_0^t x\nu e^{-\int_s^t \nu} \, ds + \int_0^t e^{-\int_s^t \nu} a(s)l(x_r) \, ds. \tag{9.2.6}$$

The following lemma can be adapted to several different situations.

Lemma 9.1. *Let $\{\varepsilon_1, \varepsilon_2, \varepsilon_3\}$ be a triple of positive numbers such that $\varepsilon_1 + \varepsilon_2 + \varepsilon_3 \leq 1 - \delta$ for fixed $0 < \delta < 1$. Assume that $|\Psi(0)| \leq \varepsilon_1 K$ and m is small enough such that $|1 - e^{-m}| < \varepsilon_2$. Assume that $l(y)$ is Lipschitz on $[-K, K]$ and satisfies $|l(y)| \leq C_l |y|$ for $y \in [-K, K]$ and for some positive constant C_l. Suppose that $a \in L^1([0, \infty))$ and $\int_0^\infty |a(s)| \, ds \leq \frac{\varepsilon_3}{C_l}$. Then the map $\mathfrak{P}$ has range in $\mathscr{S}$ and is a contraction. This implies that (9.2.1) has a unique solution in $\mathscr{S}$.*

Proof. We use the hypotheses of the theorem to bound each of the summands on the right-hand side of (9.2.6). The first term satisfies the bound

$$\left| x(0)e^{-\int_0^t \nu} \right| \leq \varepsilon_1 K,$$

since $|e^{-\int_s^t \nu}| \leq 1$. The second summand can be bounded by

$$\left| \int_0^t x\nu e^{-\int_s^t \nu} \, ds \right| \leq \|x\|_\infty \int_0^t \nu e^{-\int_s^t \nu} \, ds \leq (1 - e^{-m})\|x\|_\infty \leq \varepsilon_2 K.$$

The conditions stated above allow us to bound the third summand by

$$\left| \int_0^t e^{-\int_s^t \nu} a(s)l(x_r) \, ds \right| \leq \varepsilon_3 \|x\|_\infty \leq \varepsilon_3 K.$$

Therefore

$$|\mathfrak{P}(x)| \leq K\varepsilon_1 + K\varepsilon_2 + K\varepsilon_3 \leq (1 - \delta)K. \tag{9.2.7}$$

This implies $\|\mathfrak{P}(x)\|_\infty \leq (1 - \delta)K$ so that $\mathfrak{P}$ has the range in $\mathscr{S}$.

The contraction part follows similarly:

$$|\mathfrak{P}(x) - \mathfrak{P}(y)| \leq |x - y| \left(\int_0^t \nu e^{-\int_s^t \nu} ds + C_l \int_0^t |a(s)| \, ds \right) \tag{9.2.8}$$

$$\leq |x - y| \left(\varepsilon_2 + C_l \frac{\varepsilon_3}{C_l} \right)$$

$$\leq (1 - \delta)|x - y|.$$

Therefore $\|\mathfrak{P}(x) - \mathfrak{P}(y)\|_\infty \leq (1 - \delta)\|x - y\|_\infty$, and $\mathfrak{P} : \mathscr{S} \to \mathscr{S}$ is a contraction. $\qquad\square$

Now we consider a slightly more general equation

$$x'(t) = a(t)g(x_r) + b(t)G(x_r), \tag{9.2.9}$$

where a, b, g, G are all continuous on $[-r, \infty)$. Let us denote the right-hand side of (9.2.9) by $L(x_r)$. On the same space $\mathscr{S}$ as before, we define the new map

$$\hat{\mathfrak{P}}(x) := x(0)e^{-\int_0^t v} + \int_0^t xve^{-\int_s^t v}\, ds + \int_0^t e^{-\int_s^t v}L(x_r)\, ds. \tag{9.2.10}$$

Impose the same conditions on v as in Lemma 9.1. The only difference now is the third term. We impose the following conditions:

C1- The functions g and G are Lipschitz on $[-K, K]$. Explicitly, there are positive constants C_g and C_G such that $|g(y)| \le C_g|y|$ and $|G(y)| \le C_G|y|$ for all $y \in [-K, K]$.

C2- $a, b \in L^1([0, \infty))$, $\int_0^\infty |a(s)|\, ds \le \frac{\varepsilon_3}{2C_g}$, and $\int_0^\infty |b(s)|\, ds \le \frac{\varepsilon_3}{2C_G}$.

Lemma 9.2. *Using the notation and definitions from Lemma 9.1, under conditions C1–C2, the map* $\hat{\mathfrak{P}} : \mathscr{S} \to \mathscr{S}$ *is a contraction.*

Proof. The only difference is the third summand in the definition of $\hat{\mathfrak{P}}$. We bound it as follows:

$$\left| \int_0^t e^{-\int_s^t v}L(x_r)\, ds \right| \le K\left(C_g\frac{\varepsilon_3}{2C_g} + C_G\frac{\varepsilon_3}{2C_G} \right) = K\varepsilon_3,$$

and the rest follows similarly to the proof of Lemma 9.1. $\square$

As a consequence, we find that (9.2.9) has a unique bounded solution in $\mathscr{S}$.

9.2.1 The main inversion

Now we present a more general way of inverting (9.2.9), which starts by rewriting it as

$$\begin{aligned} x'(t) &= (a+b)g(x_r) + b(G(x_r) - g(x_r)) \\ &= -\frac{d}{dt}\int_{t-r}^t c(p+r)g(x(p))\, dp + c(t+r)g(x(t)) + bl(x_r), \end{aligned} \tag{9.2.11}$$

where $c = a + b$ and $l = G - g$. Now multiply both sides by $e^{\int_0^t v}$ and integrate as before to solve for $x(t)$. We get

$$\begin{aligned} x(t) =\ &x(0)e^{-\int_0^t v} + \int_0^t xve^{-\int_s^t v}ds \\ &- \int_0^t e^{-\int_s^t v}\left(\frac{d}{ds}\int_{s-r}^s c(p+r)g(x(p))\, dp \right) ds \\ &+ \int_0^t c(s+r)g(x)e^{-\int_s^t v}ds + \int_0^t e^{-\int_s^t v}bl(x_r)ds. \end{aligned} \tag{9.2.12}$$

Apply integration by parts in the middle line to get

$$x(t) = x(0)e^{-\int_0^t v} + \int_0^t xve^{-\int_s^t v}ds$$

$$- F(t) + e^{-\int_0^t v}F(0) + \int_0^t ve^{-\int_s^t v}F(s)\,ds \qquad (9.2.13)$$

$$+ \int_0^t c(s+r)g(x)e^{-\int_s^t v}ds + \int_0^t e^{-\int_s^t v}bl(x_r)ds,$$

where $F(t) = \int_{t-r}^t c(p+r)g(x(p))dp$.

Again, we consider the same space $\mathscr{S}$ and define the new map

$$\mathfrak{P}(x) = x(0)e^{-\int_0^t v} + \int_0^t xve^{-\int_s^t v}ds$$

$$- F(t) + e^{-\int_0^t v}F(0) + \int_0^t ve^{-\int_s^t v}F(s)\,ds \qquad (9.2.14)$$

$$+ \int_0^t c(s+r)g(x)e^{-\int_s^t v}ds + \int_0^t e^{-\int_s^t v}bl(x_r)ds.$$

We need to impose a different set of conditions on the summands on the right-hand side of (9.2.14) to get a contraction mapping of $\mathscr{S}$. First, label the terms on the right-hand side as 1–7. The new bounds on the absolute values the terms now are as follows:

D1- The first two summands are bounded exactly as before, that is, $x(0) = \Psi(0) \leq \varepsilon_1 K$ and $\|v\|_{L^1} = m > 0$, so that $1 - e^{-m} \leq \varepsilon_2$.

D2- To bound $F(t)$, assume that $\int_0^\infty |c|\,dt \leq \int_0^\infty |a|\,dt + \int_0^\infty |b|\,dt \leq \frac{\varepsilon_3}{C_g + C_G}$. Here C_g and C_G are the Lipschitz constants of the delayed functions. The third and fourth terms are each bounded above by $\varepsilon_3 K$.

D3- The definitions above produce an upper bound for the fifth term equal to $m\varepsilon_3 K$, where $m = \int_0^\infty |v|dt$.

D4- The sixth term is again bounded above by $\varepsilon_3 K$.

D5- The upper bound on b above implies that the seventh term is also bounded above by $\varepsilon_3 K$.

Therefore

$$|\mathfrak{P}(x)| \leq K(\varepsilon_1 + \varepsilon_2 + (m+4)\varepsilon_3). \qquad (9.2.15)$$

Theorem 9.2.1. *If the bounds D1–D5 above hold and there is $0 < \delta < 1$ such that $\varepsilon_1 + \varepsilon_2 + (m+4)\varepsilon_3 \leq 1 - \delta$, then the map $\mathfrak{P} : \mathscr{S} \to \mathscr{S}$ is a contraction with a unique fixed point. This fixed point is a solution to (9.2.9) that belongs to $\mathscr{S}$.*

Example 9.1. Let $\alpha, \beta > 0$ be such that

$$\dot{x} = e^{-\alpha t}g(x_r) + e^{-\beta t}G(x_r), \qquad (9.2.16)$$

where α and β are big enough so that $\alpha^{-1} + \beta^{-1} \leq \frac{\varepsilon_3}{C_g + C_G}$. This equation satisfies the previous conditions. Eq. (9.2.16) is a first-order differential equation with exponentially damped time-delayed terms. Theorem 9.2.1 implies the existence of a solution to (9.2.16) in $\mathscr{S}$.

9.2.2 Variable time delay

Here we consider $r = r(t)$ variable. We impose conditions on it later on. We want to find solutions to

$$x'(t) = a(t)g(x(t - r(t))) + b(t)G(x(t - r(t))). \qquad (9.2.17)$$

We still denote $x(t - r(t))$ by x_r. The same strategy as before gives

$$\begin{aligned} x'(t) &= (a + b)g(x_r) + b(G(x_r) - g(x_r)) \\ &= c(t)g(x_r) + bl(x_r) \\ &= \frac{cg(x_r)}{1 - r'}(1 - r') + bl(x_r). \end{aligned} \qquad (9.2.18)$$

Denote $f = \frac{c}{1-r'}$ and $f_r = f(t - r(t))$. Now we rewrite (9.2.18) as follows:

$$\begin{aligned} x' &= f_r g(x_r)(1 - r') + (f(t) - f_r)g(x_r)(1 - r') + bl(x_r) \\ &= \left(f_r g(x_r)(1 - r') - f(t)g(x(t)) \right) + f(t)g(x(t)) \\ &\quad + (f(t) - f_r)g(x_r)(1 - r') + bl(x_r) \\ &= -\left(\frac{d}{dt} \int_{t-r(t)}^{t} f(s)g(x(s))ds \right) + f(t)g(x(t)) \\ &\quad + (f(t) - f_r)g(x_r)(1 - r') + bl(x_r). \end{aligned} \qquad (9.2.19)$$

To simplify the upcoming expression, assume that $r(0) = 0$. The inversion gives

$$\begin{aligned} x(t) &= x(0)e^{-\int_0^t v} + \int_0^t xve^{-\int_s^t v}ds \\ &\quad - \int_{t-r(t)}^{t} f(p)g(x(p))dp + \int_0^t ve^{-\int_s^t v} \int_{s-r(s)}^{s} f(p)g(x(p))dpds \qquad (9.2.20) \\ &\quad + \int_0^t e^{-\int_s^t v}\left(f(s)g(x(s)) + (f(s) - f_r)g(x_r)(1 - \dot{r}) + bl(x_r) \right)ds. \end{aligned}$$

The most important condition to be imposed on r is $|r'(t)| \leq \kappa < 1$ for some κ small and positive. This condition allows us to get uniform pointwise upper bounds on the function $f = \frac{c}{1-r'}$. From the definition of $\mathscr{S}$ we impose that $r(t) \leq r$ for all $t \geq 0$. Now we impose bounds on the different terms.

E1- The first two summands are bounded exactly as before, that is, $x(0) = \Psi(0) \leq \varepsilon_1 K$ and $\|v\|_{L^1} = m > 0$, so that $1 - e^{-m} \leq \varepsilon_2$.

E2- To bound $f(t)$, assume that

$$\int_{-r}^{\infty} |f|\,dt \le (1-\kappa)^{-1} \int_{-r}^{\infty} |c|\,dt \le (1-\kappa)^{-1} \left(\int_{0}^{\infty} |a|\,dt + \int_{0}^{\infty} |b|\,dt \right)$$

$$\le \frac{\varepsilon_3}{(1-\kappa)(C_g + C_G)}.$$

This suffices to bound all the terms in the last line as well.

Theorem 9.2.2. *Define the map $\mathfrak{P}$ on $\mathscr{S}$ by the right-hand side of (9.2.20). Suppose there are constants $\{\varepsilon_1, \varepsilon_2, \varepsilon_3\}$ such that bounds E1–E2 above are satisfied and such that $\varepsilon_1 + \varepsilon_2 + \frac{2\kappa+4}{1-\kappa}\varepsilon_3 \le 1 - \delta$ for some $0 < \delta < 1$. Then $\mathfrak{P} : \mathscr{S} \to \mathscr{S}$ is a contraction. This implies that (9.2.17) has a unique solution in $\mathscr{S}$.*

9.3 Exercises

Exercise 9.1. Give the details for the proof of Theorem 9.1.3.

Exercise 9.2. Use our new inversion to show the existence of periodic solutions to

$$x'(t) = a(t)g(x) + b(t)G(t, x_L), \tag{9.3.1}$$

where $x_L = x(t - L)$, and $a(t), b(t)$ are continuous periodic functions with period $L > 0$. We assume that $g = g(t, x) = g(t + L, x)$ and $G = G(t, x) = G(t + L, x)$. We also assume that g and G are Lipschitz on $[0, L]$ (and extended periodically) with constants C_g and C_G, respectively.

Exercise 9.3. Consider the equation

$$x'(t) = -a(t)x(t) + c(t)x'(t - g(t)) + q(t, x(t - g(t))). \tag{9.3.2}$$

For $T > 0$, define $P_T = \{\phi \in C(\mathbb{R}, \mathbb{R}), \phi(t + T) = \phi(t)\}$, where $C(\mathbb{R}, \mathbb{R})$ is the space of all real-valued continuous functions. Then P_T is a Banach space endowed with the supremum norm

$$\|x\| = \max_{t \in [0,T]} |x(t)| = \max_{t \in \mathbb{R}} |x(t)|.$$

Assume that

$$a(t + T) = a(t), \quad c(t + T) = c(t), \quad g(t + T) = g(t), \quad g(t) \ge g^* > 0 \tag{9.3.3}$$

with continuously differentiable $c(t)$, twice continuously differentiable $g(t)$, and constant g^*. We suppose that $q(t, x)$ is continuous in both arguments and periodic in t. Usually, $a(t)$ is assumed to be positive, but here we only assume that

$$\int_{0}^{T} a(s)\,ds > 0.$$

Choose a positive function $v(t) = v(t + T)$. Multiply both sides of (9.3.2) by $e^{\int_0^t v}$ and integrate by parts between $t - T$ and t.

(i) Obtain a variation of parameters formula.

(ii) Use the results of part (i) to define a mapping $\mathfrak{P}$ on $P_T = \{\phi \in C(\mathbb{R}, \mathbb{R}), \phi(t + T) = \phi(t)\}$ and show that $\mathfrak{P} : P_T \to P_T$.

(iii) Impose additional conditions and prove theorems parallels to Theorems 8.7.2 and 8.7.3.

Bibliography

[1] M. Adivar, H. Koyuncuoğlu, Y. Raffoul, Existence of periodic and nonperiodic solutions of systems of nonlinear Volterra difference equations with Infinite delay, Journal of Difference Equations and Applications 19 (12) (December 2013) 1927–1939.

[2] M. Adivar, Y. Raffoul, Stability, Periodicity and Boundedness in Functional Dynamical Systems on Time Scales, Springer Nature, Switzerland AG, 2020.

[3] M. Adivar, Y. Raffoul, Existence of resolvent for Volterra integral equations on time scales, Bulletin of the Australian Mathematical Society 82 (2010) 139–155.

[4] M. Adivar, Y. Raffoul, Qualitative analysis of nonlinear Volterra integral equations on time scales using resolvent and Lyapunov functionals, Applied Mathematics and Computation 273 (2016) 258–266.

[5] M. Adivar, M. Islam, Y. Raffoul, Separate contraction and existence of periodic solutions in totally nonlinear delay differential equations, Hacettepe Journal of Mathematics and Statistics 41 (1) (2012) 1–13.

[6] R. Agarwal, Difference Equations and Inequalities. Theory, Methods and Applications, second ed., Monographs and Textbooks in Pure and Applied Mathematics, Marcel Decker, Inc., New York, 2000.

[7] R. Agarwal, D. O'Regan, Infinite Interval Problems for Differential, Difference, and Integral Equations, Kluwer Academic, The Netherlands, 2001.

[8] S. Alsahafi, Y. Raffoul, A. Sanbo, Qualitative analysis of solutions in Volterra nonlinear systems of difference equations, International Journal of Mathematical Analysis 8 (31) (2014) 1505–1515.

[9] S. Banach, Théorie des Opération Linéairs (reprint of the 1932 ed.), Chelsea, New York, 1929.

[10] R. Bellman, Stability Theory of Differential Equations, McGraw-Hill Book Company, New York, London, 1953.

[11] L. Berezansky, E. Braverman, Exponential stability of difference equations with several delays: recursive approach, Advances in Difference Equations 2009 (2009) 104310, 13 pages.

[12] M. Bohner, A. Peterson, Dynamic Equations on Time Scales, An Introduction with Applications, Birkhäuser, Boston, 2001.

[13] H. Brunner, P.J. Van der Houwen, The Numerical Solution of Volterra Equations, SIAM, Philadelphia, 1985.

[14] T.A. Burton, Volterra Integral and Differential Equations, Academic Press, New York, 1983.

[15] T.A. Burton, Stability and Periodic Solutions of Ordinary and Functional Differential Equations, Academic Press, New York, 1985.

[16] T.A. Burton, Stability by Fixed Point Theory for Functional Differential Equations, Dover, New York, 2006.

[17] F. Brauer, J.A. Nohel, Qualitative Theory of Ordinary Differential Equations, Dover, New York, 1969.

[18] E.A. Coddington, N. Levinson, Theory of Ordinary Differential Equations, McGraw-Hill Book Company, New York, London, 1955.

[19] J.M. Cushing, Integro-differential Equations and Delay Models in Population Dynamics, Lecture Notes in Biomathematics, vol. 20, Springer, Berlin, New York, 1977.

[20] R.D. Driver, Introduction to Ordinary Differential Equations, Harper & Row, New York, 1978.

[21] G. Eid, B. Ghalayani, Y. Raffoul, Lyapunov functional and stability in nonlinear finite delay Volterra discrete systems, International Journal of Difference Equations 10 (1) (2015) 77–90.

[22] S.E. Elaydi, An Introduction to Difference Equations, second ed., Undergraduate Texts in Mathematics, Springer-Verlag, New York, 1999.

[23] P. Eloe, M. Islam, Stability properties and integrability of the resolvent of linear Volterra equations, Tohoku Mathematical Journal 47 (1995) 263–269.

[24] J. Hale, H. Kocak, Dyanmics and Bifurcations, Springer-Verlag, New York, 1991.

[25] J. Hale, S. Verduyn Lunel, Introduction to Functional Differential Equations, Springer-Verlag, New York, 1993.

[26] P. Hartman, Ordinary Differential Equations, John Wiley & Sons, Inc., New York, 1964.

[27] P. Holmes, A Nonlinear oscillator with a strange attractor, Philosophical Transactions of the Royal Society of London A 292 (1979) 419–448.

[28] M. Islam, Y. Raffoul, Stability properties of linear Volterra integro-differential equations with nonlinear perturbations, Communications in Applied Analysis 7 (3) (2003) 405–416.

[29] J.L. Goldberg, A.J. Schwartz, Systems of Ordinary Differential Equations: An Introduction, Harper's Series in Mathematics, Harper & Row, 1972.

[30] W. Kelley, A. Peterson, Difference Equations an Introduction with Applications, Academic Press, San Diego, CA, 2000.

[31] W. Kelley, A. Peterson, The Theory of Differential Equations, Classical and Qualitative, Pearson Prentice Hall, New York, 2004.

[32] H. Lauwerier, Mathematical Models of Epidemics, Mathematisch Centrum, Amsterdam, Netherlands, 1981.

[33] H.J. Liu, A First Course in the Qualitative Theory of Differential Equations, Prentice Hall, Upper Saddle River, New Jersey, 2003.

[34] J. Marsden, M. McCracken, The Hopf Bifurcation and Its Applications, Springer-Verlag, New York, 1976.

[35] R.K. Miller, Nonlinear Volterra Integral Equations, Benjamin, New York, 1971.

[36] L. Perko, Differential Equations and Dynamical Systems, Springer-Verlag, New York, 1991.

[37] Y. Raffoul, Qualitative Theory of Volterra Difference Equations, Springer Nature Switzerland AG, 2018.

[38] Y. Raffoul, T-periodic solutions and a priori bound, Mathematical and Computer Modelling 32 (2000) 643–652.

[39] Y. Raffoul, Positive periodic solutions of nonlinear functional difference equations, Electronic Journal of Differential Equations 55 (2002) 1–8.

[40] Y. Raffoul, General theorems for stability and boundedness for nonlinear functional discrete systems, Journal of Mathematical Analysis and Applications 279 (2003) 639–650.

[41] Y. Raffoul, Uniform asymptotic stability in linear Volterra systems with nonlinear perturbation, International Journal of Differential Equations 6 (1) (2002) 19–28.

[42] Y. Raffoul, Periodicity in general delay nonlinear difference equations using fixed point theory, Journal of Difference Equations and Applications 10 (2004) 1229–1242.

[43] Y. Raffoul, Stability and periodicity in discrete delay equations, Journal of Mathematical Analysis and Applications 324 (2006) 1356–1362.

[44] Y. Raffoul, Periodic solutions in neutral nonlinear differential equations with functional delay, Electronic Journal of Differential Equations and Applications 2003 (102) (2003) 1–7.

[45] Y. Raffoul, Periodicity in nonlinear systems with infinite delay, Advances of Dynamical Systems and Applications 3 (1) (2008) 185–194.

[46] Y. Raffoul, Boundedness in nonlinear differential equations, Nonlinear Studies 10 (4) (2003) 343–350.

[47] Y. Raffoul, Stability in neutral nonlinear differential equations with functional delays using fixed point theory, Mathematical and Computer Modelling 40 (2004) 691–700.

[48] Y. Raffoul, Boundedness in nonlinear functional differential equations with applications to Volterra Integro-Differential Equations, Journal of Integral Equations and Applications 16 (4) (2004).

[49] Y. Raffoul, Stability in functional difference equations using fixed point theory, Communications of the Korean Mathematical Society 29 (1) (2014) 195–204.

[50] Y. Raffoul, Existence of positive periodic solutions in neutral nonlinear equations with functional delay, Rocky Mountain Journal of Mathematics 42 (6) (2012) 1983–1993.

[51] P.A. Robinson, C.J. Rennie, J.J. Wright, Propagation and stability of waves of electrical activity in the cerebral cortex, Physical Review E 56 (1997) 826–840.

[52] L. Roger, A Nonstandard discretization method for Lotka-Volterra models that preserves periodic solutions, Journal of Difference Equations and Applications 11 (8) (2005) 721–733.

[53] Y. Saito, M. Ma, T. Hara, A necessary and sufficient condition for permanence of a Lotka-Volterra discrete system with delays, Journal of Mathematical Analysis and Applications 256 (2001) 162–174.

[54] D.A. Sanchez, Ordinary Differential Equations and Stability Theory: An Introduction, W. H. Freeman and Company, New York, 1968.

[55] H. Schaefer, Uber die Method der a priori Scranken, Mathematische Annalen 129 (1955) 45–416.

[56] M. Scheffer, Fish nutrient interplay determines algal biomass: a minimal model, Oikos 62 (1991) 271–282.

[57] M.F. Scud, Vito Volterra and theoretical ecology, Theoretical Population Biology 2 (1971) 1–23.

[58] D.R. Smart, Fixed Point Theorems, Cambridge University Press, London, 1980.

[59] M.J. Smith, M.B. Wisten, A continuous day-to-day traffic assignment model and the existence of a continuous dynamic user equilibrium, Annals of Operations Research 60 (1) (1995) 59–79, https://doi.org/10.1007/BF02031940.

[60] Y. Song, C. Baker, Qualitative behavior of numerical approximations to Volterra integro-differential equations, Journal of Computational and Applied Mathematics 172 (2004) 101–115.

[61] X.H. Tang, X. Zou, Global attractivity of non-autonomous Lotka-Volterra competition system without instantaneous negative feedback, Journal of Differential Equations 192 (2003) 502–535.

[62] T. Taniguchi, Asymptotic behavior of solutions of non-autonomous difference equations, Journal of Mathematical Analysis and Applications 184 (2006) 342–347.

[63] P. Yang, R. Xu, Global attractivity of the periodic Lotka-Volterra system, Journal of Mathematical Analysis and Applications 233 (1999) 221–232.

[64] X. Yang, Uniform persistence and periodic solutions for a discrete predator-prey system with delays, Journal of Mathematical Analysis and Applications 316 (2006) 161–177.

[65] E. Yankson, Stability in discrete equations with variable delays, Electronic Journal on the Qualitative Theory of Differential Equations (8) (2009) 1–7.

[66] E. Yankson, Stability of Volterra difference delay equations, Electronic Journal on the Qualitative Theory of Differential Equations 20 (2006) 1–14.

[67] W. Yin, Eigenvalue problems for functional differential equations, Journal of Nonlinear Differential Equations 3 (1997) 74–82.

[68] X. Wen, Global attractivity of positive solution of multispecies ecological competition-predator delay system (in Chinese), Acta Mathematica Sinica 45 (1) (2002) 83–92.

[69] J. Wiener, Differential equations with piecewise constant delays, in: Trends in Theory and Practice of Nonlinear Differential Equations: Proceedings of International Conference, University of Texas, Arlington, 1982, in: Lecture Notes in Pure and Applied Mathematics, vol. 90, Dekker, New York, 1984, pp. 547–552.

[70] K. Yosida, Lectures on Differential and Integral Equations, Interscience Publishers, New York, London, Sydney, 1960.

Index